Producing iOS 6 Apps

*The Ultimate Roadmap
for Both
Non-Programmers
and
Existing Developers*

*(For Apple iPhone, iPod touch, and iPad apps,
distributed in Apple's iTunes App Store)*

UnknownCom Inc.

Producing iOS 6 Apps
The Ultimate Roadmap for Both Non-Programmers and Existing Developers

Copyright © 2012 UnknownCom Inc. All Rights Reserved.

Authored, Published and Produced by
UNKNOWNCOM INC.
3905 TAMPA RD UNIT 960
OLDSMAR FL 34677-9741

No part of this work may be reproduced, stored, or transmitted in any form, or by any means, either electronic or mechanical, including but not limited to photocopying, scanning, recording, or by any information storage or retrieval system, or otherwise, without the prior written permission of the copyright owner and the publisher. Note the "UnknownCom Inc." abbreviation of the copyright owner's/author's/publisher's full legal name "Unknown.com, Inc." was required for typesetting purposes (© 2012 Unknown.com, Inc. All Rights Reserved). The electronic (ebook) edition of this book may be uniquely formatted and/or different from the paperback edition.

LIMITATION OF LIABILITY / DISCLAIMER OF WARRANTIES:
The information in this book is provided on an "as is" basis, without warranty of any kind. The source code described in this book is available and licensed to readers, under a modified Free BSD License, on an "as is" basis without warranty of any kind, at **http://apps.unknown.com**. No warranty, or implied warranty, of any kind, is created, by sales, marketing, or promotional materials relating to this book. All code and information relating to this book are used at the reader's own risk. Readers should be aware that any references, files, web sites, and or their URL's, all of which are listed merely for convenience, and should not be construed as a recommendation or endorsement, may have changed or been deleted since this book was created. This book does not provide any tax, insurance, legal, or other professional advice whatsoever. This book is not intended to replace any professional advice from your lawyer, accountant, insurance broker, or other licensed advisor etcetera, and nothing in this book is to be construed as such.

TRADEMARKS:
Unknown.com is a trademark of UnknownCom Inc.
This book has not been authorized, sponsored, endorsed, or approved by Apple, Inc., Research In Motion Limited, or Google, Inc. UnknownCom Inc. is not affiliated with, or sponsored or endorsed by Apple, Inc., Research In Motion Limited, or Google, Inc. Apple, iPhone, iPod touch, iPad, Xcode, Dashcode, and Instruments, are trademarks of Apple, Inc. Research In Motion, RIM, and Blackberry, are trademarks of Research In Motion Limited. Google and Android are trademarks of Google, Inc. UnknownCom Inc. is not associated with any product or vendor mentioned in this book. All other trademarks are the property of their respective owners.

Rather than use a trademark symbol ™ or a registered symbol ® with every occurrence of a trademark in this book, we used the names, logos, and images, only in a nominative or descriptive editorial manner. Use in the publication of trade names, trademarks, service marks, and similar terms, even if they are not identified as such, is not to be taken as an expression or opinion as to whether or not they are subject to proprietary rights.

Paperback Edition: $29.99 US | £19.99 UK | $34.99 CN
ISBN-10: 0988337819 | ISBN-13: 978-0-9883378-1-7

Electronic (eBook) Edition: $19.99 US | £14.99 UK | $24.99 CN
ISBN-10: 0988337800 | ISBN-13: 978-0-9883378-0-0

Library of Congress Control Number: 2012918696

Printed in the United States of America

10 9 8 7 6 5 4 3 2 1

*I dedicate this book to all of our fantastic employees
who devoted themselves to conveying this previously internal-use-only
and highly confidential new-hire app development training material
to such an incredibly high quality, that we felt the need
to publish and share what has evolved into this utterly amazing
multi-part 'Tome of Knowledge' with the world.*

-- David Rajala, President, UnknownCom Inc.

ACKNOWLEDGEMENTS

Authored, Published, and Produced by: UnknownCom Inc. ("Unknown.com, Inc.")

Producer: David Rajala

Primary Contributors: David Rajala, Marc Pendleton, Erik Zimmerman, Bryan Lowe

Secondary Contributors: Jeremy Schofield, Tobias Dossin, Matthew Murphy, Ryan Brown

Additional Contributors: Unknown.com, Inc. Staff

Legal Editor: Attorney Mark Young, P.A.

Copy and Tech Editor: Bryan Lowe

Cover: Erik Zimmerman

About the Author:

UnknownCom Inc. incorporated in the State of Florida April 29th 1996 - over 16 years ago, as of this writing, and currently with unabbreviated/full legal name of "Unknown.com, Inc."

For the past several years, UnknownCom Inc. has capitalized on Apple iOS (iPhone, iPad & iPod Touch), Mac OSX, and Android mobile app development. UnknownCom Inc. is an Apple iOS, OSX, Safari; and, Google Android, Research In Motion Blackberry, and Microsoft Windows developer. Key members of UnknownCom Inc. have been involved in producing **hundreds of mobile apps** over the years, **including several #1-ranked apps** - which combined, have been **downloaded millions of times** over the years - either under, by, or for various employers and or clients, as well as UnknownCom Inc.'s own stand-alone products (some of which confidentially in an *"unknown"* capacity, with others distributed both directly, as well as indirectly, by partners and/or affiliates, etc.).

As of this writing, just a few of the more well-known mobile apps proudly contracted and distributed by UnknownCom Inc. include: Hollywood movie studio **LIONSGATE's official *SAW 3D Jigsaw Your Voice***, for fans of their terrifyingly famous **Saw** horror movies. As well as **Johnny Brennan's official *The Jerky Boys Prank Caller*, and *The Jerky Boys Pinball***, *for fans of his hilariously famous comedy routines. (And for those old Jerky Boys fans who may be interested: Mr. Brennan's latest work for over 10 years now has been in doing some of the hilarious voices for the Family Guy TV show on Fox. According to Mr. Brennan, Mort Goldman, Sol Rosenberg's long lost cousin, has really taken off and has an incredible fan base...)*

About the Contributors:

David Rajala, UnknownCom Inc.'s founding President, originally fell in love with computers with the acquisition of a coveted Commodore 64 back in the 1980's. Seeking most any computing job available in those early pioneering days, he found a cutting-edge role in a few startups as a network designer and installer thereby earning his first computer-industry certification, the Novell CNE. Later, upon entering the larger enterprise of Corporate America, Mr. Rajala excelled as a Network Engineer, while obtaining numerous (now-expired) certifications for computer operating systems and networking solutions, including, but not limited to, the following: Novell Master-CNE and Enterprise-CNE; Microsoft MCP, MCSE, MCT; Cisco CCNA, CCDA, CCNP, CCDP; and others. During his long stay with Corporate America, Mr. Rajala also taught official Cisco networking classes (as a Cisco Certified Systems Instructor) for several years, then deciding to blaze his own trail, left the corporate world of cubes in order to begin working and teaching as an independent contractor and IT consultant under the "Unknown" brand label.

Later, as CEO and co-founder of a prestigious adult learning and education-oriented software company and as President of its instructor-led training division, Mr. Rajala proved influential in growing and partnering the business with Pearson's Cisco Press imprint in order to provide practice tests and a network simulator application within their prestigious official Cisco Press books. His efforts proved instrumental in partnering with Cisco Systems, which ultimately granted the company its prestigious official status as a Cisco Learning Partner (CLP). Upon selling his share of the business in 2005 and again under the "Unknown" brand label, Mr. Rajala provided several years of IT consulting services focused on unique systems integrations and networking jobs such as Apple Xserver and Xserve-RAID fibre channel configurations for high-end Apple Final Cut Studio shops.

Mr. Rajala later returned to the software industry and co-founded as CEO of a unique online-oriented marketing and software company. ***That is, until Apple's launch of the original iPhone.*** Finding love at first sight of the Apple iPhone, Mr. Rajala believed it to be the most amazing handheld pocket computer ever conceived that, as an added bonus, just so happened to be able to make phone calls as well. With his passion ignited, he refocused the direction of the company to developing mobile apps shortly after the launch of Apple's iTunes App Store. The company established itself as one of the early App Store developers with several #1-ranked apps as well as millions of combined total app downloads! Mr. Rajala exited his stay with that early mobile apps company in 2010, and later sold his share of it as well.

Immediately upon leaving that initial software and app development company, Mr. Rajala has been solely dedicated and entirely focused on the business of UnknownCom Inc.'s mobile apps development for primarily Apple's iOS and Google's Android platforms as well as the ongoing writing and publication of this book.

Mark Young, P.A. (www.myoungpa.com) represents clients in intellectual property matters and technology transactions and disputes. He earned a B.S. degree, with honors, in aerospace engineering from Polytechnic University in 1988; a Juris Doctorate degree from the University of Dayton in 1994; and, while working full-time as a patent attorney, an LL.M. degree, with high honors, in intellectual property law, from the John Marshall Law School in Chicago in 1995. He is a member of the Florida and Illinois bars and admitted to practice before the U.S. Patent and Trademark Office, U.S. District Courts for the Middle District of Florida and Northern District of Illinois, and Court of Appeals for the Federal Circuit. Since 1994 he has represented many large companies, small businesses and individuals in a wide array of patent, trademark, copyright, transactional and litigation matters.

Prior to practicing law, Mr. Young worked for several years as an engineer on various research and development projects. He has experience in numerical modeling, embedded systems and propulsion system design. From 1997 through 2001, Mr. Young served as an adjunct professor of Intellectual Property Law at the Florida Coastal School of Law, an ABA accredited law school in Jacksonville, FL.

Marc Pendleton hails from Miami where he earned his B.S. in Computer Engineering from Florida International University. After a few years in the corporate world, he decided to pursue his passion for making games and mobile apps and joined the Unknown team of developers. Drawing from his background as a computer engineer coupled with an unbridled passion for research, Mr. Pendleton has graciously "brain-dumped" his pertinent mobile apps knowledge into this book with the hope that it will help others on their path to creating their mobile apps.

Erik Zimmerman is a jack-of-all-trades. In his youth, he discovered the technological awe of the computer. Destined to learn how they worked, he took one apart and examined it piece by piece. Since then he has been programming, building, and loving computers both inside and out. Mr. Zimmerman has also expanded his expertise into the fields of the fine arts, visual arts and music composition. He is a one-man app-making army.

Bryan Lowe, a seasoned and well-certified Information Security Professional (CISSP, CISM, CRISC, CCSK), eloquently and enthusiastically serves as an effective Information Technology consultant to Fortune 500 senior management teams and successfully communicates many different concepts to a broad range of both technical and non-technical audiences. He is an accomplished and highly respected advisor who works directly on security systems design, procedural documentation, product support guides, and enterprise-wide security policy & compliance development. Holding degrees in both English and Education, Mr. Lowe complements his technical strengths with essential team-building skills sharpened by exercise in four global corporations, the Department of Defense, numerous classrooms, a small college, two newspapers, several community volunteer activities, and his ongoing personal home improvement projects.

Jeremy Schofield, an experienced mobile software developer, has been creating apps for various platforms since the inception of Apple's App Store, Google's Android Marketplace, and Research In Motion's Blackberry App World. Before focusing on mobile apps, Mr. Schofield's outstanding skills as a .NET programmer enabled him to provide valuable insights to this tome's Objective-C "mini-book".

Matthew Murphy and Ryan Brown, two of the most exceptionally skilled Unity 3D game programmers on this particular plant, provided many contributions to this book. Their dedication to improving their programming skills is surpassed only by the many hours of commitment that they devote to crushing both friends and co-workers alike in competitive online computer games. They call it "research".

Tobias Dossin, the ultimate man-of-many-hats regarding his office activity, is a very well rounded and technically accomplished individual who performed a good deal of research for this book.

Needless to say, a tremendous amount of vision, teamwork and experience has all come together and contributed to the creation of this book – what we internally call our ***Tome of Knowledge*** – which we now share with you, our cherished and valued reader.

Please note that "UnknownCom Inc." is merely an abbreviation of and for the full legal name of "Unknown.com, Inc." as required for typesetting purposes.

Contents at a Glance

Contents at a Glance .... *ix*
Contents in Detail .... *xiii*
Foreword .... *xxv*
Preface .... *xxvii*
Introduction .... *xxxi*

SECTION I:
GETTING STARTED 1

Chapter 1: Turn Your App Idea into a Plan for Success *3*
Chapter 2: App Monetization Techniques *11*
Chapter 3: Getting Started With In-App Advertising *15*
Chapter 4: Advertising Aggregation *19*
Chapter 5: Critical Business Decisions *23*
Chapter 6: Intro to the Apple Developer Program *37*
Chapter 7: Agreeing to Contracts and Setting-Up Financial Information *43*
Chapter 8: Reserving an App Name *49*
Chapter 9: Preparing for Development *57*
Chapter 10: Checklist – Transitioning to Development *67*

SECTION II:
CHOOSE A DEVELOPMENT OPTION 69

Chapter 11: Introduction to iOS Development Options *71*
Chapter 12: Cross-Platform Development Options *75*
Chapter 13: Apple Recommended Development Method *77*
Chapter 14: Overview of Non-Apple Development Options *79*
Chapter 15: Development Options Summary *83*

SECTION III:
APPLE SUPPORTED DEVELOPMENT METHODS (VIA XCODE) 85

Chapter 16: Introduction to Xcode *87*
Chapter 17: Introduction to Objective-C *107*
Chapter 18: Foundation Framework *129*
Chapter 19: App Features, Structure, and Lifecycle *139*
Chapter 20: iOS Design Basics *157*

Chapter 21: Data Storage and Handling *181*
Chapter 22: Important Frameworks *195*
Chapter 23: User Input *201*
Chapter 24: Instruments *223*

SECTION IV:
ALTERNATIVE DEVELOPMENT OPTIONS *255*

Chapter 25: Unity 3D *257*
Chapter 26: ShiVa3D *275*
Chapter 27: PhoneGap *287*
Chapter 28: MonoTouch *293*
Chapter 29: Marmalade *305*
Chapter 30: Adobe Flash Builder *319*
Chapter 31: Adobe Flash Professional *331*
Chapter 32: Cocos2d (via Xcode) *343*
Chapter 33: Corona SDK *365*
Chapter 34: GameSalad *377*
Chapter 35: Titanium Studio *389*
Chapter 36: MoSync *397*

SECTION V:
SUBMITTING YOUR APP *407*

Chapter 37: App Submission Overview *409*
Chapter 38: Preparing For Submission *413*
Chapter 39: Submitting Apps Using Xcode *431*
Chapter 40: Submitting Apps Using Application Loader *437*
Chapter 41: Bonus Chapter of Gotchas! *443*

SECTION VI:
AFTER THE UPLOAD *451*

Chapter 42: Post Upload Marketing *453*

AFTERWORD *461*

Appendix A: Apple Contact Info Quick Reference *463*
Appendix B: Apple iOS 6 New Features *465*
Appendix C: Additional Reading *467*
Appendix D: App Review Websites *469*
Appendix E: Marketing Tools & Websites *471*

Appendix F: Useful Services & Websites... 473
Appendix G: Graphics, Images & Icons... 475
INDEX 477

Contents in Detail

Contents at a Glance ix
Contents in Detail xiii
Foreword xxv
Preface xxvii
Introduction xxxi

    How to Use this Book: The Sections xxxii
    Section I: Getting Started xxxii
    Section II: Choose a Development Option xxxv
    Section III: Apple Supported Development Method xxxv
    Section IV: Alternative Development Options xxxvii
    Section V: Submitting Your App xxxviii
    Section VI: After the Upload xxxix

SECTION I:
GETTING STARTED 1

Chapter 1: Turn Your App Idea into a Plan for Success 3

    1.1 Starting Your Own App Business 6
        1.1.1 SBA's Step 1: Write an App Business Plan 6
        1.1.2 SBA's Step 2: Get Business Assistance and Training 7
        1.1.3 SBA's Step 3: Choose a Business Location 7
        1.1.4 SBA's Step 4: Finance Your Business 7
        1.1.5 SBA's Steps 5 and 6: Determine the Legal Structure of Your Business; and; Register a Business Name ("Doing Business As") 7
        1.1.6 SBA's Step 7: Get a Tax Identification Number 8
        1.1.7 SBA's Step 8: Register for State and Local Taxes 8
        1.1.8 SBA's Step 9: Obtain Business Licenses and Permits 8
        1.1.9 SBA's Step 10: Understand Employee Responsibilities 9
    1.2 Getting to Know the Process 9
    1.3 Pre-Development Market Research 9

Chapter 2: App Monetization Techniques 11

    2.1 Market Research and Analysis 11
    2.2 Monetizing on Free Apps 12
        2.2.1 In-App Advertisements 12

	2.2.2 In-App Purchases and In-App Subscriptions 12
2.3	**Monetizing on Paid Apps** 13
2.4	**App Store Monetization Conclusion** 13

Chapter 3: Getting Started With In-App Advertising *15*

3.1	**Advertising Overview** 15
3.2	**Advertising Networks** 16
	3.2.1 Apple's iAd Mobile Advertising Platform 16
	3.2.2 Google Admob 16
	3.2.3 Millennial Media 16
	3.2.4 Jumptap 17
	3.2.5 Smaato 17
	3.2.6 Greystripe 17
3.3	**Advertising Services Conclusion** 17

Chapter 4: Advertising Aggregation *19*

4.1	**Aggregators vs. Ad Networks** 19
4.2	**AdWhirl** 20
4.3	**MobClix** 20
4.4	**Built-in Ad Aggregation** 21
4.5	**Summary** 21

Chapter 5: Critical Business Decisions *23*

5.1	**Legal Beagle: It's Getting Out Of Hand** 24
5.2	**Using Good Judgment and Common Sense** 24
5.3	**The Trademark Tango** 26
5.4	**Trademark Scan Database** 26
5.5	**Trying to Avoid Trademark Trouble** 26
5.6	**Playing with Trademark Fire** 27
5.7	**Trademark Summary** 28
5.8	**The Copyright Clash** 29
5.9	**Open Source Isn't Always Free** 29
5.10	**Open Source License Agreements** 29
5.11	**No Safety in Numbers** 30
5.12	**Copyrights Last a Long Time** 31
5.13	**Mobile Devices Don't Void Copyrights** 32
5.14	**Think Globally** 32
5.15	**Rights of Individuals** 32
5.16	**Misconception of Free Apps** 33
5.17	**Legal Barbeque and Disclaimer** 33
5.18	**From Our Lawyer, Mark Young, P.A.** 33

| | 5.18.1 COPYRIGHT FAIR USE | 34 |
| | 5.18.2 TRADEMARK FAIR USE | 35 |

Chapter 6: Intro to the Apple Developer Program ... 37

6.1	Conquering Apple Developer Paperwork	38
6.2	Determine the Account Type	38
6.3	Create an Apple ID	39
6.4	Potential Signup Delay	40
6.5	Agree and Agree to Agreements	40
6.6	Paying for the Program	40
6.7	Signup Walkthrough Summary	41

Chapter 7: Agreeing to Contracts and Setting-Up Financial Information ... 43

| 7.1 | iTunes Connect | 43 |
| 7.2 | Developer Portal Contracts | 47 |

Chapter 8: Reserving an App Name ... 49

8.1	Creating an App ID through the iOS Provisioning Portal	50
8.2	Creating a New App Page in iTunes Connect	52
8.3	Chapter Overview	56

Chapter 9: Preparing for Development ... 57

9.1	Requesting a Development Certificate	57
9.2	Adding a Device For Development	60
	9.2.1 Adding a Device Manually Through the iOS Developer Portal	60
	9.2.2 Adding a Development Device Using Xcode 4	62
9.3	Creating Generic Provisioning Profiles	62
	9.3.1 Manually Creating Provisioning Profiles	63

Chapter 10: Checklist – Transitioning to Development ... 67

SECTION II:
CHOOSE A DEVELOPMENT OPTION ... 69

Chapter 11: Introduction to iOS Development Options ... 71

11.1	From Web Apps to iOS Apps	71
11.2	Game-Focused Alternatives	72
	11.2.1 Independent Authoring Tools	72
	11.2.2 GUI-Based Development	72
11.3	Advantages of Using Development Alternatives	73
11.4	Disadvantages of Using Development Alternatives	73

Chapter 12: Cross-Platform Development Options **75**

Chapter 13: Apple Recommended Development Method **77**

Chapter 14: Overview of Non-Apple Development Options **79**

    14.1 Unity 3D **79**
    14.2 ShiVa3D **79**
    14.3 PhoneGap **79**
    14.4 MonoTouch **80**
    14.5 Marmalade **80**
    14.6 Adobe Flash Builder **80**
    14.7 Adobe Flash Professional **80**
    14.8 Cocos2D **80**
    14.9 Corona SDK **81**
    14.10 GameSalad **81**
    14.11 Titanium Studio **81**
    14.12 MoSync **81**

Chapter 15: Development Options Summary **83**

SECTION III:
APPLE SUPPORTED DEVELOPMENT METHODS (VIA XCODE) 85

Chapter 16: Introduction to Xcode **87**

    16.1 Main Feature Overview of Xcode **88**
        16.1.1 Interface Builder 88
        16.1.2 Simulator 89
        16.1.3 Instruments 89
        16.1.4 Other Notable Features 90
    16.2 Xcode Installation **90**
    16.3 Xcode's Windows **92**
        16.3.1 Editors 93
        16.3.2 Navigator View 94
        16.3.3 Debug view 97
        16.3.4 Utilities view 98
        16.3.5 The Organizer Window 100
        16.3.6 Hello World! 102

Chapter 17: Introduction to Objective-C **107**

    17.1 Objective-C Basic Types **108**
        17.1.1 Boolean Types 108
        17.1.2 Foundation Framework Data Types 108

17.2 Objective-C Classes 108
17.2.1 Class Interface 109
17.2.2 Class Implementation 109
17.2.3 Instance Variables 110
17.2.4 Methods 110
17.3 Objects 111
17.3.1 Messaging 112
17.3.2 Selectors 113
17.3.3 Properties 114
17.3.4 Instantiation 116
17.3.5 Destructors 119
17.4 Memory Management 120
17.4.1 Reference Counting 120
17.4.2 Garbage Collection 121
17.5 Root Class 122
17.5.1 Introspection 122
17.6 Protocols 123
17.6.1 Formal protocols 123
17.6.2 Categories 124
17.7 Exception Handling 125
17.8 Compiling Objective-C with C++ 127

Chapter 18: Foundation Framework 129

18.1 Data Types and Storage 131
18.1.1 NSData 131
18.1.2 NSValue 131
18.1.3 NSNumber 131
18.1.4 NSString 132
18.1.5 NSMutableString 132
18.1.6 NSRange 132
18.1.7 NSDate 132
18.2 Utilities 132
18.2.1 NSTimer 132
18.2.2 NSCalendar 133
18.2.3 NSFormatter 133
18.2.4 NSDateFormatter 133
18.2.5 NSNumberFormatter 133
18.2.6 NSUserDefaults 133
18.3 Collections 134
18.3.1 NSArray 134
18.3.2 NSMutableArray 135

    18.3.3 NSSet 135
    18.3.4 NSMutableSet 136
    18.3.5 NSDictionary 137
18.4 Summary **138**

Chapter 19: App Features, Structure, and Lifecycle 139

19.1 Views **139**
    19.1.1 Interface Builder 140
    19.1.2 Creating interfaces manually 147
    19.1.3 View's Life Cycle 149
    19.1.4 Delegates and Data Sources 150
19.2 The Application Life Cycle **150**
    19.2.1 Local and Remote Push Notifications 151
19.3 Basic Debugging **153**
19.4 Summary **155**

Chapter 20: iOS Design Basics 157

20.1 Designing Views **157**
    20.1.1 Displaying images... 157
    20.1.2 Selecting elements from a list 158
    20.1.3 Displaying web content 160
    20.1.4 Displaying alerts 161
    20.1.5 Controls 164
    20.1.6 Table Views 166
20.2 Controllers **169**
    20.2.1 Navigation Controller 169
    20.2.2 Tab Bar Controller 171
20.3 App Example **173**
    20.3.1 Creating a Navigation-based Application 173
    20.3.2 Creating a table view with various section 174
    20.3.3 Editing in a Table View 177
    20.3.4 Selecting cells in a Table View 178

Chapter 21: Data Storage and Handling 181

21.1 File and Network I/O **181**
    21.1.1 Reading Data from a File 181
    21.1.2 Writing Data to a File 182
    21.1.3 File Management with NSFileManager 182
    21.1.4 Network Access to File Input and Output 183
    21.1.5 Managing XML Data 183
21.2 App Settings **184**

21.2.1 Settings Data Types 184
21.2.2 Contents 184
21.2.3 App Settings Icon 185
21.2.4 Quick Example 185
21.3 SQLite **186**
21.3.1 Deploying the Database 186
21.3.2 Getting the Results 188
21.3.3 Usage and Implementation Considerations 189
21.4 Core Data **190**
21.4.1 The Anatomy of Core Data 190
21.4.2 Example Project 191

Chapter 22: Important Frameworks **195**

22.1 Address Book Framework **195**
22.2 Core Data Framework **196**
22.3 Core Location Framework **197**
22.4 Event Kit Framework **197**
22.5 AVFoundation Framework **197**
22.6 Core Audio Framework **198**
22.7 OpenAL Framework **198**
22.8 Media Player Framework **198**
22.9 Core Animation Framework **198**
22.10 Core Graphics Framework **199**
22.11 UIKit Framework **200**

Chapter 23: User Input **201**

23.1 Touch **201**
23.2 Multi-Touch **203**
23.3 Gestures **206**
23.3.1 Tap Gesture 207
23.3.2 Swipe Gesture 208
23.3.3 Pinch Gesture 208
23.3.4 Pan Gesture 209
23.3.5 Rotate Gesture 210
23.3.6 Long-Press Gesture 211
23.4 Camera **212**
23.5 Microphone **215**
23.6 Core Location **219**
23.6.1 Positional Data 219
23.6.2 Compass Heading 220
23.7 Core Motion - Gyroscope **220**

Contents in Detail   xix

23.8 The Accelerometer 221

Chapter 24: Instruments 223

24.1 Instruments for the iOS Simulator and iOS Device 224
    24.1.1 Allocations 224
    24.1.2 Leaks 229
    24.1.3 Activity Monitor 231
    24.1.4 Time Profiler 232
    24.1.5 Automation 237
24.2 iOS Simulator-Specific Instruments 241
    24.2.1 Zombies 241
    24.2.2 Threads 243
    24.2.3 File Activity 244
24.3 iOS Device-Specific Instruments 244
    24.3.1 Energy Diagnostics 244
    24.3.2 System Usage 245
    24.3.3 Core Animation 245
    24.3.4 OpenGL ES Driver 247
    24.3.5 OpenGL ES Analysis 249
24.4 Custom Instruments and Building Your Own Instrument Template 251

SECTION IV: ALTERNATIVE DEVELOPMENT OPTIONS 255

Chapter 25: Unity 3D 257

25.1 Installation 257
25.2 Unity 3D Walkthrough 261
    25.2.1 Creating a new project 261
    25.2.2 Creating your first game 262
    25.2.3 Writing your first script 265
25.3 Publishing Builds 270

Chapter 26: ShiVa3D 275

26.1 Installation 275
26.2 The "Hello ShiVa" Example 277
26.3 Building for Distribution 283

Chapter 27: PhoneGap 287

27.1 Installation 287
27.2 Hello World! 288
27.3 Building for Distribution 292
    27.3.1 PhoneGap Build Service 292

Chapter 28: MonoTouch ... 293

28.1 Installation ... 294
28.2 Example ... 295
28.3 Building for Distribution ... 301

Chapter 29: Marmalade ... 305

29.1 Installation and licensing for using Marmalade ... 305
    29.1.1 Installation for Mac OS X ... 305
    29.1.2 Installation on Windows 7 ... 307
29.2 The "Hello Marmalade" Example ... 309
29.3 Building for Distribution and Deploying to a Device ... 315
    29.3.1 Deploying to an iOS Device ... 316
    29.3.2 Building for Distribution ... 317

Chapter 30: Adobe Flash Builder ... 319

30.1 Installation ... 319
30.2 Walkthrough ... 320
30.3 Building for Distribution ... 327

Chapter 31: Adobe Flash Professional ... 331

31.1 Adobe Flash Installation ... 331
31.2 Adobe Flash Walkthrough ... 334
31.3 Deploying to an iOS Device and Building for Distribution ... 341

Chapter 32: Cocos2d (via Xcode) ... 343

32.1 Installation ... 343
32.2 Hello World! ... 346
    32.2.1 Create a New Project ... 347
    32.2.2 Creating a Sprite Sheet ... 350
    32.2.3 Creating an Animated Sprite ... 351
    32.2.4 Responding to Touches ... 357
    32.2.5 Make a Main Loop ... 358
    32.2.6 Building for iOS Distribution ... 362
32.3 Cocos2d Summary ... 363

Chapter 33: Corona SDK ... 365

33.1 Installation ... 366
33.2 Hello World ... 367
    33.2.1 Understanding the Workspace ... 368
    33.2.2 Writing the Code ... 370
    33.2.3 Running the Simulator ... 374

33.3 Building for iOS Distribution 375
    33.3.1 Compiling 375
    33.3.2 Adding to a Device 375
    33.3.3 Publishing 375
33.4 Advertising 375

Chapter 34: GameSalad 377

34.1 Installation 377
34.2 Hello, World! 377
34.3 Testing on an iOS Device and Distribution 383

Chapter 35: Titanium Studio 389

35.1 Installation & Setup 389
35.2 Hello World! 390
35.3 Testing on a Device and Building for Distribution 394

Chapter 36: MoSync 397

36.1 Installation 397
36.2 Example 399
36.3 Building for Distribution 405

SECTION V: SUBMITTING YOUR APP 407

Chapter 37: App Submission Overview 409

37.1 Xcode 4 App Submission Overview 410
37.2 Application Loader Overview 411
37.3 Submission Methods Overview 411

Chapter 38: Preparing For Submission 413

38.1 Avoiding Instant Rejection 413
    38.1.1 Locating the Metadata 413
    38.1.2 Potential Metadata Issues 415
38.2 Top Technical Reasons for Rejection 415
    38.2.1 Incorrect image sizes used for device 416
    38.2.2 Incorrect, missing, or misplaced imbedded device desktop icon .. 418
    38.2.3 Mismatch between iTunes Connect image and device's icon 418
    38.2.4 Feature or API used for incompatible device type 418
    38.2.5 Mismatch between App Store and reality 419
38.3 Walkthrough on Creating a Distribution Certificate 419
38.4 Generating and Importing Profiles into Xcode 423
38.5 Verifying Items and Set as "Ready for Upload" 425

    38.5.1 Set up In-App Purchases in iTunes Connect. 426
    38.5.2 Each In-App Purchase needs to have the following metadata: 426
    38.5.3 Activate Game Center in iTunes Connect 426
    38.5.4 Set up Leaderboards 426
    38.5.5 Set up Achievements 426
    38.5.6 Set up iAds in iTunes Connect. 427
38.6 Finding, Verifying and Tuning Metadata 427
38.7 Configuring App Build Settings 428

Chapter 39: Submitting Apps Using Xcode *431*

39.1 Walkthrough on Submitting Apps with Xcode 431
39.2 Viewing Crash Reports 434
39.3 How to add a new version of your app in iTunes Connect 436

Chapter 40: Submitting Apps Using Application Loader *437*

40.1 Walkthrough on Submitting Apps with Application Loader 437
40.2 Viewing Crash Reports 440
40.3 How to Add a New App Version in iTunes Connect 441

Chapter 41: Bonus Chapter of Gotchas! *443*

41.1 Team Admins: There Can Be Only One! 443
41.2 App Transfers: No Can Do! 443
41.3 Don't Scam the Ranking System 445
41.4 Don't Spam the App Store 446
41.5 Sexual Content is Not Permitted 446
41.6 Make an App Name Reservation 446
41.7 Resolving Rejection 446
41.8 Easter Egg Apps, Misleading Descriptions, etc. 446
41.9 Following Apple Guidelines 447
41.10 Expedited Reviews 447
41.11 Device Limits, Use Safari, Keyword Restrictions 447
41.12 UDID Myths (aka "The good old days") 448
41.13 Reinstalling Xcode if needed 448
41.14 Illegitimate Content 448
41.15 Overview of Gotchas! 449

SECTION VI:
AFTER THE UPLOAD *451*

Chapter 42: Post Upload Marketing *453*

42.1 What now? 453
42.2 Post-upload Marketing review 454

42.2.1	Utilize Social Media	454
42.2.2	Review your competitors	454
42.2.3	Review your own App Page in iTunes Connect	454
42.2.4	Promotional Codes	455
42.2.5	Set an App Release Date	455
42.2.6	Use Your Own Website	455
42.2.7	Video Impact	456
42.3	**Press Releases**	**456**
42.4	**Advertising Campaigns**	**457**
42.5	**Sticking to your Business Plan**	**458**

AFTERWORD *461*

Appendix A: Apple Contact Info Quick Reference *463*

Appendix B: Apple iOS 6 New Features *465*

Map Kit	465
Social Integration (Facebook)	465
Passbook and Pass Kit	465
Reminders and Event Kit	465
In-App Purchase	465
Game Center and Game Kit	465
Camera	466
iCloud	466
Webkit and Safari	466

Appendix C: Additional Reading *467*

Appendix D: App Review Websites *469*

Appendix E: Marketing Tools & Websites *471*

Appendix F: Useful Services & Websites *473*

Appendix G: Graphics, Images & Icons *475*

INDEX *477*

Foreword

Way back in 1994, when domain names were government-funded and freely registered to the public with a simple email to the InterNIC, Mr. Rajala first registered the infamous "unknown.com" domain name far before the Internet was fashionable and prime for explosion. After making two exceptionally humorous observations: (1) that the majority of early network operating systems would crash if their internally-reserved keyword "unknown" was used within its configuration, and (2) that the "unknown.com" domain name was the internet's bit-bucket for a tremendous amount of unwanted global email. The idea of the "Unknown" brand name was born.

***With longtime roots in education**, Unknown.com, Inc. continues to this day to provide free educational information and guides on its website for learning: (1) how to properly read "spoofed" spam email headers, and (2) how to properly look up IP address ownership from erroneous Gmail "attempted access" notices; among others.*

*Unknown.com, Inc. is also known for providing assistance to end-user victims who have been "spoofed" to help determine the **real** information for them to report to their ISP and the Internet Crime Complaint Center (IC3), located at: www.ic3.gov - which is a partnership between the United States FBI and NW3C.*

It is all part of the good we do, and just an example of our ongoing contributions to the community.

Preface

I personally wish to thank you for your purchase of our multi-part **Tome of Knowledge**, otherwise known as ***Producing iOS 6 Apps: The Ultimate Roadmap for Both Non-Programmers and Existing Developers***. We have made every effort to make this a useful and practical complete guide to building mobile Apple iOS applications, or "apps" for short, and to help you get them onto the Apple iTunes App Store as efficiently as possible. **This is, effectively, several books in one!** It's not merely a business/legal/sales/marketing/promoting your app book, nor is it merely an app programming book, but rather, a combined and comprehensive guide to all things required to create and produces apps – and to help turn your idea into reality.

It is not an easy task to go from a great idea into a real plan to build and successfully launch an app for sale on Apple's App Store. Based on real world experience, this book is for audiences that range from beginners without programming experience to even the most experienced programmers who want to build an app for the Apple iPhone, iPod touch, and iPad. All you need is the desire to take your idea to fruition. With this extensive tome's help, you will be able to turn that idea into a real app with real value worth paying for, to the millions of iPhone, iPod touch, and iPad users worldwide!

Along with countless thousands of others, when I bought my first iPhone upon its original release, I instantly realized that the device was not just a phone – **it is a mini handheld computer, that just so happens to make phone calls!** The moment that thought struck me, I realized, it is nothing less than awesome with a hugely profitable future. Even that original model had capabilities far beyond the "traditional smartphone" (which all had limited functionality and tedious interfaces at the time). The original iPhone was indeed, a powerful and unique handheld computer. Until the release of the iPhone, I had only dreamed of having a phone that "knew" where it was via GPS, had instant access to information on the Internet, had the ability to "know" the orientation of the phone, and had hundreds of other cool little features. From this global realization, nearly a million "Apps" are now available for the iPhone, iPad and other devices based on the Apple iOS operating system.

This book, affectionately dubbed the 'Tome of Knowledge', contains many resources not normally seen in a traditional programming book.

For example, in part, this major work contains an extensive guide for existing programmers to jump right into Xcode 4.5 and iOS 6 development. It even includes step-by-step walkthroughs, complete with numerous screenshots for guiding you through various topics! Many developers we have hired over the years already had some kind of more-mainstream C++ or Java experience, or even Microsoft's Visual Studio .net coding. They just wanted to jump right into Apple's Xcode, but they were perplexed with Apple's Objective-C. Others could pick up the Objective-C syntax easily, but were then stumbling on the Xcode IDE or the iTunes Connect app upload processes. Therefore, we have shared all relevant parts of our own internal new-hire employee training, previously highly guarded as company secret, but now shared with you our reader. If you are in that category, then this book is most certainly of very high interest to you. This extensive part of the book is effectively a mini-book in and of itself.

Another example, in part, is that this major work also contains extensive "how to" and "walkthroughs" on getting the app done and into the App Store, regardless of extensive programming knowledge. Similar to the other major sections, it too contains numerous step-by-step walkthroughs, complete with screenshots to guide you through many of its various topics. It also contains serious legal issues to avoid specific to the App Store from our attorney, and other more exciting topics such as promoting and monetizing your apps. These parts are, again, somewhat mini-books in and of themselves as well. And finally, as if that wasn't enough, this book is also a collection of our years of experience, knowledge, and lessons learned (with more than a little frustration) experienced from making useful utilities, silly apps to get a laugh, lucrative games of skill, and other custom software for business and personal use. This is a comprehensive guide based upon the collective experiences from a dedicated team, which has been a part of literally hundreds of iOS applications since Apple's App Store launched. It is their collective knowledge and experience – from our valued employees, both past and present - that is at the heart of this book.

With the evolution of the iPhone and other iOS based products changing the consumer market space, you have fortunately chosen a great time to turn your idea into reality. Depending on your skill level and budget, there is a wide variety of development tools and environments available to build and test applications. You do not have to be a hardcore programmer to make great iPhone or iPad Apps anymore! Within this book, we explore many of those tools in detail, and thoroughly explain how to best utilize them to build your own apps.

The chapters regarding Apple's App Store submission process are especially important, as there are many stages. Often, there are weeks of time waiting for approval before acceptance into Apple's program as a developer, and again long periods of time after submitting the app while waiting for Apple's approval for it to appear for sale. The contributors – a dedicated team of developers – have devoted a lot of detailed material to help you avoid many common developer mistakes, and make the highest and best use of your most precious resource: time. Each chapter reflects the experience acquired from our wasting valuable time submitting, having rejected, and re-submitting, many of our own apps to the App Store.

This multi-part Tome of Knowledge begins from a practical perspective from a "real world" experience of guiding you through everything, starting with the roadmap of creating an App Store specific business plan then leading directly into creating the developer account with Apple. Because the process is very involved, there are chapters devoted to each important stage. Many chapters also cover material that will help you make the practical business decisions vital to successfully submitting and publishing your work. These detailed guides will save you time, much frustration, and bring immediate value and power to your development effort.

For ease of navigation, this extensive tome is divided into six easy to follow sections that begin with the vital first steps which must be completed in order to meet Apple's requirements, and then proceeds with information on the iOS development tools that are the best for your needs and budget. It covers the most widely used development environments (there are many options to Apple's Xcode and Objective-C nowadays!), complete with emphasis on helping you choose the best tools for your level of skills and programming knowledge, if any. These guides target a wide range of people from those who are not programmers, to those highly skilled programmers looking for tools they can easily and most efficiently use to develop their app – as quickly as painlessly as possible.

In summary, this book provides everyone – both the programmers and non-programmers alike – with the ultimate app roadmap: **From an experienced app company's team, who decided to "tell all" and share everything it knows with you, our valued reader**. Now, finally, our internal secrets have been revealed, and happily provided to you, within in this extensive 'Tome of Knowledge'.

We sincerely thank you for your support, and wish you all of the best on your own roadmap to "h**App**iness" and success!

        Best Wishes,

        David Rajala
        President
        Unknown.com, Inc.

Introduction

Encompassing the core development framework of the Apple iOS 6 SDK and Xcode 4.5, *"Producing iOS 6 Apps: The Ultimate Roadmap for Both Non-Programmers and Existing Developers"* makes every effort to provide the widest possible range of mobile application, or "app",-creators with the core understanding, and recommended tools, all with easy to follow step-by-step examples. For the experienced, the book also includes a very detailed Objective-C jumpstart guide to help existing programmers easily begin preparing native apps quickly. This book was written with the intent as the must-have guide to producing apps for the Apple family of devices: the iPhone, the iPad and the iPod touch. ***This is effectively two books in one inclusive compendium.***

Originally generated as an internal-only, and highly-confidential, training guide for onboarding new employees into the company's mobile app development lifecycle and methodology. Serious book production began shortly after a major app release party when a beer-fueled intern asked to use the training material as a source for his Master's thesis. Additional publishing motivation came from the realization that a large number of iOS app development books shamelessly spamming today's market merely contain such hilarious paraphrases similar to: *"Sketch out your app. Hire a contractor for tens of thousands of dollars to do absolutely all of everything for you. Get ready to 'get rich quick'! The end."* How complete is a guide that skips all details, misleads with overly unrealistic statements, or even lacks the guidance on how to do it yourself? Simply horrific! The following week, all the Unknown development staff started learning to write prose in addition to code, and began working towards bringing these confidential internal training materials into something to be proud of.

It is from all of the team's combined years of experience that this complete guide and reference to absolutely-every-single-little-thing required to deliver your app idea into the Apple iTunes App Store as quickly, painlessly, and profitably, as possible.

This book covers the following topics:

- Performing market research and analysis for a successful app and a solid business plan.
- Using Advertising: iAd Mobile, Google Admob, Millennial Media, Jumptap, Smaato, Greystripe, AdWhirl, MobClix, etc.
- Making critical business & legal decisions: Trademark, Copyright & Open-Source Freeware.
- Walking through the Apple Developer Program through the creation of an Apple ID and other Developer Portal accounts required for iTunes Connect and the online App Store.
- Authoring in iOS 6, Xcode 4.5, Objective-C and Third-Party tools as Unity 3D, ShiVa3D, PhoneGap, MonoTouch, Marmalade, Adobe Flash Professional, Cocos2D, Corona SDK, GameSalad, Titanium Studio, MoSync, etc.
- Designing views, interfaces, images, controls, objects, classes and user input touch and gestures.
- Managing memory and dealing with data types, databases and storage.
- Submitting your app and marketing via social networks and various media outlets.

- Including a BONUS chapter of "real-life" GOTCHA items and how to avoid them.

Thank you very much for your kind consideration of the hard work and earnest efforts put forth in producing this work.

How to Use this Book: The Sections

This book contains an enormous amount of information!

The overall "flow" of this book contains related Chapters into Sections, presented in a logical progression of what YOU need to be successful. Many of the Sections are available as a reference guide to provide information on not only how to plan, design, create and upload the app successfully; but also to provide detailed material to help decide to use the Apple Development tools or use one of the many top non-Apple Development Platforms.

Next, extensive materials are provided, to help select and use the development platform of your choice: Apple or Non-Apple.

This book has thoughtfully divided its content areas into the P³ groups, in a total of 6 sections, for easy reading and quick reference - As shown in the figure below:

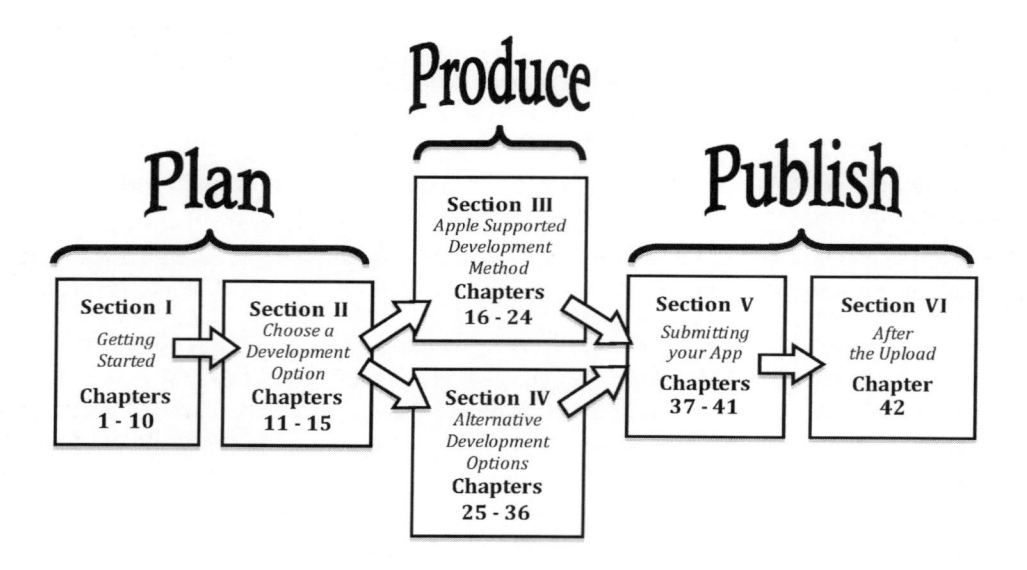

```
                        Produce

      Plan                                        Publish

  Section I     Section II    Section III    Section V    Section VI
   Getting       Choose a    Apple Supported  Submitting     After
   Started     Development    Development     your App    the Upload
               Option         Method
  Chapters     Chapters       Chapters        Chapters     Chapter
   1 - 10       11 - 15        16 - 24         37 - 41       42

                             Section IV
                            Alternative
                            Development
                              Options
                             Chapters
                              25 - 36
```

Section I: Getting Started
Chapter 1: Turn Your App Idea into a Plan for Success

Best described by the old proverb, the reality of success in the mobile App Store market, "Those that fail to plan – plan to fail".

xxxii   Introduction

This tome begins with a reality check in the first Chapter: Turn Your App Idea into a Plan for Success. While there are hundreds of competing books that focus only on app programming, or just basic marketing, those other books almost entirely neglect the App Store market and its unique issues; the app's pre-launch, release, and post release marketing.

Many of those other books have also commonly neglected critical business issues specific to working with the Mobile market and specifically the App Store, such as their failing to offer focused business plan suggestions and addressing legal issues relating directly the App Store. In Getting Started, these vital topics are covered as well as other important areas (all very important, but often ignored). This book strives to fill all of those gaps, and more.

Essentially, whenever building any new business product, service, or decision, it is always a good idea to begin first with a plan of action – for no matter what happens. In the world of developing and publishing iOS apps, it is especially important to think ahead to ensure that your idea is *technically possible,* and that the ultimate goal of monetization is achieved in the most effective way possible; while at the same time, having a logical and pre-planned exit strategy to eliminate emotions in the unfortunate event of failure. This is covered in great detail.

Chapter 2: App Monetization Techniques

Since you are reading this book, it is safe to assume you want to make money. There is a wide variety of ways to make money in the Mobile App Marketplace explored here.

This chapter explores ways to earn money from offering Free Apps, Using In-App Advertising, the wide variety of ways to provide income from offering In-App Purchases and In-App Subscriptions - as well as just selling the app outright.

Chapter 3: Getting Started With In-App Advertising

A continuation of Chapter 2, embodied here are discussions and ever growing ways that companies provide advertising for and within your app.

Additionally, explore the increasing importance of Advertising Networks and highlight several specific services available to you including: Apple's iAds, Google's Admob, Millennial Media, Jumptap, Smaato, and Greystripe

Chapter 4: Advertising Aggregation

Yes! Continue with understanding alternate ways to make money! In this chapter, a focus on the advertising options available and, an exploration of common mobile techniques: Aggregators vs. direct Ad Networks. Highlighted are several of most established aggregators, including: AdWhirl, MobClix, and the ways to use built-in Ad Aggregation.

After this chapter: You will know a lot about in-app advertising!

Chapter 5: Critical Business Decisions

There is no escape from the legal issues that have to be addressed, especially now with hundreds of thousands of apps and hundreds of thousands of lawyers! Please note that this book is not giving legal or other professional advice. The contributors are merely offering some hypothetical scenarios, based on what may or may not have been experienced.

For example, the author of this book, Unknown.com, Inc., did indeed follow its own advice, and hired an intellectual property attorney (Mark Young, P.A.) for a complete review of this entire book! Mr. Young shares some of his own contributions, as well.

With disclaimers in mind, this book bravely shares the contributors collective experiences combined with many others. These developers' unfortunate hypothetical situations you may want to avoid are shared, and some general suggestions about using good judgment and basic common sense are made. It cannot be stressed enough how important it is to respect and protect not only your intellectual property rights, but those of others as well.

Chapter 6: Intro to the Apple Developer Program

This chapter is where the fun begins! It starts with exploring and showing you how to conquer the Apple Developer paperwork. This includes exploring the reasons behind why it is so very important to determine the Account Type (e.g. individual vs. company), how to create a separate Apple ID for the developer program (important for many reasons), and other key topics. There are also areas that are important to avoid potential Signup Delays, the Agree and Agree to Agreements and of course instructions on paying for the Program with a useful signup Walkthrough Summary.

Chapter 7: Agreeing to Contracts and Setting-up Financial Information

Unfortunately, there is a lot of paperwork in any new venture: but it is not so bad, especially if following this roadmap to guide you through it. Once approved for Apple's iOS Developer Program, this book guides you throughout the additional steps required, so that your developer account can be ready to accept payments or distribute paid apps.

Once all of those steps are completed, the account is ready to start submitting apps for distribution and to earn revenue.

Chapter 8: Reserving an App Name

Before beginning development and creating necessary assets such as graphics, it is wise to first search for competitors to ensure you are not using a confusingly similar name (as discussed in Chapter 5), and to reserve that app name within Apple's iTunes App Store. This helps to ensure that the name is available and that another developer has not or hopefully will not take that name before you are ready to release your own app. Nothing is worse than spending time and money on competitive analysis, name research, resource creation, and its promotion and marketing, etcetera: and then realizing that the desired app name is already in use.

The concept of reserving an app name seems simple, but the process required to accomplish that task, starting with creating an App ID through the iOS Provisioning Portal, and creating a new App Page in iTunes Connect, requires careful thought.

Chapter 9: Preparing for Development

An extensive chapter, providing detailed explanations and precise guidance throughout requesting a development certificate, creating a development provisioning profile, adding a device for development, and ultimately testing and deploying your app.

Chapter 10: Checklist – Transitioning to Development

This chapter provides a brief summary of Section I, complete with a checklist for you to make sure that you have learned what is needed to move on to Section II.

Section II: Choose a Development Option

This section helps guide you through the process of deciding on which development platform to use, based on your existing experience and coding or scripting language knowledge.

Chapter 11: Introduction to iOS Development Options

In this chapter, you are introduced to several different types of development options which should help you gain some insight as to whether or not you want to create your app in what's referred to in this book as the Apple provided and officially-supported method, using the official Apple Xcode IDE and Objective-C programming language, or one of the several other third-party options to create an app that are also covered in this book.

Chapter 12: Cross-Platform Development Options

This chapter provides advantages and disadvantages to cross-platform development (e.g. write an app once, and have it work one both Android and iOS without any further time or effort), and why it's an important factor in the decision between the different development options. If you are seeking to develop for multiple platforms and not just iOS – all at the same time for maximum efficiency and monetization - then this chapter is a must read.

Chapter 13: Apple Recommended Development Method

This chapter covers Apple's officially-supposed Xcode, and why it is considered by many to be the preferred development platform for iOS app development. A brief overview of some of the major features of Xcode is discussed as well.

Chapter 14: Overview of non-Apple Development Options

This chapter lists many of the popular development option alternatives to Apple's provided and officially supported Xcode. Along with each third-party alternative, is a brief overview that discusses the supported platforms, approximate costs, and programming languages (if any!), required to create an app within that particular solution. As an important aside, some third-party options do not require programming (or very simple scripting), thereby enabling programmers and non-programmers alike equally competitive opportunities on the App Store.

Chapter 15: Development Options Summary

This chapter contains a quick-reference chart that helps summarize the information provided in the previous chapter, to better help you decide which development option is right for you at a glance. The development option choice is such a major decision, and a huge commitment of your time, money, and resources, that it is definitely worthy of your time.

Section III: Apple Supported Development Methods

Section III covers much of what you will need to start developing and creating Apps, using Apple's officially supported Objective-C and Xcode development tools. This entire section is more of a mini-book, geared towards those with at least a little existing programming experience. If you have no interest in the Xcode development environment, or in the Objective-C programming language syntax, then you can just skip right to the next Section.

Chapter 16: Introduction to Xcode

This chapter will be your first look at Apple's Xcode IDE. Some of the main features of Xcode will be explained in this chapter. There will even be a brief walkthrough on how to install Xcode as well as a brief "Hello World" first app! After this chapter, you should be able to make your way around the Xcode IDE well enough to start building apps.

Chapter 17: Introduction to Objective-C

In this chapter, you will be introduced to some of the basics of Objective-C; the programming language you will be using to create "native" iOS apps. If you are a programmer that already knows an Object-Oriented Programming language such as Java, C++, or C# then many of these concepts will be familiar to you already. However, you will notice the syntax is much different, so this chapter provides an important crash-course and jumpstart guide, to quickly understand this unique programming language.

Chapter 18: Foundation Framework

This chapter builds on the prior Objective-C chapter, and contains an overview of some of the main classes that are part of the Foundation Framework. It consists of Data Types and Storage classes, Utility classes, and Collection classes.

Chapter 19: App Features, Structure, and Lifecycle

This chapter focuses more on the structure and lifecycle of an iOS app by going into concepts like views, data source, and delegates. Also explained in this chapter is the process of creating an interface for the app either manually or with Xcode's Interface Builder.

Chapter 20: iOS Design Basics

This chapter builds from what you learned in the previous chapter and goes into more detail on designing views and structuring your app with different view controllers. Also contained in this chapter is an example app that takes much of what you learned in the previous chapters and puts it to use building a simple app.

Chapter 21: Data Storage and Handling

This chapter contains information that you will need if you want to store and retrieve any data for your app. It covers file and network I/O, app settings, SQLite, and Core Data.

Chapter 22: Important Frameworks

This chapter reviews some important frameworks, and some examples of their functionality are shown. Frameworks included in this chapter are the: Address Book, Core Data, Core Location, Event Kit, AVFoundation, Core Audio, OpenAL, Media Player, Core Animation, Core Graphics, and UIKit Frameworks. Some of the important frameworks are reiterated, while others are introduced to lead-in for discussion in upcoming chapters.

Chapter 23: User Input

This chapter will teach you all about user input. Whether you want certain touch functionality such as multi-touch or gestures, or you want to have camera and microphone functionality in your app, this chapter will show you how. Also included in this chapter is information on positional data, compass, gyroscope, and accelerometer functionality.

Chapter 24: Instruments

Is something wrong with your app and you cannot figure it out? Is your app running slow and you want to optimize it? Then you might be interested in this chapter, which covers many of the high tech instruments that come with Xcode. Using these instruments you will be able to view important memory related information pertaining to your app, such as the amount of memory allocations or if there are any memory leaks. Other important uses for these instruments are if you want to check file activity, system usage, or if you want to see if there are bottlenecks in certain parts of your app. NOTE: Viewing app Crash Reports is covered further in Chapter 39.

Section IV: Alternative Development Options

Section IV contains several different alternatives to the previously discussed Apple officially provided and supported method of app development, as found in Section III. Many of these third-party tools have different specialties and purposes. Some are aimed towards 3D game design, some aimed towards multi-platform functionality, while others just make it easier to start making iOS apps quickly with straightforward scripting knowledge, and even solutions for non-programmers. NOTE: Additional non-programming app-creation solutions are covered in the Appendices.

Chapter 25: Unity3D

This chapter walks through Unity 3D installation and creating a simple game. Afterwards it will show you how to build the app for the desired platform (e.g. iPhone iOS; and yes, it is a cross-platform compiler, so Android or others is also possible). After completing this chapter, you should be familiar with the Unity Editor, some simple scripting and well on your way to making your own games. The Unity Editor can be downloaded and run on a XP SP2 or Later; Mac OS X: Intel CPU + OS X 10.6 or later.

Chapter 26: ShiVa3D

In this chapter, you will learn about ShiVa3D and be walked through the process of installing it, creating your first app, and building something for distribution on the App Store. In order to use ShiVa3D you will need a computer with Windows to run the editor and when building the app for distribution you will need a Mac with OS X 10.6 or higher.

Chapter 27: PhoneGap

In this chapter, you will learn about PhoneGap while being walked through the process of its installation, creating you first app, and building it for distribution on the App Store. To create iOS apps with PhoneGap, you will need an Intel-based computer with Mac OS X 10.6 or higher. Other platforms such as Blackberry and Windows Phone may require Windows.

Chapter 28: MonoTouch

In this chapter, you will learn about the MonoTouch SDK, the Mono framework and how to use them to create iOS apps using C# and .NET libraries. *Existing Microsoft .NET developers may like this option.*

Chapter 29: Marmalade

This chapter covers the cross-platform development tool Marmalade. You will be guided through the process of installing Marmalade on either Mac OS X or Windows 7 (depending on your preference), and then walked through your first app; afterwards you will be shown how to deploy it to an iOS device and how to build the app for distribution on the App Store.

Chapter 30: Adobe Flash Builder

This chapter will walk you through using Adobe Flash Builder to create your first iOS app using ActionScript and MXML.

Chapter 31: Adobe Flash Professional

This chapter focuses on the Adobe Flash Pro CS5.5 or CS6 platform, utilizing Adobe Air. You will learn how to create your first iOS app and get it ready for deployment on the App Store.

Chapter 32: Cocos2d (via Xcode)

This chapter walks through installing and creating an app with the Cocos2d framework as designed for building 2D games. In addition, you will learn about "sprite sheets," and how to create an animated sprite in Cocos2d.

Chapter 33: Corona SDK

Within the Corona SDK, you will find a complete walkthrough detailing how to create your first app with Corona SDK as well as instructions on how to build the app for distribution. This chapter also includes information on the advertising services supported in Corona SDK.

Chapter 34: GameSalad

This chapter will cover the GameSalad game or multimedia app creator. This is a unique chapter since **this tool requires absolutely no programming whatsoever** to create an app, such as a game. You will be guided, in detail, through the process of creating a game - and testing it on an iOS device - and then distributing it on the App Store.

Chapter 35: Titanium Studio

In this chapter you will be walked through the process of installing and setting up Titanium Studio - and creating your first iOS app with it. Then, you are guided through the process of testing the app on a device - and how to prepare it for distribution through the App Store.

Chapter 36: MoSync

This chapter focuses on the MoSync SDK, which can be used to create cross-platform apps with HTML5 and JavaScript, C/C++, or a combination of both. You will be walked through the process of creating your first app with MoSync using C/C++. *Existing web developers may like this option*.

Section V: Submitting Your App

Section V covers the process of submitting your iOS app to Apple for review, and ultimately public worldwide distribution through the App Store.

Chapter 37: App Submission Overview

This chapter is an overview of the different options for uploading an app to the App Store. Some apps can be uploaded directly through Apple's Xcode 4.x (v4.5 required for iOS 6), while others will require the use of Apple's latest Application Loader tool (both require OS X).

Chapter 38: Preparing For Submission

This chapter focuses extensively on the steps required for preparing your new creation for the app submission process. Some of the important topics covered here include: avoiding missteps that can cause instant rejection, various technical reasons for rejection, provisioning profiles, verifying the app's metadata, etcetera.

Chapter 39: Submitting Apps Using Xcode 4

This chapter contains a walkthrough on submitting apps with Apple's Xcode 4.5, and some of the best practices to do so. Also, even after the app has been approved for distribution: view crash reports for your existing apps, so you know which ones to fix first as time goes on, depending on the severity.

Chapter 40: Submitting Apps Using Application Loader

This chapter contains a walkthrough on submitting and uploading apps to the App Store with Apple's Application Loader tool. This is same tool used for uploading apps built with some of the third-party development opinion alternatives to Apple's Xcode, and or apps otherwise incompatible with Xcode.

Chapter 41: Bonus Chapter of Gotchas!

This bonus chapter covers a wide variety of "gotchas" to avoid. Learn from real-world and theoretical scenarios of painful mistakes. Just a few: There can be only one Team Admin, app-transfers between two developer accounts is not supported at this time, etcetera. The big trouble any developer can get in to with scamming the ranking system, or spamming the App Store, are also covered here for you to help avoid.

Section VI: After the Upload

Chapter 42: Post Upload Marketing

In this chapter, you will learn about what you should do once your app is uploaded to the App Store. Content covered in this chapter consists of information on sticking to your business plan, marketing your app, creating a press release, and setting up an advertising campaign.

SECTION I

GETTING STARTED

In this first section, an overview on how to develop a working plan to bring your app idea into a successful existence is provided in detail. Provided are suggestions for you to consider when developing your own app-focused business plan. This planning stage is very important, as it allows you the opportunity to methodically plan for the development and life of your app.

One factor of success that many app developers fail to take into consideration is the existing market of "what is hot" and "what is not". Suggestions are offered on the use of tools for you to perform some product research prior to investing the time and effort on an app to help ensure it's not for a dying niche. **Additional resources are listed in the Appendices.**

One of the main reasons most people wish to develop an app for the iPhone or iPad is to make money. Being an app developer corporation ourselves, this is clearly understood: Some best practices around utilizing advertising networks and in-app purchasing is included. Different advertising models and how to best choose the right blend to monetize your app is also included.

From experience, a good deal of content and several theoretical scenarios on potential legal issues (grimly dubbed "Legal Landmines" whenever seeing someone side-blinded by them) that developers may encounter and want to avoid is included. So important, in fact, that a full chapter devoted to the subject, in which offers insights into the topics of copyrights and trademarks. As if that wasn't thorough enough, the perspectives on these topics are provided directly from the Unknown.com, Inc. lawyer, Mark Young, P.A.

Finally, after taking all this planning and research into consideration, you are walked through the requirements and essential steps necessary to set up your Apple developer account – ***as quickly as possible***. Upon the completion of this section of the book, you will be functionally ready and well prepared to start the development of your app.

Turn Your App Idea into a Plan for Success

It is absolutely essential to create an app business plan. Your plan does not have to be a 1,000-page corporate nightmare that requires an MBA and an accountant to decipher. The plan is just a written review of your app idea, answering many questions.

Perhaps, two of the important reasons to have an app business plan - and to sticking to it - may include scenarios from a positive side, as well as a negative side.

On the positive side, say the app becomes successful and is downloaded a few thousand times with good reviews. This can cause anyone to become so excited, that they fail to perform many of the strategic steps recommended in order to maximum its exposure - which may have brought the app to a higher status - which could have, perhaps, resulted in the app being downloaded a million times instead!

On the other hand, perhaps the app has not done so well no matter what is done, and it's simply not going to be successful for whatever reason. The app's business plan (which should contain, among other things, when it was previously decided up-front to halt the investment and cut losses in the event of failure), can be absolutely instrumental in overcoming anyone's burning desire to continue to burn through resources and money trying to make the app successful until bankruptcy.

In other words, whether it's letting go and cutting losses, or sticking to the post-release checklist in the event of success to maximum revenue, it is hard for anyone in a time of crisis to think clearly - that's just human nature. Either way, the business plan will greatly help to remove the emotion from the logical process.

Does your app, or something similar, already exist? Go look! How will your app stand out and succeed (or, depending on your attitude, "crush and destroy them" as determined by your personality type)? Even with the best ideas, it is amazing what is popular and sells, versus the outstandingly high quality apps that outright fail miserably. It is critically important to investigate what the customer wants (or even better, thinks they need), and ensure that your app fills that missing hole.

Did you check both App Stores? Apple's "iOS" (mostly mobile) iTunes App Store, abbreviated simply as "App Store", is for devices running Apple's iOS. Apple's "OS X" (desktop/laptop) iTunes Mac App Store, abbreviated simply as "Mac App Store", is for Macintosh computers running Apple's newer OS X. Some apps may be made for *both* platforms, by using specialized products called *cross-platform compilers* and other methods as mentioned in this book. So, whenever performing competitive analysis, do not forget to check *both* stores – they're separate!

Your app needs to not only be good as a standalone computer program, but more importantly –it needs to fill a need - that meets the Apple guidelines - and is technically possible to develop. It is amazing how many great developers (and large companies) never check to see if there are similar apps that already exist and how well they are doing prior to blindly entering the fray.

If there are similar apps to your idea, do not fret. Just download them – all of them – and learn from them. Competition is inevitable. It is often a signal that your idea is a good one. You need to learn how to make your app fit into the current list of existing apps, and of course, how to make it stand out.

The lack of any competition or similar apps is potentially a warning signal. If your idea is not "out there," then it might not be commercially viable, or more often than not, it's because it isn't permitted under Apple's guidelines (and everyone who has submitted such an app was rejected). Find out why. The answers may lead to an even better avenue for your idea to work around the issues, and to be successful. One of the best chances of any product's success is always to be first to market!

These are the first few questions to ask before spending time and money on an idea that may be great on paper, but not so great in the marketplace. Most people get hung up on the paperwork, talk the idea to death to everyone they know, but then never get started on the actual project. Don't get hung up on total paperwork overkill – by its very nature, paperwork is for most folks outright discouraging. Find a balance for yourself that only you can answer – a business plan that is as long or as short as you believe will be beneficial to you - so now, Journey Onward!

Everything, ranging from great inventions and products to new services and apps, starts with an idea. Once you have an idea for your app, it is important to analyze what it would take to create your app, your potential competition and your potential market. But you also need to look into development time, cost (research, development, marketing), and your anticipated Return On Investment ("ROI").

One of the first and most important decisions you should make, prior to signing up for any accounts anywhere (including but not limited to a separate checking account for your new venture), is deciding if you are going to conquer the App Store as an individual or under your own company.

> ***Note that nothing in this book provides any tax, legal, insurance, or other professional advice, nor does it attempt to replace your licensed advisor, and should not be construed as such.***

This book is not intended to replace any professional advice of any such professional whatsoever, including but not limited to your lawyer, accountant, insurance broker, etcetera, and nothing in this book is to be construed as such. For obvious liability reasons, this book isn't attempting to replace your bar-certified attorney, your certified public accountant, your licensed and registered insurance broker, or other licensed professional, etcetera. You should always consult a lawyer or accountant if you face any legal or accounting issues, such as the legalities and taxes of establishing your business and the promoting of your app in your geographic location.

You should also discuss professional, general, errors and omissions, and intellectual property liability insurance options, along with other options as they may pertain to you, with your duly licensed and authorized insurance broker and underwriter. Experience has shown Apple's sensitivity to the regional legal issues as extreme, for they clearly have a lot to lose. You have been warned.

It is absolutely critical to decide if you are going to sign up all of your accounts as a business or as an individual in advance, because all of the relevant account types require you to sign up as either a business or an individual up-front, and most cannot be changed later. Specifically, the primary Apple Developer account that you will need to distribute any app on the App Store, cannot be switched between individual or company status after it is initially set up! Thankfully, setting up a new business and its separate checking account is a relatively fast and straightforward process in most states.

The U.S. Small Business Administration website (http://www.sba.gov/) offers a well-organized resource of documents and guides for starting your own business and, if so desired, also offers resources for getting out of the business and planning your exit strategy - whether that be selling the business for a huge profit, or calling it quits as painlessly as possible if things do not work out.

The SBA website has articles on everything from helping to guide you in forming your business, to selling your business. Before you undertake any activities, the one thing all sources stress is to reach out for some expert professional advice from accountants, lawyers, bankers, insurance agents, tax experts, and other professionals as appropriate.

Not to worry! Feel free to bookmark and pause for a moment on the business plan right here, and jump ahead to review Chapters 5 and 6, to reflect on some of these topics discussed at great length. Those two Chapters provide advantages and disadvantages, as specific to the App Store, and should help you, along with your licensed professional advisor, decide on your new app venture being a company or individual business status. Then, come back here to Chapter 1, with that decision made, so you can be in the correct mindset for forming your business plan.

Additionally, the remaining getting-started chapters, specifically Chapters 2, 3, and 4, will all assume that you've already pre-made the decision in advance, on whether to sign up everywhere under individual or business account types.

> **Return here after reviewing the App Stores issues pertaining to company vs. individual account types in Chapters 5 and 6.**

1.1 Starting Your Own App Business

Assuming you've now at least skimmed chapters 5 and 6, begin with visiting one of the top authorities in the United States for starting your own small business: The U.S. Small Business Administration ("SBA"). Specifically, the SBA's article #2815, "Follow These 10 Steps to Starting a Business", located at:

http://www.sba.gov/sba-direct/article/2815

Of course, this is not some random generic or basic getting-started business book! Within each of those 10 SBA steps, ***the pointed comments will draw your attention directly to the specific App Store issues***. In addition, to every SBA "step" is provided commentary on each item to help place you in the mindset for thinking of them ***as relevant to the App Store.***

IMPORTANT: Even if you do not plan on forming your own company for your App Store venture, you will still find this information extremely helpful for your individual goals.

1.1.1 SBA's Step 1: Write an App Business Plan

So important, in fact, it is the focus of this first chapter as well. Here's the link:

http://www.sba.gov/content/templates-writing-business-plan

NOTE: Notice there's even a link to a wizard-oriented business planner, to help guide you through creating the business plan. To help with the "business planner", jot some of your ideas down, and focus on the body of the document to get started.

- First, the description of your business. You're already done with that step, because you know it's going to be distributing Apps on the App Store! See? That was easy!

- Marketing. This is covered within this book, as specific to the App Store. For now, just think of some basic budget of both time and resources you are willing to spend.

- Competition. Again covered in this book, determining your competition is important to any new venture. Take a quick look through the App Store now, searching for random apps that may be similar to your idea, and jot down a few competitors that you may be up against.

- Operating procedures. As pertinent to the App Store, a great start would be how you are going to obtain assets for your app idea. For example, will you be programming it solo, or simply using one of the many app resources mentioned within this book to create the app entirely by yourself?

- Personnel. If you decide to have others work on the app for you, will you contracting it all out, or are you thinking of hiring employees?

- Business insurance. A very important decision indeed! After reading some of the issues in Chapters 5 and 6, you may want to consider an insurance policy covering issues specific to the App Store. In addition to an ordinary "general liability" policy, which may also be required by landlords, it's equally if not even more beneficial to obtain an "intellectual property" policy as well. If shopping around for an insurance policy, some of the more relevant coverage's to the App Store may include, but probably are not limited to: trademarks, copyrights, and patents. Sometimes these are covered by what's known as "errors and omissions" policies, but quite often they are not. Be sure to ask the underwriters you are working with those specific questions!

- Financial data and Supporting Documents. Only you will know the budget you have in mind for your new venture on the App Store. One of the important things you should do in this section is to make the decision now, up front: in a worst-case scenario, how much are you willing to lose before you decide to cut losses and walk away? Deciding this in advance, before getting started, is reiterated several times throughout this book. It truly will help remove the emotional attachment to letting go of your hard work, if that unfortunate need arises.

NOW, that you've jotted down some notes as relevant to the App Store, you are ready to tackle the SBA's guided "business planner", as located at:

http://web.sba.gov/busplantemplate/BizPlanStart.cfm

Okay! With your Step #1 Business Plan in hand, review the remaining SBA steps with a focus towards the App Store!

1.1.2 SBA's Step 2: Get Business Assistance and Training

This is nothing less than astonishingly fantastic - free new business owner training! You've already shown the Entrepreneurial Spirit mentioned in their video, proven by the fact that you've picked up this book! One of the contributor's favorite catch phrase is, "If it's free, it's for me!" So how could you go wrong with free training, courtesy of the SBA?

1.1.3 SBA's Step 3: Choose a Business Location

You may already know the answer to that: The App Store!

1.1.4 SBA's Step 4: Finance Your Business

If you are thinking really big, then to make your vision a reality it may need some financial help. This can help you get started.

1.1.5 SBA's Steps 5 and 6: Determine the Legal Structure of Your Business; and; Register a Business Name ("Doing Business As")

This decision is, quite frankly, absolutely critical and a grouped together here as they are so closely related.

The topic of forming your own legal company business entity under the guidance of licensed professionals is discussed at great length, and its importance as relevant to the App Store is reiterated several times throughout this book. For example, as discussed in later chapters, Apple does NOT currently accept App Store Developers who wish to register as sole proprietorship's under a simple "Doing Business As" (DBA) registered aliases of individuals. Also note that, depending on the State you register your legal business entity (e.g. S-Corp), there may or may not be a separate requirement for a DBA registration if you are already a bona-fide corporation. Many online websites provide inexpensive legal services for forming an LLC, corporation, etc. Be sure they're credible!

Perhaps you've seen television commercials for inexpensive flat-rate legal documents from places such as, by way of example only, Legal Zoom, located at www.legalzoom.com. However, often times, a lawyer's advice is to ask a lawyer BEFORE getting yourself into something that you do not fully understand. In relation to this scenario, it would be best to ask a lawyer in your local area for advice if you are unsure how to proceed in your area.

1.1.6 SBA's Step 7: Get a Tax Identification Number

This is fantastic advice, especially since you may not want to go giving out your own personal social security number to all of the revenue generation sources that will be covered in upcoming chapters! Not just the App Store, but also all of the other websites out there, who need a tax number for reporting your revenue. For one thing, this is not just a privacy concern; it's also an identity theft concern. According to some sources, identity theft is the #1 crime in the United States. NOT giving out your social security number, birthday, etc., may help you to protect yourself in this regard. Additionally, your own company having its own Tax ID (e.g. EIN) can be beneficial for tax reasons, as should be discussed with your own accountant.

1.1.7 SBA's Step 8: Register for State and Local Taxes

Certain aspects of licensing may or may not be necessary, depending on your local state laws, and the avenues in which you plan to monetize your work in the App Store. For example, Apple's iTunes Connect payment system, automatically charges local state sales tax and pays it themselves, as part of the service that they provide to United States (US) developers! Apple also shields developers from many customer service issues, because as you will find, your customers will typically contact Apple iTunes support first. However, if you are looking to sell any Apps directly to end-users, by way of example only Google Android apps through the Google Marketplace under your own direct-sales account, then you'd almost certainly need a sales tax license to collect and pay those State sales taxes manually yourself. Either way, there may be other tax or license issues to consider as well. Regardless, this is where your accountant should help guide you for your local area.

1.1.8 SBA's Step 9: Obtain Business Licenses and Permits

Some new business owners erroneously believe that by forming a C-Corp, LLC, or similar legal entity, that they do not have to obtain additional local licenses or pay local taxes for their county, etcetera. Depending on your area, this may or may not be true, but can be as straightforward as going to the local branch office, telling them what your business type is, paying another tax license fee, and walking out with a paid certificate on the spot. Of course, every State and County is different, so again, you will want to ask your accountant.

1.1.9 SBA's Step 10: Understand Employee Responsibilities

This may not be on your immediate agenda, but it's good to know as your company grows! Note that there are many reputable payroll service companies out there today, which run full payroll services, send all tax forms and can even file your tax returns on your behalf, thereby essentially providing complete payroll services. Perhaps that's a bit more expensive, but it would allow you to focus more on what you do best: producing Apps!

That's it! Now, you are really getting somewhere! In essence, your app business plan should have been created with app Production, Publishing, and Profiting (P³), through Conceiving, Creating, and Capitalizing (C³) on the opportunities you've seen from your App Store research in mind. The remainder of this chapter, and upcoming getting-started chapters (specifically Chapters 2, 3, and 4), will all assume that you've already pre-made the decision in advance, on whether to sign up your new App Store and related accounts under individual (e.g. sole proprietor) or company (e.g. business LLC, S-corp, etc.): ***Account Types.***

1.2 Getting to Know the Process

In order to distribute an app through Apple's App Store, Apple must review and approve the app for distribution. Developers are required to abide by Apple's Human Interface Guidelines found at ...

https://developer.apple.com/appstore/guidelines.html

Direct link:
https://developer.apple.com/library/ios/#documentation/UserExperience/Conceptual/MobileHIG/Introduction/Introduction.html

... as well as Apple's Terms of Service at: http://www.apple.com/legal/itunes/ww/.

If an app does not conform to these guidelines, Apple will turn it away. Before beginning development, read through their guidelines to be certain that your idea satisfies them; otherwise you risk rejection. For example, within Apple's documentation, it states that apps which promote illegal activities will be rejected. Once you are certain that your idea meets Apple's requirements, the next step is to analyze its technical feasibility. This is where it is important to research and explore the features and technical limitations of Apple's devices.

Once your idea meets Apple's expectations and is technically possible to create, it is now time to do some initial market research. There are various methods to conducting market research: browsing the App Store, utilizing web search engines such as Google, or using social media tools like Facebook and Twitter. Looking at available, popular apps will give you much insight. Read customer reviews on similar products – often you can see where apps are lacking and where their strong points are. You will gleam the elements of the missing niche – and you can create your plan around filling that niche. **Additional tools and resources to help guide you, some of which can provide extensive competitive detail, are contained in the Appendices towards the end of this book.**

1.3 Pre-Development Market Research

Conducting market research will teach you about your target audience. You will be able to create your app around the expectations and needs of your target market.

After your feasibility of success is established, and initial market research is complete; with business plan in hand, your app can now be crafted. Take into consideration your target market, original idea as well as the features and limitations of the devices themselves. With all these things in mind, you are now ready to create what's traditionally called an app "Features, Design and Specification Document", which will make tackling the project much easier down the road.

Before diving in to "mockup" tools for helping to draft an app interface, first decide on other factors which may affect the interface design itself: App monetization, covered in the following Chapter on app monetization techniques.

App Monetization Techniques

Monetization techniques available on Apple's iTunes App Store are constantly evolving. It is important to consider all options once you have an idea and before beginning development. Monetization can be tricky; it differs on a per app basis and requires constant monitoring, research, competitive analysis, and tweaking to find the best fit for each of your products. The different types of revenue-generating options available through the App Store include paid apps, in-app purchases, in-app subscriptions, "lite" and full versions, as well as in-app advertisements. Promoting your products using outside advertising campaigns is also an effective method to increase your customer base and revenue.

2.1 Market Research and Analysis

The first step to deciding on a monetization method for your app, or a combination thereof, is research. Once you have an idea for a product, it is necessary to again perform more competitive analysis. This time, take note of the quantity of similar apps that are already available on the store, the price or advertising usages within those apps, their features, ratings, and end-user feedback. This will give insight as to what is already available, for what price, as well as what the end-user ultimately wants and expects in an app of a conceptually-similar nature. If you find that your idea is unique, with no competitors, you may want to consider going the "paid" route as opposed to free. If you will be competing for market share and most of your competitors' apps are in the $0.99 to $1.99 price range, it may be advantageous for you to price your app as free and integrate banner advertisements or in-app purchases. It is important to note that prices can be changed at any time, however there is a huge penalty: when changing a free app to a paid app, or vice-versa, all user reviews, download ranks, and ratings will be reset! What? That is correct: Your app will start over again as if newly uploaded, as far as rankings are concerned, if you switch between free/paid or paid/free. Clearly, you will want to decide this up front, and hopefully never have to change it!

In general, free apps get more downloads than paid apps. This is due to the large number of apps available on the App Store and the idea that the end user wants to spend as little of their savings as possible. Why would a user pay $0.99 for an app that does the exact same thing as a free equivalent? In some cases, it may be useful to provide an option to the user by offering the ability to remove in-app advertisements through the use of an in-app purchase. If you have a large number of competitors, but your product is superior, then there are a number of different approaches you can take. These range from offering BOTH free "lite" and paid "full" versions of your app, incorporating in-app purchases/subscriptions, or combinations of the techniques.

2.2 Monetizing on Free Apps

After you conduct your market research and decide which approach to take, there are several ways you can additionally monetize your app: Utilizing in-app purchases, in-app subscriptions, or in-app advertisements. In many circumstances, it is beneficial to use a combination of these methods.

Always check the latest Apple guidelines for the latest rules, as there have been some controversy surrounding some of these more savvy tactics.

2.2.1 In-App Advertisements

There are different advertising networks you can choose to work with, both as detailed later in this book as well as others. After you sign up and create an account with one of them, you will receive an SDK (Software Development Kit) for integration that will enable your app to serve advertisements. Chapter 3 describes the varying advertising networks in more detail. While specifics vary between advertising networks, generally, a banner ad is 320 x 50 pixels in portrait mode and 480 x 32 pixels in landscape mode for non-retina iPhone and iPod devices, and 768 x 66 pixels in portrait mode or 1024 x 66 pixels in landscape mode for a non-retina iPad device (please see the complete charge of resolutions later in this book). Ad networks may also offer interstitial ads, which appear in between screens or at set intervals. Every time the ad appears, as well as each time the user clicks on the ad, a fee is earned.

Different ad networks offer different campaigns, which result in varying *effective cost per impressions* also known as eCPM. The ad network calculates this by multiplying the CPC, or *cost per click*, by the CTR, or *click through rate*, and multiplying the result by whatever number they use. In general terms, the more users you have viewing the ads and possibly clicking on them, the more revenue you will likely generate. It is important to note that if you choose the advertising-based approach, it is a good idea to integrate more than one ad network to account for unfilled or empty ad requests, using an *Ad-Aggregator*. This shall be covered in more detail in Chapter 4.

One note of interest, which have become somewhat main-stream for the App Store here in the year 2012: Many savvy developers now have their own internal advertising systems. This is to in effect build their own mini app-store, within their own apps which can be enabled or disabled remotely, usually for promoting their other apps within their own mini-store. For example, if they get lucky and have a top-ranked app, then they may move a percentage of banner advertising revenue over to self-promotion from their top-ranked app to their other apps; at the expense of losing out on a bit of banner-ad revenue.

2.2.2 In-App Purchases and In-App Subscriptions

Other ways to monetize free apps are in-app purchases or in-app subscriptions. In-app purchases are commonly used for unlocking content, or purchasing virtual items. One way to use the in-app purchase model is to offer the game for free but require a purchase in order to advance beyond a specific level. In-app purchases are useful in these sorts of situations because when the user becomes engaged, they will be more willing to spend money to continue playing (e.g. by unlocking new levels) or add to their overall in-game experience (e.g. by purchasing virtual goods). In-app subscriptions are useful for apps that provide services, such as a newspaper or radio station subscription. In-app subscriptions have the option of being one-time or auto-renewable.

Many savvy developers now provide the ability, using the in-app purchase features, to simply turn off their banner advertisements within their free apps, in exchange for a one-time or even subscription-based fee (e.g. no ads for $0.99 per year).

2.3 Monetizing on Paid Apps

Paid apps tend to do well if you have a unique idea or are selling something of value that is difficult to obtain elsewhere. Users will usually pay for something that provides valuable information, or for something that is completely new and groundbreaking. When faced with a list of possible apps to download, users will gravitate more towards the free or low cost apps. This is where your market research comes in handy.

Developers have the ability to choose the price they wish to charge for the app, ranging from free to US$999.99. Choosing a price for your app depends on several factors: competitor pricing, market research results, the amount of content available within the app, whether or not the content is accessible via other means, as well as desirability.

It is also possible to combine monetization methods. As discussed, a common way is to offer an in-app purchase to remove advertisements from a free app. Another, often more effective way, is to offer virtual goods or special levels as in-app purchases. These two examples offer revenue from both advertisements as well as from in-app purchases.

> **TIP:** Some of the Top-Grossing Apps on the App Store are using a savvy combination of all techniques, with many being "Free" downloads! Clearly, of the 3 charts (Free, Paid, Top Grossing), the one to watch ongoing for maximizing revenue and finding trends is Top Grossing!

Another possible monetization combination is to offer an in-app subscription, say for 30 days, to an online product belonging to one of your advertisers. During that 30 day period, your user has enhanced game play, no advertisements, or another bonus. Once the 30-day subscription ends, then the perk disappears. If they renew the subscription, the perk continues. If you follow the in-app subscription model, be sure to talk with your advertiser about being compensated for the one time subscriber who renews. In the subscription business, the renewal is crucial – make sure you discuss how you will benefit from the customer who converts from a one-time purchase to a reoccurring one.

2.4 App Store Monetization Conclusion

To reiterate, there are a variety of approaches to monetize your development efforts. It is a good idea to look into all of your options and to devise a plan before submitting your app for distribution on The App Store.

Advertisers are your friend. They want to help you. They want your app to be successful, and be downloaded as many times as possible, so they can display as many ads as professionally looking as possible. Ask them for help; almost always, you will find most of them more than willing to bend over backwards to help you.

Researching the App Store, researching your target market, and an honest evaluation of your proposed app from potential advertisers, will help you maximize your potential revenue.

Getting Started With In-App Advertising

Developers like you can monetize their apps through in-app advertising. In-app advertising services are available through a number of different companies in various formats and styles. Now it's time to roll up your sleeves and dig in to see how it's done!

3.1 Advertising Overview

As discussed in Chapter 2 there are several different advertising networks that offer developers the ability to sign up for an account, configure their app to serve advertisements, and generate revenue from those advertisements. This chapter will go over some of the current top advertising networks for developers. It is important to note that different advertising networks offer different services and advertising types, such as banner ads, video ads, rich media ads, and interstitial advertisements. Each of these advertising types are often available in a variety of sizes and in a variety of payout rations based on different criteria such as click thru or impressions. Generally, banner advertisements have the lowest pay ratio but they are the most consistent, while full screen interstitial advertisements and video advertisements have higher pay ratios, yet are not always available. There is a variance between the advertising networks so it pays to request information from a number of sources before making a decision.

When choosing the advertising networks you wish to incorporate into your apps, it is important to evaluate the different ad formats and types that are offered. Full screen interstitial and video advertisements are intended to display when transitioning between views. Banner advertisements can be configured to display whenever the app is running or only in specified views. Rich media advertisements are advertisements that engage the user in such a way that encourages user interaction and is not just a static image or text. Depending on the type of app you are developing, it may be useful to integrate a combination of the varying ad formats supported to help analyze what works best for your specific app. Advertising techniques should be addressed on a case-by-case basis and vary from app to app.

It is important to note that most of the advertising networks mentioned in this chapter offer services for developers as well as for advertisers. Services for developers include in-app advertising, while services for advertisers include creating and running paid advertising campaigns. The focus of this chapter is on the app-sales developer side of things, but it is good to know that advertising services are available through these companies as well and can be useful for promoting apps and increasing your user base. This is especially useful if you have an app do exceptionally well, and want to have your own banner ad for your app appear in competing apps.

All of the advertising networks mentioned in this chapter have their own developer library, or Software Development Kit (SDK), that must be incorporated into your app in order for ads to be served. The relevant information on how to incorporate these SDKs found at their respective websites. If you wish to serve advertisements from multiple advertising networks, you must use an advertisement aggregation service, which is discussed in Chapter 4, or implement your own solution. Note that various advertiser's SDK's may or may not be compatible with some of the development options mentioned within this book. Compatibility with your chosen development method option may very well be the deciding factor for your advertising solution.

3.2 Advertising Networks

In this section, you will investigate some of the top advertising networks that offer in-app advertising services for iOS. There are others, but these are some favorites. Please review the appendices for more information.

3.2.1 Apple's iAd Mobile Advertising Platform

The iAd advertising platform is Apple's own advertising platform for mobile devices. This platform allows developers to serve ads within their app once iTunes Connect is configured and the iAd framework is included in the apps project. Apple's iAd advertising service is available in the banner format on iPhone, iPod, and iPad devices running iOS 4.0 and above. At the time of writing this book, iAd advertisements running on iPad devices with iOS 4.3 and above have the ability to serve full screen interstitial advertisements. If the target iOS SDK is lower than 4.0, it is good practice to weak-link the library, making it optional. This is so that devices running a version of iOS older than 4.0 (e.g. the original iPhone model which supports maximum iOS version 3.13), will function as normal, but will not receive ads from the iAd platform. Remember, the lower the iOS version the better, in order to have the most device and therefore customer coverage as possible.

3.2.2 Google Admob

Google offers an in-app advertising solution through their Admob service available at http://www.admob.com. To serve advertisements from Admob, sign up for an account using a Google Account through the Admob website. Once the account has been created and you have logged in, add the app in which you wish to serve ads. This will then allow you to download their SDK as well as provide you with a unique identifier, called Publisher ID, which is required to serve advertisements. Details on integration are located on Admob's website. Admob currently offers only banner advertisements for iPhone, iPod and iPad devices. At the time of this writing, Admob offers full screen interstitial advertisements to a select group of publishers. Admob gives the developer the ability to serve advertisements from other Google advertising services such as AdSense and Adwords if an Admob ad is not available when requested by your app.

3.2.3 Millennial Media

Millennial Media is another advertising service that offers in-app advertising for iOS developers. Millennial Media's website is located at the URL http://www.millenialmedia.com. Signing up for a developer account through this website is required to utilize their in-app advertising service. Once an account has been created, apps can be added and the SDK can be downloaded for integration into apps. Millennial Media offers banner advertisements, rich media advertisements, and full screen video advertisements. They also give developers the ability to use alternative, third-party advertising networks to populate unfilled ad requests.

3.2.4 Jumptap

Jumptap provides developers with an in-app advertising service, which is available through their website: http://www.jumptap.com. Signing up for an account through Jumptap's website is required to gain access to their services. Once an account has been created, apps can be added to the account and the SDK can be downloaded for app integration. Jumptap offers rich media ads, banner ads, as well as full screen interstitial and video ads for all iOS devices.

3.2.5 Smaato

Smaato is another advertising solution for iOS developers available through their website: http://www.smaato.com. Signing up for a developer account through their website is required to utilize their in-app advertising service. After an account is created, apps can be added and the SDK can be downloaded for app integration. Smaato offers developers full screen video and interstitial ads as well as banner ads for all iOS devices.

3.2.6 Greystripe

Greystripe is an advertising network that has partnered with Adobe and offers in-app advertising solutions for iOS apps. Greystripe's website is located at http://www.greystripe.com. In order to gain access to their developer services, you must establish an account through their website. Once your account has been created, apps can be added and the SDK can be downloaded for app integration. Greystripe offers a variety of full screen advertisements as well as banner ads. Due in part to its Adobe partnership, Greystripe makes it easy to integrate advertisements within iOS apps created with Adobe's Flash product.

3.3 Advertising Services Conclusion

There are many advertising networks available to iOS developers; however, they all follow the same basic process: signup for an account, implement the advertising network's SDK into your project, configure app specific settings through the advertising networks website, and sit back while revenue is generated.

NOTE: It is always recommended to download the latest version of the advertisers SDK to address recently-updated iOS compatibility issues, and iOS 6 is no different. If you have a problem, the first thing to try is a newer version of the SDK, if available

Advertising Aggregation

4.1 Aggregators vs. Ad Networks

Advertising aggregation is a handy method that can be used to maximize revenue from in-app advertisements. In a nutshell, advertising aggregation combines the use of multiple advertising networks within one app. Instead of serving advertisements from one ad network, advertising aggregation allows you to serve ads from more than one network.

There are several reasons to use ad network aggregation, the most important being an increase in revenue. Depending on what method is used, it is possible to achieve extremely high fill rates. Fill rates are the percentage in which ads are served to your apps. While average fill rates vary between different ad networks, reaching a 100% fill rate with one ad network rarely happens. This means your app is requesting an ad, and it is given nothing: a blank nothingness. This is the last thing you want happening to your app. This most often happens due to the fact that an advertisers' ad inventory is low or the ad network is experiencing a large number of requests. Depending upon on the network, fill rates can drop after being served a certain number of advertisements. With ad aggregation, when an ad network SDK reports that it is not able to fill an ad, the ad aggregator will immediately attempt to fetch an ad from a different network, bypassing the preset wait time before attempting to fetch a new ad. Nice!

In addition to achieving higher fill rates, ad aggregators also make it possible to prioritize ad networks as well as allocate percentages of ads served from specific ad networks without uploading a new version of the app. Depending on the ad aggregator, this can be done through a web portal. Allocating percentages and prioritizing ad networks is useful when an ad provider is offering a promotion or when higher eCPM's are on specific networks. In some cases, an ad network paying a high eCPM will have a low fill rate. In this situation it is usually beneficial to allocate and prioritize the majority ads served from the ad network that is paying high due to the fact that if an ad is not served, the ad aggregator will attempt to fetch a replacement from the next prioritized ad network. Using this method, the user will be served a new ad from a different ad network; if the ad network that is paying higher is not able to fill a request then the end user will see a fresh advertisement.

> **TIP:** *Constantly monitoring which ad networks are paying what, and using ad aggregators to nimbly change between ad networks on-the-fly, is critical to maximizing your app's ad revenue.*

4.2 AdWhirl

AdWhirl is a popular ad aggregation service. Both the server and the client aspects of the system are open source and free to use. Due to the open source nature of the project, it is possible to host an entire AdWhirl solution privately.

AdWhirl makes it possible utilize several ad networks at once as long as the individual ad network SDKs or libraries are included in the app. By default, AdWhirl offers control over advertising within app through their website. For AdWhirl to work properly, you must sign up the supported ad networks, such as Admob, and configure AdWhirl to serve advertisements from the various ad networks through a web interface. It is also important to know that for AdWhirl to work properly, you must include the proper adapter classes, for each ad network, which can be found in the AdWhirl client project. These adapter classes can also be found in each of the ad network libraries available from their respective websites. AdWhirl serves standard banner ads; however, it can also serve other ad types as well. More details can be found in the AdWhirl documentation located at http://www.adwhirl.com.

AdWhirl functions somewhat differently than other ad aggregation services in the sense that developers are paid directly by the respective ad networks; AdWhirl does not collect funds from these networks or charge fees whatsoever.

Extensive documentation and further instructions are available through their website including instructions on how to host private instances of the AdWhirl system.

AdWhirl remains extremely popular, partially due to the fact that if any one ad network or partner disappears or fails to pay you for whatever reason, then you can simply log in to the AdWhirl control panel and change your app's ad ratio to 0% for the now-defunct ad network, instantly resuming ad revenue on another ad network without delay.

4.3 MobClix

Mobclix is another popular advertising aggregation solution. Mobclix functions in the same way as AdWhirl by allowing developers the ability to serve ads from multiple ad networks. To use MobClix, you must sign up through their website and implement their SDK into the developer's project. Mobclix's ad providers offer a variety of ad sizes, varying from standard banner ads to full screen video ads.

Mobclix differs from AdWhirl through automation allowing you to sign up for additional ad networks and by collecting and consolidating payment from all ad networks used. So instead of receiving payments from each individual network as with AdWhirl, a single, consolidated check is received from Mobclix. Mobclix also has the ability to serve house ads, or advertisements created by the developer.

Mobclix offers different payment options. They include Net 15, Net 30, and Net 90. Net 15 guarantees payment within 15 days from month end and charges a 10% fee. Net 30 guarantees payment within 30 days from month end and charges an 8% fee. Net 90 guarantees payment within 90 days after month end and charges no fee. Regardless of which payment option you choose, a minimum of $100 in earnings must be accumulated before payment is processed.

For detailed information and instruction on how to sign up and implement Mobclix's SDK, please visit their website at http://www.mobclix.com.

4.4 Built-in Ad Aggregation

Increasingly more ad networks are offering built in ad aggregation services.

One such company, as mentioned in the previous chapter, is Millennial Media (http://www.millennialmedia.com). Millennial Media is a growing ad network that has recently added the support to serve advertisements from other ad providers such as Admob, Amobee, Jumptap, and Mojiva.

Another such ad network, Google's Admob (http://www.admob.com), also has recently added the ability to serve Google AdSense, Adwords, or Double Click advertisements when an Admob advertisement is not available.

Both of these services require configuring options through their respective control panels.

4.5 Summary

With low fill rates and constantly changing eCPM values, it can be difficult to maximize monetization with just one ad network. By using multiple ad networks through an ad aggregation service, you are able to maximize revenue and minimize unfilled ad requests as well as change ad network allocation on the fly. This gives you the power to constantly tweak and improve advertising solutions through a web interface without having to manually update your published app.

Critical Business Decisions

As fully detailed in Chapter 1 of this book, and just to reiterate its importance again, the U.S. Small Business Administration website (http://www.sba.gov) offers a well-organized resource of documents and guides for starting your own (App Store) small business. To reiterate one last time, if you've decided on signing up for accounts either as an individual or as a company, there is a wealth of valuable information there: If you have any interest of taking your app development plans to a professional level, you may want to add the SBA.gov website to your research stack.

> ***Note that nothing in this book provides any tax, legal, insurance, or other professional advice, nor does it attempt to replace your licensed advisor, and should not be construed as such.***

This book is not intended to replace any professional advice of any such professional whatsoever, including but not limited to your lawyer, accountant, insurance broker, etcetera, and nothing in this book is to be construed as such. For obvious liability reasons, this book isn't attempting to replace your bar-certified attorney, your certified public accountant, your licensed and registered insurance broker, or other licensed professional, etcetera. You should always consult a lawyer or accountant if you face any legal or accounting issues, such as the legalities and taxes of establishing your business and the promoting of your app in your geographic location.

You should also discuss professional, general, errors and omissions, and intellectual property liability insurance options, along with other options as they may pertain to you, with your duly licensed and authorized insurance broker and underwriter. Experience has shown Apple's sensitivity to the regional legal issues as extreme, for they clearly have a lot to lose. You have been warned.

With all that said, the "legal issues" are a huge can of worms, often never opened, or at best skimmed, glazed over, or completely dodged in nearly all comparable Apple iOS developer books. In other words, it is difficult to locate any resources to help avoid "legal landmines" - which relate *directly* to the App Store.

This book will bravely share some unfortunate theoretical situations which you may want to avoid, and makes general suggestions regarding the use of good judgment and common sense.

But to be absolutely clear, this book **is not** providing any legal advice. The author and contributors of this book are geeks that have absolutely no legal expertise or training whatsoever.

However, the author and contributors of this book did follow their own advice, and hired intellectual property attorney Mark Young. Mr. Young was gracious enough to permit the usage of his name, and edited all of the relevant portions of this book for accuracy. Mr. Young even provided his own contributions towards the end of this chapter!

5.1 Legal Beagle: It's Getting Out Of Hand

One of the tactics frequently used by some ruthless companies to remove their competition from the marketplace is not through innovation of new and better products, but rather by attacking their competition through the legal system. ***The App Store is no different.*** With hundreds of thousands of Apple iOS apps now available for download from the App Store, literally millions in local currency ***per month*** are up for grabs. Sadly, money always gets the attention of the lawyers.

> **Reminder:** There are many free and open-source code-bases, images, graphics, sounds, and other content, which you may be tempted to pull from the Internet and then modify for your use, ***but use caution***: Many free and or overly cheap sources have invalid licenses, are scams, or may even contain outright illegally pirated content ***sold under the guise of being legitimate!*** While <u>this book is not providing any legal advice whatsoever</u>, note that, in theory, even if you have a "receipt" from paying for so-called "pirated" or otherwise unlicensed content in your app, when it comes to intellectual property rights and or copyright law, ***paid receipts MAY NOT HELP your defense!*** Please be sure to validate the license, its legality, and the credibility of the sourced creative, of any code, image, sound, or other content that you use or otherwise incorporate into any project. Just because you see content on a "Sharing" site does not mean the real copyright holder knows it's there or gave permission! When in doubt, ask a lawyer. If you see amazing free or bulk-rate content offerings, just remember the old adage: *Deals that seem "too good to be true", probably are. While imaginary and theoretical, this is all entirely possible.*

5.2 Using Good Judgment and Common Sense

The author and contributors of this book have some rather unfortunate and unpleasant stories told via situational scenarios as experienced by their colleagues. These are shared with you in the hopes that you are able to avoid placing yourself in similar scenarios.

As some situations will show, using good judgment and common sense will help to avoid conflict. Unfortunately, experience has also shown that occasionally developers fail to use good judgment or common sense in a desire to turn a quick profit in the now highly competitive App Store.

If common sense and various laws are ignored, you will eventually receive an ugly legal notice; perhaps, similar to this ***actual email*** (edited to remove contact info and sensitive details):

From: App Store Notices <AppStoreNotices@apple.com>
Date: [Your_future?]
To: [Your_email]
Subject: Apple Inc. (our ref# APP123...)

Dear [Your_name],

Please include APP123 in the subject line of any future correspondence on this matter.

On [date], we received a notice from [angry third-party] that [angry third-party] believes your application named "Your_app" infringes [angry third-party]'s intellectual property rights. In particular, [angry third-party] believes you are infringing their [copyright/trademark/patent].

You can reach [angry third-party] through [their lawyer's name] (phone: 123, email: oh_no@why_is_this_happening_to_me.omg). Please exchange correspondence directly with [angry third-party].

We look forward to receiving written assurance that your application does not infringe [angry third-party]'s rights, or that you are taking steps to promptly resolve the matter. Written assurance may include confirmation that your application does not infringe [angry third-party]'s rights, an express authorization from [angry third-party], or other evidence acceptable to Apple.

Under our terms of agreement, Apple may remove your application from the App Store at any time. You may remove your application using the steps provided below, for example, while you make any necessary changes to your application.

Visit iTunes Connect at http://itunesconnect.apple.com

1) Access your app in the Manage Your Applications module.

2) Click on the "Rights and Pricing" button from the App Summary Page.

3) Click on the "Deselect All" button to uncheck all App Store territories.

4) Click on the "Save Changes" button.

We look forward to receiving confirmation from you within 5 days.

Thank you for your immediate attention.

iTunes Music Marketing & IP Legal | Apple | 1 Infinite Loop | Cupertino | CA | 95014 | AppStoreNotices@apple.com

[Attachment:]

Content Provider Name: [Whoever's mad at you...]

This request was sent to you via the following iTunes Connect Contact Us pathway:
Contracts and Legal -> Rights Infringement

Link to Disputed Content: [Your_app]
Disputed Developer: [You!]
Available Cease and Desist letter: [(You do not want one of these...)]

Whoa! Now that's an email that you do not want and one that will surely ruin your day.

5.3 The Trademark Tango

Common sense: Don't Use Trademarks of third-parties anywhere in your app, app description, search keywords, etc. without written permission!

Also note that marketing strategies such as, "If you liked #1 ranked app XYZ, then you will love this app!" whereby XYZ is the name of a competing app, are generally not permitted on iTunes. So-called "fair use" does not apply to this scenario, since this is apparently against an Apple policy. This is presumably to try and help avoid some trademark conflict issues, by avoiding the topic altogether. *Please see lawyer's comments on Fair Use towards the end of this chapter.*

If your despicable goal isn't to make money, but rather cause a huge stir, waste countless thousands on legal fees, and "do the tango" arguing with lawyers and judges in your country, then ignoring trademark law and registered trademarks is probably one of the fastest ways to achieve that goal. So fast, in fact, that you may even trigger an automated instant-rejection response from Apple immediately upon upload! Not a fun time.

5.4 Trademark Scan Database

Over time, it is speculated and appears that Apple has built a database of trademarks and their holders who have asserted their legal rights against developers. This is presumably done when the owners send legal notices to Apple, complaining about other apps on the App Store claimed to be violating their intellectual property rights. These legal notices usually have ugly demands for the app to be immediately taken down and removed from iTunes. These notices also may include threats of, or actual lawsuits for, damages resulting from infringing their legal rights.

Sometimes, these legal notices are sent to Apple many months after the app first appeared on the App Store. This is the case when a trademark holder discovers that an app is attempting to confuse – by using a similar sounding/looking name - or outright using the company or product name.

It does not matter if you see another competing app on the store using the unauthorized trademark of a third-party. They may appear to be getting away with it month after month, but once discovered, the app will disappear from the store and the developer will have a very bad day. In general, the longer the app has been on the store, the larger the sum of potential damages that can be demanded by the trademark holder.

However, if you are legitimately uploading an app with a third-party trademark and you have permission for using such trademark, simply be sure to plan for this verification process and corresponding delay. It is a best practice to keep a paper trail outlining and authorizing your usage.

5.5 Trying to Avoid Trademark Trouble

At a bare minimum, there are a few things you should try to do to help yourself avoid a potential naming conflict, **before** you get too excited about using any particular company or product name for your apps. Nothing is more devastating than being attacked by lawyers for an honest mistake after all of your hard work making an app! Even if you cannot afford to "do it right" and "play it safe" by hiring a lawyer to professionally guide you and to perform trademark searches on your behalf, which is highly recommended, then at a bare minimum: Practice using some common sense and at least review and follow the brief list of referenced sites *before* getting your heart set on a name.

1. Scan the official government trademark websites for the countries you plan to sell your app, for any registrations or pending applications for the exact or "confusingly similar" trademarks for use in connection with the same, similar or related goods or services. In the USA, for example, this is the United States Patent and Trademark Office (USPTO), using its Trademark Electronic Search System (TESS), located at: http://www.uspto.gov.

2. Scan the major search engines for any other company, goods or services and other product names that are either the same as, or even "confusingly similar" to, whatever company or app name you are thinking of using.

3. Scan the internet registrars for any domain names that may be "parked" or simply not showing up on regular search engines. This is referred to as a "Whois" search. A Whois will help you determine if someone else is already working on a product or company under the name. This is also sometimes called doing a "Whois" lookup search.

4. Apple also tries to help everyone avoid conflict. When your app is uploaded, all of its Metadata is scanned for potentially conflicting trademarks in the Apple database. Only those developers who have complained to Apple in the past are probably in the Apple trademark scan database, so do not rely on this alone! If any existing previously known third-party trademarked term is found, Apple's system may automatically reject or place your app on hold for using the trademark of a third-party, pending formal written verification of permission to use it.

The use of anything which is even vaguely similar to anything owned by Apple isn't allowed. Basically, if Apple thinks its customers may in any way be confused into thinking your app was created or approved by Apple, then it generally will not be permitted on the App Store.

When searching for existing names (trademarks): Keep a date stamped written log of all of your search efforts, and consider other things you can do such as making a checklist and emailing your scan results to yourself as further evidence. This will help, in case of conflict, to show that you made, at a bare minimum, honest effort to avoid conflict in advance of using any name.

In other words, even if all of your existing trademark search efforts come up clean, it does not automatically mean the mark is safe to use. There is always risk in not using a lawyer to do it right, but sincere scanning efforts *in advance* could come in handy if you do inadvertently step on someone else's trademark toes. There are no guarantees, but everyone is more likely to be forgiving if you can prove it was an honest mistake!

5.6 Playing with Trademark Fire

Of course, there are always many exceptions to every rule. Some weird situations you may have heard of in this regard revolve around subtle trademark details. It almost goes without saying, that these are the unfortunate situations where some developers have gotten into the most trouble. Therefore, these scenarios are where you *really* want to get a lawyer's advice, in advance, as some big companies out there only want to protest their name and do not care about such details:

1. In theory, it ***might*** be possible to use an app name that is similar to an existing completely unrelated third-party product name. For example, you might have your heart set on "XYZ" as the perfect computer video game app name, but then discover it's perhaps already being used for a brand of tissues. Unless the third-party product name is truly famous, the third party may have no legitimate basis to object.

2. In theory, "fair use" ***might*** permit your mentioning another trademark within your app description, for comparative marketing or to indicate compatibility. Since Apple does not seem to permit this on its App Store as a matter of its own policy anyway, this is somewhat a moot point. ***Please see lawyer's comments on Fair Use towards the end of this chapter.***

3. In theory, "fair use" ***might*** also permit use of a term that happens to be another's trademark where that term is used in a descriptive sense to fairly describe a feature, function or characteristic of an app. For example, describing your app as a "flashlight" if your app enables and holds the device's flash on, thereby literally providing a flashlight-type of feature, might constitute a descriptive fair use, even if the term "flashlight" is another's trademark. Despite a valid "fair use" defense, there is still the possibility that the developer of a prior app called "flashlight" will object.

Think of those options as similar to playing with fire – sometimes, trademark holders are only interested in protecting their brand names at all costs, and it does not matter what you think you know; they might legally come after you to try to scare you away. Is it worth paying a lawyer countless thousands per month for years on end, just to argue: Probably not.

5.7   Trademark Summary

Trademark rights in the United States (US) derive from use of a mark in commerce and extend as far as the owner's trade. A mark can be a word, combination of words, color combinations, graphic image, combinations of the foregoing, or other subject matter that identifies and distinguishes a party's goods or services. Registration is not required for trademark protection in the US, though it is advisable as it provides important legal advantages including favorable presumptions of ownership and validity as well as nationwide constructive notice and use. In some foreign countries, registration is a prerequisite for trademark protection.

Trademark laws protect against infringement and dilution. In general, infringement occurs when there is a likelihood of confusion due to similarities between a latecomer's and the owner's marks, goods and trade channels. Dilution, which protects only truly famous marks, may occur if the latecomer's mark is similar enough to evoke, in the minds of consumers, the owner's mark, even if the latecomer's mark is used on goods that are entirely unrelated to the owner's goods.

It behooves a business to adopt a distinctive mark that avoids infringement and dilution and does not provoke a senior trademark owner. It is embarrassing, costly and time consuming to change a product's mark, re-brand a website, and make amends with co-workers and business partners in response to a takedown notice or other infringement allegation. Additionally, a distinctive mark can be protected, federally registered and constitute a valuable asset.

A thought to ponder is why you would even want your app to be called something so confusingly similar to anything else. Confusion harms both parties and consumers. It prevents a product from standing out in a crowd. It also limits the value of the name and your ability to protect it.

In sum, by researching the availability of a trademark in advance, a developer can avoid costly and embarrassing pitfalls. Concomitantly, a distinctive trademark not only helps avoid conflict, but it also may become a valuable brand that is entitled to a meaningful scope of protection.

5.8 The Copyright Clash

Common sense: Don't Use Copyrighted Materials of third-parties anywhere in your app without written permission! By way of example only, the app code, graphics, sounds, videos, etc.

In addition to the third-party trademark holder database, Apple also has a similar verification routine for various famous and other top-ranked apps. In other words, if you try to upload a "rip-off" app that is nearly identical to another commonly downloaded app, even if it's called something different, then be prepared for accusations of violating the competitor's intellectual property rights, in particular, the competitor's copyright. While this scenario does not always result in an instantaneous automatic rejection upon upload, it will most likely lead to a rejection notice as soon as a human reviewer has manually inspected your app. It is very difficult or impossible for a latecomer to explain away striking similarities with another's earlier work.

5.9 Open Source Isn't Always Free

Many developers falsely assume open source projects are "okay" to use within their apps, when in fact, using certain open source within your project is almost always a violation of its license. This is why, if you read your Apple developer agreement, Free and Open Source Software (or FOSS for short), has its own huge separate section. In essence, just because some software out there is "open" source, it does NOT automatically mean its "free" source to use: Perhaps, not even within a "free" app.

In other words, "Open" and "Free" are essentially **unrelated**.

5.10 Open Source License Agreements

There are many different kinds of license agreement for free and open source software. Some are good, but many are very bad. For example, GNU is a very common open source license, but it's also extremely limiting and effectively prohibits the use of GNU-licensed open source code within your free or paid mobile device app. Other licenses, such as MIT, are friendlier to both paid and free mobile device use. Be sure to carefully read the project's specific open source license agreement that you are considering using.

Some open source licenses, such as the GNU license, require licensees to provide free copies of their derivative works in source code for others to use, modify and redistribute. While such conditions are favorable to the open source community, they make it very difficult for a business to use such code as a foundation for its app.

Open source code is impossible to audit. The typical open source project is a grassroots efforts comprised of contributions from many sources. It creates many opportunities for contributors to introduce infringing code, while making the source of the contribution almost impossible to track.

Another noteworthy gotcha on using open source projects is the ***feeling*** many third-party developers have when they release their software under a license they either do not fully understand or did not think entirely through. Open source contributors sometimes do not realize the entire project they're releasing under a certain license may legitimately be used exactly as-is, meaning including the entire look and feel, any sounds included, all graphics and everything. This is why occasionally you may see dozens of the exact same app on the app store without ever being taken down; it's possible they're all based on the same open codebase. However, this can be dangerous.

> ***TIP:*** *Always change the graphics, sounds, and look and feel of open source projects that you may use for your app. Why? Sometimes, contributing developers to open source projects become hostile when an "exact copy" of their work is distributed; not realizing the license under which they released the project permits said exact copies. Again, regardless of right or wrong, is it worth paying lawyers just to argue about it? Probably not. Another very good reason is the fact that you do not want to be falsely accused of copyright infringement or spamming the app store with rip-offs of other apps. Making yours unique may help avoid those issues.*

5.11 No Safety in Numbers

Often, developers erroneously think if there are already dozens of apps on the app store violating a policy or outright breaking the law, then that makes them okay. ***They're not okay.*** The moral of the story: It's irrelevant if you see many apps on the App Store which appear to be infringing on third-party intellectual property rights.

Your mom probably told you that just because you see someone else doing something wrong; it does not make it right. The same applies here. It's quite common for Apple to discover a series of "competing" rip-off apps which have somehow slipped pass the Apple reviewers checks and verifications for violating third-party intellectual property rights, but once discovered they tend to immediately remove all such apps without notice from iTunes distribution in one big swoop. This has happened many times in the history of the app store, so be warned: Apple has aggressively identified such developers, and has been known for deleting entire developer accounts for repeat offenses.

In the best case scenario, you may receive an unfriendly demand notice to correct the violations and re-upload a new app, with a reminder that if Apple is sued because of your app, then you have to pay all of their legal bills. Can you even imagine the cost of paying for multi-billion revenue publically traded corporation's legal bills?

In short, you may eventually see a long list of rip-off apps on iTunes which have slipped past the Apple reviewers, but do not be tempted to join the insanity! It's virtually guaranteed: they're all going to be abruptly removed one day – and potentially sued the next.

5.12 Copyrights Last a Long Time

Copyright is a form of intellectual property protection provided to authors of original works of authorship fixed in a tangible medium of expression. Apps are one example of such works. Registration is no longer a prerequisite for protection, but provides important legal advantages including the right to pursue statutory damages and attorney's fees for post-registration infringements. Likewise, a copyright notice is no longer required, but is advisable as a deterrent to infringers and to foreclose any defense of innocent infringement.

Copyright lasts a long time. For works created in 1978 or later, copyright protection subsists upon creation of the work and lasts for a term equal to the life of the author plus 70 years or, if an anonymous or pseudonymous work, or a work for hire, 95 years from publication or 120 years from creation, whichever is first.

The United States (US) Copyright Act and copyright laws of other countries generally give the owner of copyright the exclusive right to do and to authorize others to do certain things, including reproducing the work, preparing derivative works based upon the work, distributing copies of the work, performing the work publicly, in the case of such works as music and motion pictures, and displaying the copyrighted work publicly, in the case of such works as works of art. Doing any of these things during the copyright term without permission from the copyright owner risks infringement.

No this is not legal advice and yes there are many exceptions. Unfortunately, there are countless subtle details which affect copyrights and its timeframes, and naturally every country is different.

Suffice it to say, the myths and urban legends that copyrights are valid for only the year the software displays in its notice or that software must be updated every year to show the current copyrighted year, or that software copyrights expire in 10 or 20 years, are all for the most part completely false.

Developers have also occasionally falsely assumed that an "older" piece of software (or its contents), say for example made 10 or 20 years ago, are somehow now "public domain" and free to copy. This is a horribly dangerous and foolish assumption to make!

Just imagine this unfortunate hypothetical situation: A developer thinks they're doing the "community" a favor and ports a popular 1980's arcade game to the iPhone without written permission. The developer thought nothing was wrong with this action, because after all the game was created and published over 30 years ago, and not only is the game no longer available in arcades but the original video game company had also since gone out of business. Unfortunately, this hypothetical developer was rather unhappy one day to receive a DMCA takedown notice from Apple due to this fictional arcade app, because Apple had in turn been served legal notice from the law firm retained by the video game company who had purchased the copyrights to this old 1980's arcade game without said developer's knowledge. The developer was even more unhappy to later be served a lawsuit from the new copyright holder of that 1980's video game, and eventually learned the hard way that not only was their "free" app removed from the App Store, but that they are now also responsible for damages plus Apple's and the video game company's legal fees. **While imaginary and theoretical, this is all entirely possible.**

5.13 Mobile Devices Don't Void Copyrights

Common sense: Don't Clone Old or Third-Party Desktop Software without Written Permission

In most countries, the owner of the copyright of an original software application has the right to prevent others from producing ported versions of a particular piece of software. "Ported" being converted or translated editions of substantially the same work, such as the software code and the app's other content being moved from one device platform to another. A copyright owner has the right to prevent others from creating derivative works during the entire copyright term. Yes, the broad legal protection and lengthy timeframe of copyrights, usually includes all of the app's overall sounds, graphics, and other forms of creative expression as well – not just code.

Many developers have made false assumptions about ***what they think they know*** about the law, sometimes from numerous websites that post false information, and have been badly burned.

For example, some developers have made the false assumption that they can make a derivative work of some existing third-party desktop software, because in their misinformed minds the copyright was for the original desktop version which somehow does not apply to mobile ported versions.

In any case, those misinformed developers are eventually hit with a DMCA takedown notice, and a potential lawsuit, for infringing the copyright holders' intellectual property rights.

Again, if you've obtained written permission from the copyright holder for your mobile app, then simply plan for a potential delay when such written evidence is requested by Apple.

5.14 Think Globally

Remember, by default, your app is available worldwide. Prior to enabling the selling of your app, scan each country carefully in the iTunes Connect control panel in order to ensure that you are not infringing on any intellectual property rights of any kind in any country. Just because a trade name is available in one country, it does not automatically imply it's available in all countries around the world.

Apple's App Store is a bit unique in regards to worldwide sales. For example, you've been selling the PC edition of your app for the past 10 years on your global ".com" internet website. Clearly, foreign countries could have found the PC edition of your software as it was offered to them all that time in the past. Oddly, this can be somewhat irrelevant to ugly legal issues you may encounter on the App Store! Basically, foreign companies or other entities, residing in those specific countries, may have the legal ability to complain to their "local" Apple location, in their particular country, and have your app taken down off that specific country's App Store.

5.15 Rights of Individuals

Personality Rights. Likeness Laws. Right of Publicity. Whatever you want to call it, Apple, Inc. is based in the State of California, and California has some of the strongest laws to protect individual identities, i.e., names and likenesses, from exploitation. In essence, this means that if you do not have specific written permission to do so, then do not use it. Any app, which so much as invokes the thought of a famous person, such as a Hollywood superstar or a professional athlete (even if you do not show a person's actual face or real name!), might violate someone's legal rights. For example, if you use the likeness of a celebrity without written permission, Apple will most likely reject the app from their App Store. Politicians are public figures, which may be subject to commentary and ridicule without recourse. The First Amendment protects such speech; however, Apple's policies mostly forbid any such apps on the App Store regardless.

5.16   Misconception of Free Apps

Many developers falsely assume that there is nothing wrong with using the intellectual property of others in a "free" app. **This is incorrect.** Free versus paid is irrelevant in regards to legal claims of infringing other parties' intellectual property rights. By way of example only, that includes patents, trademarks and copyrights. It is believed that most previous third-party legal takedown notices sent to Apple are stored, and searched for automatically upon upload of every app, in order to auto-reject all app uploads containing content in previous takedown complaints. It's usually less painful to learn of your app's auto-rejection from Apple before your app can hit the market and cause any "damages" to a third-party, than it is to receive the formal legal takedown notice directly from that third-party.

A specific example would be a so-called "fan app," meaning an app uploaded to the App Store by a well-meaning enthusiast or supporter of a particular movie. Imagine you are a fan of a movie about a young wizard, and put the name of that movie's young wizard into your app name. Even if the app is free, and the lack of commercial gain is probably irrelevant, the movie studio who owns the rights to that young wizard's name may still send you a legal takedown notice, file a lawsuit, etc.

5.17   Legal Barbeque and Disclaimer

Don't be one of those roasted at the next legal barbeque. Paying a real lawyer in your local area for advice up front is always a lot cheaper than paying for one after trouble hits. And do not forget: *The author and contributors of this book are not lawyers, and are not providing any legal or other advice in this book whatsoever.* These are only vague outlines of unfortunate situations some developers have learned the hard way, which have been shared with you here for your entertainment in the hopes of your avoiding learning these lessons the hard way yourself. In other words, hopefully the information contained in this book (provided "merely for your entertainment value"), does somehow help you think about at least trying to avoid similar missteps or conflict. After all, the objective is to take joy in sharing your creative work with the world – on the App Store – and hopefully make some money in the process!

5.18   From Our Lawyer, Mark Young, P.A.

Once more, the author and contributors of this book did indeed following their own advice and hired intellectual property attorney Mark Young, P.A. for a complete review of this entire book! Mr. Young is a real, actual, living, intellectual property lawyer, who's well versed in, and currently practicing copyright, trademark, and patent law.

Mr. Young has graciously provided written permission to use his name and law firm, along with his own contributions. Additionally, Mr. Young edited all of the legal sections of this book for overall correctness.

In an overabundance of caution, both the contributors to this book and its publisher have outright obtained permission, or otherwise substantiated the ability, to use the screenshots and other content cited within this book for educational use.

Please note, however, that nothing contained in this book is providing any of Mr. Young's legal, professional, or other opinions on any particular subject matter whatsoever, including, but not limited to, his following contributions.

5.18.1 COPYRIGHT FAIR USE

The fair use doctrine under copyright law limits the exclusive rights held by a copyright owner. Whether a use is fair in particular circumstances depends upon the application of various factors, including:

(1) the purpose and character of the use, including whether such use is of a commercial nature or is for nonprofit educational purposes;
(2) the nature of the copyrighted work;
(3) the amount and substantiality of the portion used in relation to the copyrighted work as a whole; and
(4) the effect of the use upon the potential market for or value of the copyrighted work.

Applying the factors, the United States (US) Supreme Court has ruled that home videotaping of copyrighted television programs for more convenient time-shifting purposes and 2 Live Crew's rap parody of the well-known song by Roy Orbison, "Oh, Pretty Woman" each constituted a fair use. In contrast, The Nation magazine's publishing of excerpts of a purloined copy of the manuscript of President Ford's memoirs was not a fair use. While the aforementioned cases seem clear, persons planning to rely on a fair use defense should beware. Application of the fair use doctrine is somewhat imprecise and different courts have come to different conclusions on seemingly similar cases. For example, the Ninth Circuit Court of Appeals rejected a fair use defense where the well-known "Cat in the Hat" poem of Dr. Seuss served as the basis for a fully rewritten poem telling the story of the O.J. Simpson criminal trial. Just as one should not gamble their life's savings on a spin of a roulette wheel, one should avoid situations where freedom from copyright liability may depend upon viability of a fair use defense.

In an influential decision arising in a quite different context—reverse engineering (and thus copying) a computer program embodied in a game console—the Ninth Circuit Court of Appeals in *Sega Enterprises, Ltd. v. Accolade, Inc.* upheld the defense of fair use. The defendant's purpose was to create an "intermediate copy" which served as the basis for analysis of the program and ultimately for the design of compatible video games. Reverse engineering was the only practicable way for the defendant to discover unprotected elements (e.g., ideas and methods of operation) that were embedded in the copyrightable "object code" that operated the game console.

The Digital Millennium Copyright Act (DMCA) was added to the Copyright Act in 1998, after the Sega case. One of its purposes is to ensure that "technological protection measures"—such as scrambling or encrypting digital versions of recordings, films and books—are not circumvented without proper authorization. Such technological protection measures typically ensure that the copyright-protected work will not be copied, stored or transmitted to others. Raising this prohibition, in *Universal City Studios, Inc. v. Corley*, eight major motion picture studios that had incorporated in their DVD versions of their copyrighted films a Content Scramble System (CSS), which prevented making copies of the DVDs, or playing them on devices lacking licensed decryption technology, or transmitting them on the Internet, sued a Norwegian teenager who reverse engineered CSS and devised a computer program to circumvent it (DeCSS). The defendant posted DeCSS online. The court readily found the defendant to have violated section 1201(a)(2) ("trafficking") of the DMCA, by offering and providing on a website circumvention software that was "primarily designed" for the purpose of circumventing the CSS "technological measure that effectively controls access to a work" protected by copyright.

5.18.2 TRADEMARK FAIR USE

The law does not prohibit the use of another's trademark on or in connection with the sale of one's own goods or services so long as such use is not deceptive. Under appropriate circumstances, for example, a trademark can truthfully be used on repaired, damaged, deteriorated, or repackaged goods sold by someone other than the trademark owner if it is made clear that the goods are in fact repaired, damaged, deteriorated, repackaged, unguaranteed or otherwise varying from the trademark owner's standards. One may also lawfully use the trademark of another in truthful comparative advertising. Additionally, under the doctrine of nominative fair use, one may lawfully use another's trademark to truthfully refer to a particular product for purposes of comparison, criticism, point of reference and the like, provided that the use does not "suggest sponsorship or endorsement by the trademark holder." For example, a developer of an add-on may rightfully indicate that the add-on is compatible with another's software by referencing the trademark name of the software. Furthermore, under the doctrine of descriptive fair use, non-deceptive descriptive use of a term that another has claimed as a trademark for a similar product may be permissible under appropriate circumstances, provided that the use does not "suggest sponsorship or endorsement by the trademark holder." Moreover, a well-known trademark normally may be lawfully referenced as part of a parody or satire, as long as the parody or satire pokes fun at the trademark brand and no likelihood of consumer confusion results.

One important caveat--while the law may permit use of another's trademark as discussed above, trademark owners and Apple may take a more restrictive view. What good is a developer's knowledge that its use of another's trademark is lawful, if it will be forced to fight a protracted and prohibitively expensive legal battle to defend such use? Many trademark owners, especially well-funded sophisticated trademark owners, jealously guard their trademarks from any perceived encroachment. Likewise, although the law may allow use of another's mark in certain circumstances, a private entity such as Apple may prohibit such use to preserve decorum and avoid conflict. Thus, Apple may prohibit comparative advertising and other lawful types of fair use. A prudent developer will avoid unnecessary controversy.

Intro to the Apple Developer Program

Welcome to Apple's iTunes App Store universe! Think of the App Store as its own unique world, somewhat a "mini-internet" and not directly connected to traditional search engines or marketing efforts. As you will discover within these pages, the App Store and the Apple policies which surround it are quite unique.

As of this writing, the App Store is the one and only legitimate way to sell apps for the iPhone, iPad, and iPod touch. Currently, the only way to publish apps to the App Store is by obtaining an account in the Apple iOS Developer Program. Many would-be developers who have a great app idea burnout while searching extensively for the fragmented information required to develop and publish an app. Too many often give up on the process and those great, revolutionary ideas never get to see the light of day.

In the shortest possible timeframe, and most concise way possible, this book guides you through:

- Deciding on the type Apple Developer Account which is right for you
- Signing up for an App Store Developer Account
- Using common sense: Some common yet unfortunate legal landmines and scenarios to avoid
- Avoiding app rejection and other ugly Gotchas!
- Checking for available app names
- Reserving your own app name
- Deciding on the free vs. paid app revenue models
- Understanding, signing up, and incorporating some common third-party advertiser accounts and media ads technologies to generate revenue from your app
- Creating an app based on your idea
- Avoiding any programming efforts
- Providing an extensive and thorough quick-start coding reference, for existing programmers
- Uploading your app to the App Store

- Marketing and promoting your app

This book **does not** try to teach you how to be a programmer: because you do not have to be one in order to have a great app on the App Store! But, this book **does** have a huge resource guide for assisting existing programmers to immediately jump straight into productive iOS development.

To get started, this introductory chapter will guide you step-by-step on:

- Deciding which Apple Developer account type is right for you, and
- Signing up for the App Store Developer Program.

6.1   Conquering Apple Developer Paperwork

Creating and uploading an app is sometimes easier for some folks than navigating through all of the different systems, legal documents, and tax forms encountered while trying to create the account – just to get started.

In a nutshell, your goals are in multiple phases and places: First, create an Apple ID login on one system (http://appleid.apple.com accounts), and apply for the Apple developer program membership on a second system (http://developer.apple.com portal). After waiting for and obtaining approval, agreeing to several contracts, and paying for access to the program in a third system (Apple iTunes App Store). Finally, after waiting for payment processing, agree to several more contracts, and enter your banking and other information in a fourth system (iTunes Connect).

But do not worry – this chapter will guide you precisely through every single step!

6.2   Determine the Account Type

Before you jump into creating an account, you may want to first speak with a legal and tax advisor about limiting your tax and personal liability exposure. It is generally wise to keep your business affairs separate from your personal affairs. Some developers may minimize their tax and personal liability exposure by commercializing an app through a limited liability company or corporation. ***The potential tax and liability benefits could be substantial.*** This book does not provide any tax, insurance, legal, or other professional advice whatsoever, and nothing contained herein should be construed as such. (Please accept some apologies here for the ongoing repetitive legal disclaimers mumbo-jumbo.)

In short, this is an important decision – one that you need to put some serious thought into.

There are three categories of Apple Developer Account types to choose from (although really only two options for the vast majority of folks), each with their own advantages and disadvantages, depending on your geographic location and situation:

Individual Account Types:

Advantages:	Disadvantages:
Nearly-instant Apple approval	Potential for personal social security number disclosed to Apple(if US paid apps)
Potentially lower overall annual total cost of ownership	Greater potential for personal legal liability
Potentially easier income tax returns	Only one login

Company Account Types:

Advantages:	Disadvantages:
Potentially reduced personal legal liability	Potential for added annual expenses (e.g. professional legal and tax advice)
Potentially reduced overall taxes (e.g. write offs)	Delayed signup during Apple's manual review and approval process
Multiple team account logins	

If choosing a Company Account Type, please note: Don't waste your time forming a simple Doing Business As (DBA), Fictitious Business or Trade Name – because Apple will not accept them for a developer account. If you want to create a company developer account (either to try to limit your legal liability and or for tax reasons), then you will need an Apple-approved "bona-fide" business entity. That means it has to be legitimate and legally formed, as recognized by Apple. If your company is based in the United States, and you do not feel like trusting Apple and others with your personal social security number, then one possible Apple-approved alternative is for your business to obtain its own Employer Identification Number (EIN) issued by the Internal Revenue Service (IRS). Please ask your accountant, and see Chapter 1 for more information.

Enterprise Account Type:

Advantages:	Disadvantages:
Additional internal private app deployment flexibility	Cannot sell apps on the public App Store

Note: Enterprise accounts are only available to extremely large corporations, and those specific enterprise-issues are not detailed per se in this book.

6.3 Create an Apple ID

You most likely do not want to use any existing Apple ID for your Developer Account. There are many reasons for this, but primarily, it's for technical reasons such as iTunes Connect reporting, and accounting reasons such as sales reports and corresponding bank deposits. You can setup a new Apple ID at http://appleid.apple.com

Apple ID's are very valuable: As a developer, they are your app's gateway to the App Store universe. Be sure to use a very long and difficult password, and answer all of the security questions with long and tough answers to help avoid your account being hacked!

6.4   Potential Signup Delay

Please Note: If you choose to sign up as a company, for legal and or tax reasons, then you must plan in advance for the corresponding delay. In some states a new limited liability company or corporation may be established in a matter of hours. In other states, the process may take several days. However, Apple's verification process can take days or even weeks.

6.5   Agree and Agree to Agreements

No matter what type of account you setup, there will be multiple phases of contractual agreements that you will be required to agree to in order to sell your app on the App Store.

From: **Apple Developer Support** <noreply-iphonedev@apple.com>
Subject: Apple Developer Program Enrollment Update
To: YOU

Apple Developer Program Enrollment Update

Dear YOU,

You can now continue the Apple Developer Program enrollment process by reviewing and agreeing to the Program License Agreement. You must click through this agreement in order to purchase or complete your enrollment in an Apple Developer Program(s).

If you need further assistance, please contact us.

Best regards,

Apple Developer Support

Figure 6-1: iTunes Enrollment Update

6.6   Paying for the Program

If you choose to create a company account, you will not be able to pay right away. Developers with company accounts are only permitted to pay after their corporate documents have been manually reviewed, verified, and approved. This is akin to a background check of sorts, to confirm the responsible party in the event of a lawsuit, and other legal and tax issues.

Eventually, you will receive an email stating that you have been accepted to the Apple Developer Program for iOS. Within that welcome email, there will be a special link to click on which will bring you to the App Store in order to submit payment. As of this writing, the fee is US$99.00 per program per year.

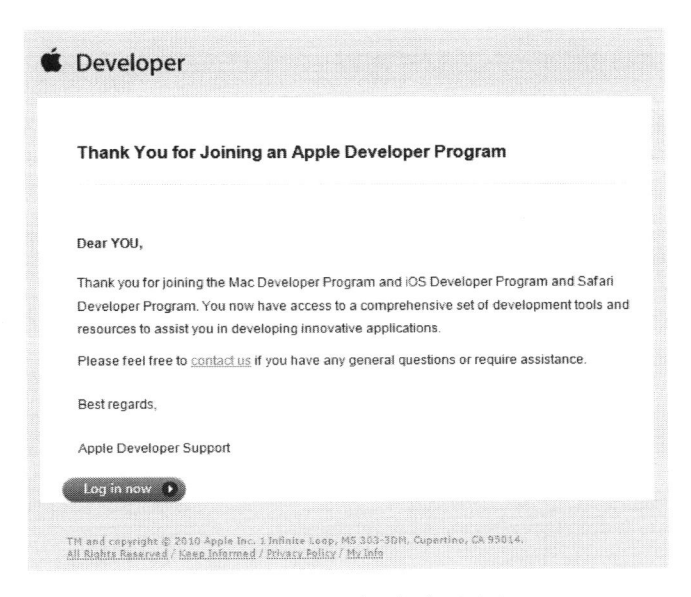

Figure 6-2: iTunes Thanks for Joining

6.7 Signup Walkthrough Summary

With your Apple ID already created and ready to go:

1. Go to http://developer.apple.com/programs
2. Choose iOS Developer Program, for iPhone, iPod touch, and iPad.
3. Click *Enroll Now*, then click *Continue*

   If you followed the suggestion of creating a separate Apple ID, and assuming you are not an existing Apple Developer, you will select the "I have an Apple ID I would like to use for my enrollment in an Apple Developer Program."

4. Choose Individual or Company

If you choose *Individual*, then your personal name will appear as the seller for your apps in the App Store. Payment, account activation, and app uploads will most likely be permitted immediately.

If you choose *Company*, then be prepared to submit your corporate documents to Apple for manual review, verification, and delayed approval. This is presumably done to ensure every developer on the App Store is legitimate and held accountable for everything uploaded for sales to the general public. In other words, Apple wants to know exactly who they're paying and who to point to, in the event of any disputed apps.

Agreeing to Contracts and Setting-Up Financial Information

Once approved for Apple's iOS Developer Program, additional steps need to be taken before the developer account is ready to accept payments and be used to distribute apps. After Apple has accepted the developer's app for access to the Developer Program, there are agreements and financial information that needs to be approved and configured. The agreements requiring attention are located in two different places: iTunes Connect and the iOS Developer Portal. Financial, tax and contact information must be configured through iTunes Connect while developer agreements must be reviewed and accepted through the iOS Developer Portal. Once these steps are completed, the account you created is ready to begin submitting apps for distribution and earning revenue.

7.1 iTunes Connect

The iTunes Connect system processes sales, taxes, contracts with Apple, as well as several other mission critical items such as registering new apps, which will be covered in more detail in Chapter 8. Before the developer can begin selling apps on the App Store, contracts must be agreed to; tax and banking information, as well as contact information must be configured and verified. These steps are required.

1. Log into iTunes Connect (http://itunesconnect.apple.com) with your Team Admin username and password. If you have questions about the use of Team Admin, please review Section 41.1 (Team Admins: There Can Be Only One!) for more information.

Figure 7-1: iTunes Connect Login Screen

2. Select the module, "Contracts, Tax, and Banking", which is shown in Figure 7-2.

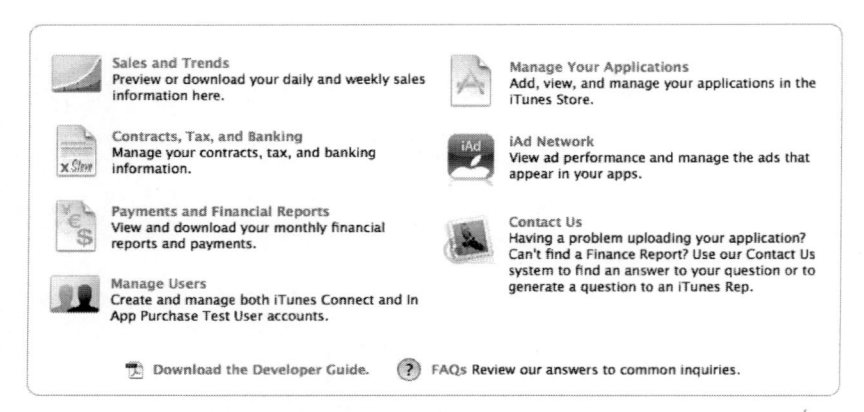

Figure 7-2: iTunes Connect Control Panel

3. Read and agree to the contracts displayed in Figure 7-3.

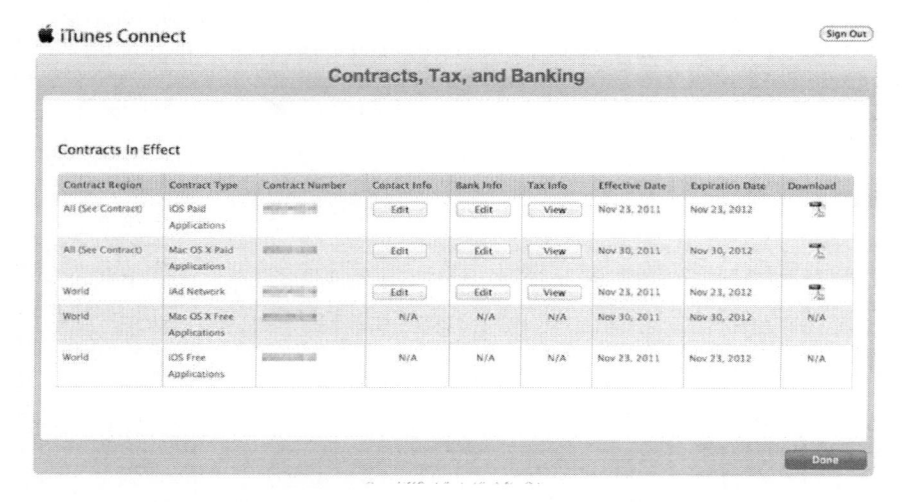

Figure 7-3: iTunes Connect – Contracts, Tax, and Banking

4. Enter contact information by clicking on the "Edit" button under the "Contact Info" column. This will display a page where the contact information can be entered, shown in Figure 7-4.

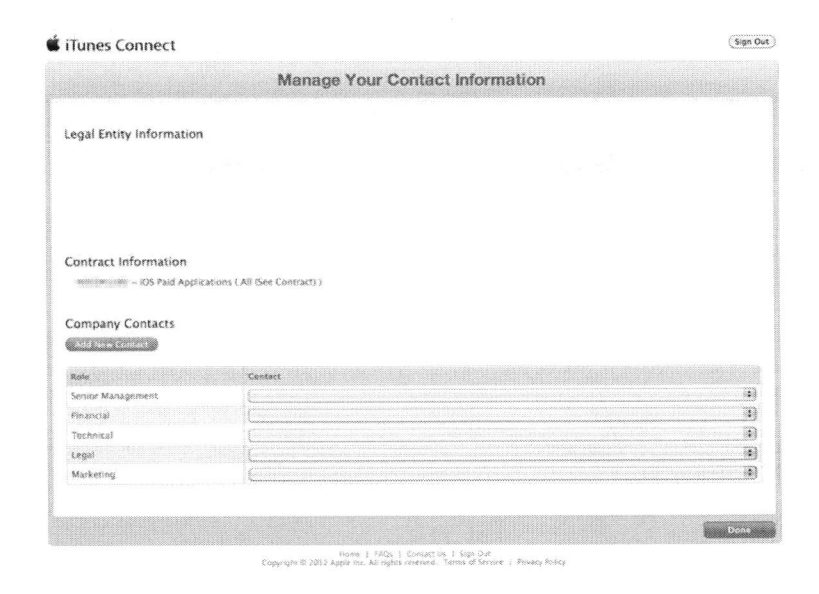

Figure 7-4: iTunes Connect – Contracts, Tax, and Banking – Contact Info

5. Select "Add New Contact", as shown in Figure 7-4, and enter the appropriate information when the pop-up appears - Figure 7-5.

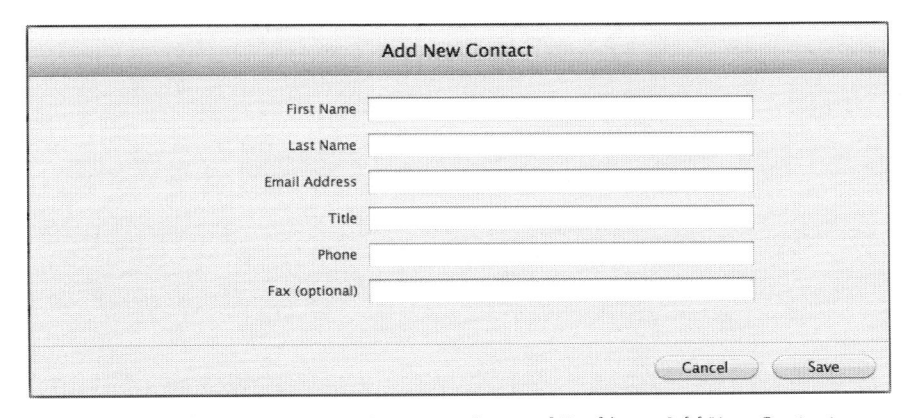

Figure 7-5: iTunes Connect – Contracts, Tax, and Banking – Add New Contact

6. After adding the required contact information, press the "Edit" button under the "Bank Info" column from Figure 7-6.

Chapter 7: Agreeing to Contracts and Setting-Up Financial Information   45

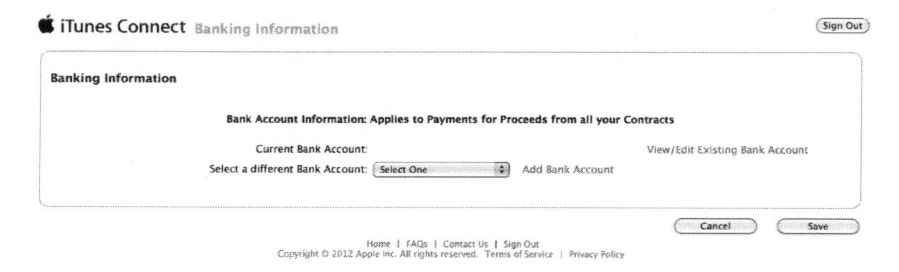

Figure 7-6: iTunes Connect – Contracts, Tax, and Banking – Banking Info

7. Click on the "Add Bank Account" link shown in Figure 7-6 and enter the appropriate information when requested.
8. After the correct banking information has been entered, press the "Edit" button under the "Tax Info" column – see Figure 7-7.

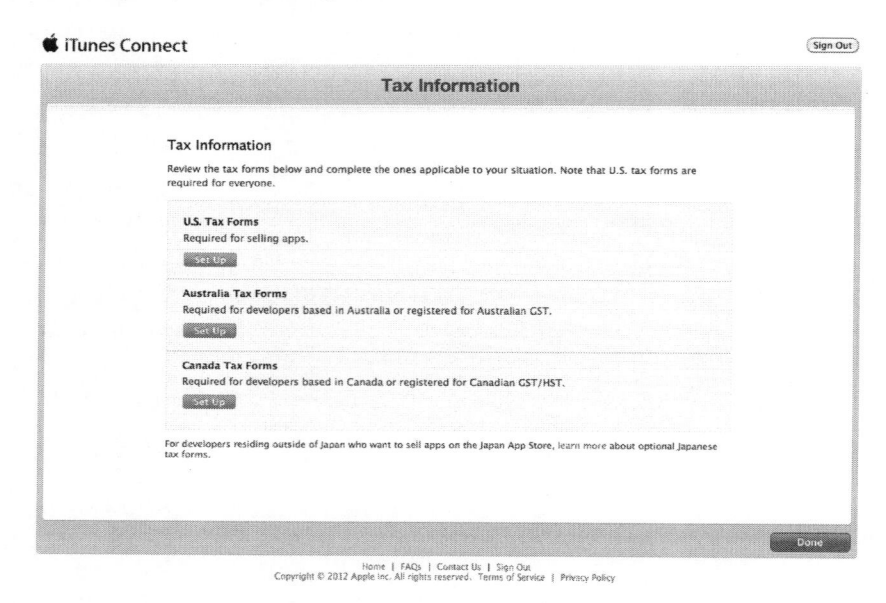

Figure 7-7: iTunes Connect – Contracts, Tax, and Banking – Tax Information

9. Click the button labeled "Set Up" under the appropriate tax form section displayed in Figure 7-7. When presented fill out the appropriate tax forms. In Figure 7-8 the U.S. tax information page is shown.

46   Chapter 7: Agreeing to Contracts and Setting-Up Financial Information

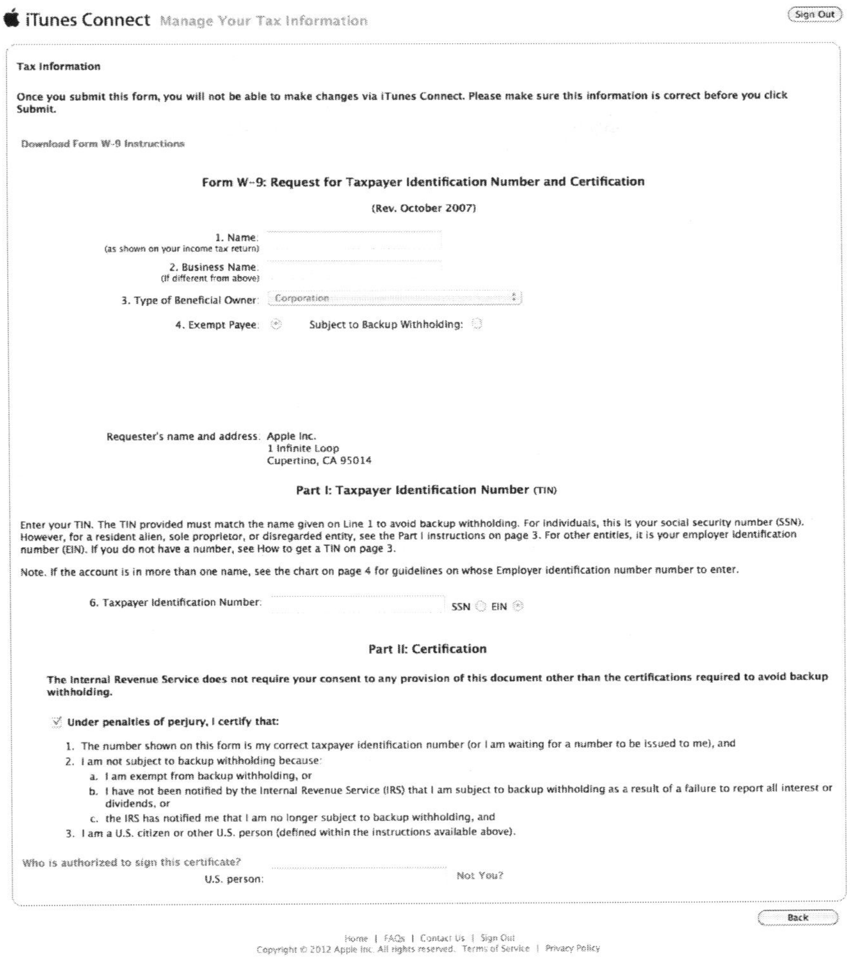

Figure 7-8: iTunes Connect – Contracts, Tax, and Banking – US Tax Information

This concludes the setup and configuration of the required contracts, contact, banking, and tax information accessible through iTunes Connect. Once this information is completed, the account is ready and able to sell apps through the App Store.

7.2   Developer Portal Contracts

The Apple Developer Portal provides access to developer related materials, such as documentation, beta releases of software and operating systems, and the provisioning portal. The developer portal will be discussed in more detail in Chapters 8 & 9; the purpose of this section is to guide developers to important contracts that need to be read and accepted in order to participate in the developer program. It is also important to keep current with developer agreements to ensure access to the latest developer materials.

Chapter 7: Agreeing to Contracts and Setting-Up Financial Information   47

1. Log in to the developer portal (http://developer.apple.com) with your team admin user name and password and select "Member Center."

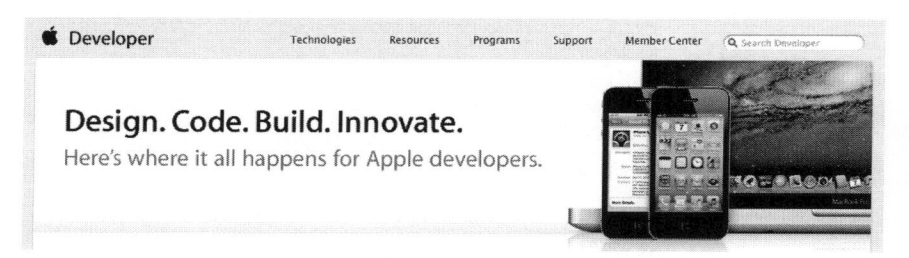

Figure 7-9: Apple Developer Portal

2. Within the Member Center, select "Your Account", shown in Figure 7-10.

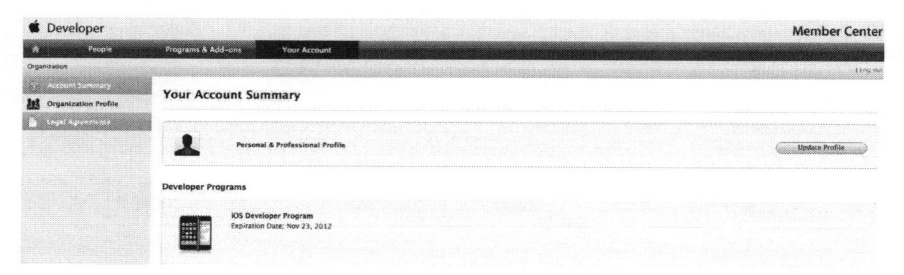

Figure 7-10: Apple Developer Portal – Member Center – Your Account

3. Select "Legal Agreements" on the left-hand side as depicted in Figure 7-11. From here, agreements can be reviewed, accepted or rejected. It is necessary to agree to the iOS Developer Program License Agreement to activate the developer account and distribute apps.

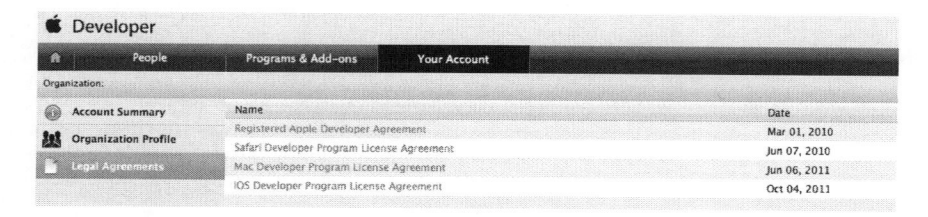

Figure 7-11: Apple Developer Portal – Member Center – Your Account – Legal Agreements

This concludes the section on legal agreements within the Apple Developer Center. Remember, it is important to keep up to date on agreements to ensure access to the latest developer information.

Reserving an App Name

Before beginning development and creating necessary assets such as graphics, it is wise to first scan for similar names and trademarks already in use to avoid (as discussed in earlier chapters), then to reserve the app name within Apple's iTunes App Store. This ensures that the name is available and that another developer will not take it before you are ready to release your app. Nothing is worse than spending time and money on resource creation, promotion, and marketing and then realizing that the desired app name is already in use. To help ensure that this does not happen, it is recommended to consult a lawyer for trademark scans and reserve the non-conflicting app name with Apple before beginning any development. The steps necessary to reserve an app name are also required to submit your app for distribution on the App Store.

> *Note that nothing in this book provides any tax, legal, insurance, or other professional advice, nor does it attempt to replace your licensed advisor, and should not be construed as such.*

Reserving the app's name consists of two main parts: creating an App ID through the iOS Provisioning Portal and creating a new app page within iTunes Connect.

It is important to note that Apple does not permit developers to reserve app names for an indefinite period of time. If you have not submitted an app within roughly 90 days, Apple will send an email notification to you stating that the app name will expire and become available for others to use unless action is taken. This is meant to prevent developers from "name squatting" (similar to domain name squatting). Essentially this prevents developers from reserving names in the hopes that an individual or company will require the reserved name and be forced to purchase it from the holder.

> **TIP:** *Before an app name is reserved, as discussed in earlier chapters, it is strongly recommended to help avoid potential conflict with others, by performing extensive similar-name scans, and consulting a lawyer for trademark checks. Again: Trademark searches can be performed through the USPTO website: http://www.uspto.gov/trademarks/index.jsp*

Protecting yourself and your intellectual property is covered in more detail in Chapter 5: Critical Business Decisions. The purpose of performing such checks is to ensure there will be no future conflicts and it will allow the developer to protect their own content once the app is released. As discussed in previously, the last thing a developer wants is to create an app whose name conflicts with an existing trademark. If there is a conflict, it is very likely that you will be served with a

DMCA takedown notice (or worse!) once the app goes live, requesting that the app be removed from sale and or possibly face legal consequences.

Once the app name is reserved and there are no potential trademark conflicts, the developer may proceed with development, marketing, and securing intellectual property rights relating to their app.

8.1   Creating an App ID through the iOS Provisioning Portal

The first step to preparing your app for distribution and reserving an app name is creating an App ID through the iOS Provisioning Portal. App IDs are unique identifiers that can be assigned to apps and are also used as unique identifiers that can be assigned to in-app purchases.

1. Begin the process of creating an App ID by logging into the Developer Portal, specifically known as the iOS Dev Center (http://developer.apple.com/ios) and browsing to the "Provisioning Portal" section, as seen in Figure 8-1.

Figure 8-1: iOS Dev Center

2. Once in the "Provisioning Portal" section, select "App IDs" from the menu on the left, shown in Figure 8-2. Please note that on the App IDs landing page within the iOS Provisioning Portal, Apple provides links to useful documentation that explains App IDs in great detail.

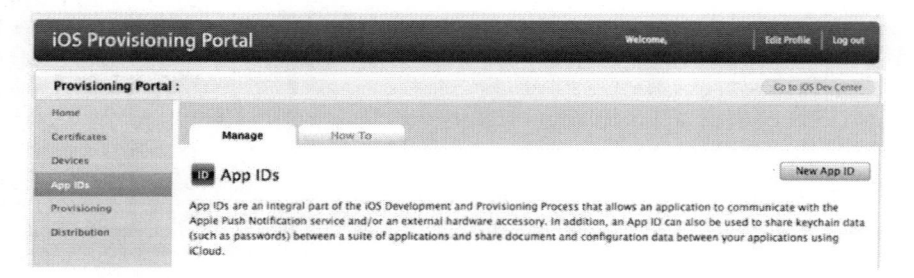

Figure 8-2: iOS Provisioning Portal

3. From the App IDs landing page within the iOS Provisioning Portal, press the button labeled "New App ID" as pictured in Figure 8-2. Pushing this button displays a web page where an App ID can be created for your iOS app.

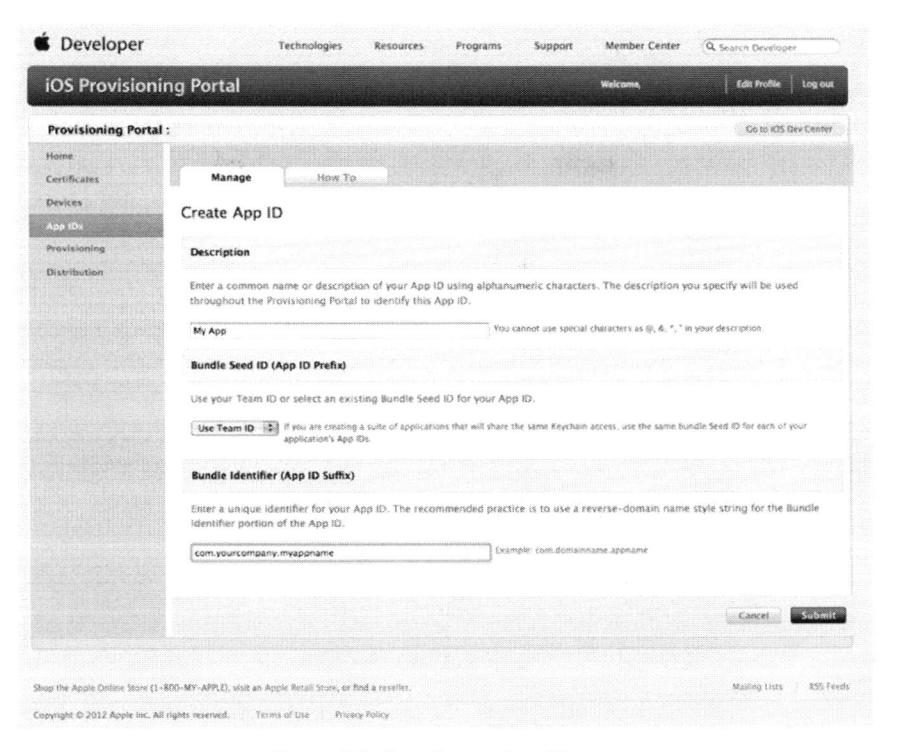

Figure 8-3: Creating an App ID

4. For the first field, entitled "Description," enter a name or description for the App ID. This description is used to identify your App ID within the iOS Provisioning Portal and is not visible to the public. This will often be the name of the app.

5. The next field, entitled "Bundle Seed ID (App ID Prefix)," is for selecting a Bundle Seed ID for your App ID. A Bundle Seed ID is the App ID Prefix used to uniquely identify Bundle IDs. By default "Use Team ID" is selected. When "Use Team ID" is selected, a randomly generated common Bundle Seed ID is used for the Bundle Seed ID. In most cases, the option to "Use Team ID" for the Bundle Seed ID is sufficient for the app. An exception to this would be if you were creating a suite of multiple apps in which the desired functionality would be to share an identical service or services across the app suite, such as Apple Push Notifications, In-App Purchases, or Game Center.

6. The final field required, entitled "Bundle Identifier (App ID Suffix)," is a unique identifier for your app. It is recommended to use a reverse-domain name style string as the Bundle Identifier, such as "*com.yourcompanyname.theappname*".

7. Once all fields have been completed and verified to be correct, press "Submit." This finalizes the creation of your App ID.

This covers the steps required to create an App ID from within the iOS Provisioning Portal for a single app. You can now create a new app page within iTunes Connect, which will link to the your newly created App ID and thus allowing you to reserve an app name.

8.2 Creating a New App Page in iTunes Connect

After the App ID has been created through the iOS Provisioning Portal, the next step to reserving the app name is to create a new app page in iTunes Connect.

1. Log into iTunes Connect (http://itunesconnect.apple.com) with your administrator username and password. Select the "Manage Your Applications" module from the menu as seen in the Figure 8-4.

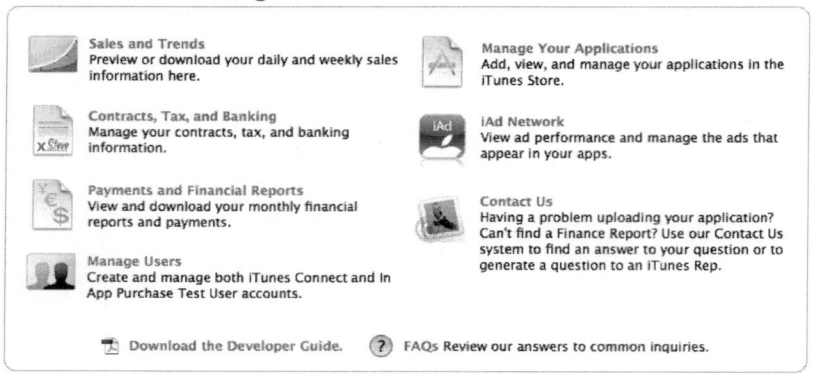

Figure 8-4: iTunes Connect – Select "Manage Your Applications"

2. Once in the "Manage Your Apps" module, select "Add New App" from the top left corner of the screen, as shown in Figure 8-5.

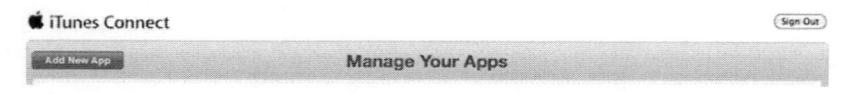

Figure 8-5: iTunes Connect – Select "Add New App"

3. On the following screen, see Figure 8-6, you are required to specify if it is an iOS Application or a Mac OS X Application. Select "iOS App."

Figure 8-6: iTunes Connect – Choose "iOS App"

Chapter 8: Reserving an App Name

4. Select the default language for the app and the app name, SKU number, and Bundle ID created previously in Section 8.1. Choose a default language for the app and enter the desired app name. The SKU is a unique identifier you will use internally to associate with the app and will not be visible to end users. The Bundle ID is selected from the drop down list of available Bundle IDs created through the iOS Provisioning Portal. In Figure 8-7 "My App – com.yourcompany.myappname" is selected for the Bundle ID. Press the button labeled "Continue" when finished. If no error messages are displayed, the app name is available.

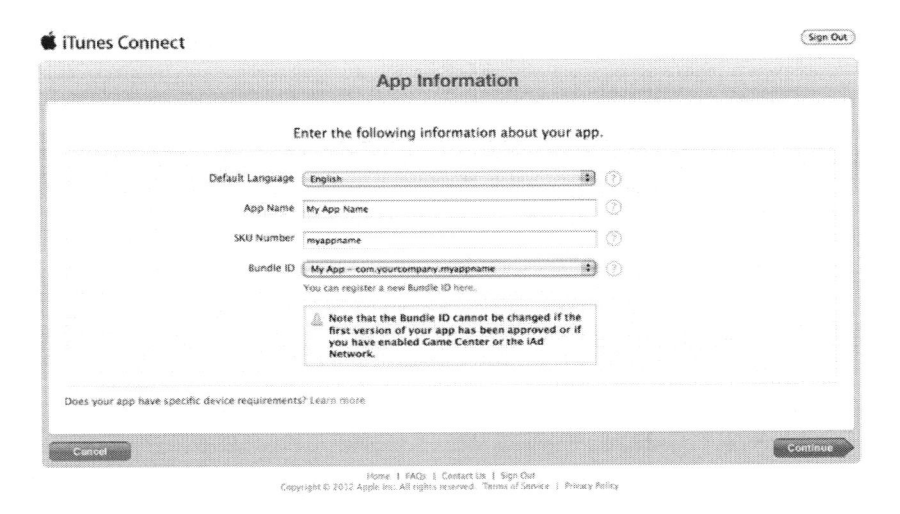

Figure 8-7: iTunes Connect – Enter App Information

5. The next screen, Figure 8-8, prompts the user for more information about the app such as availability date, price tier, as well as choice selections for which countries the app will be available. This can be modified at a later date, prior to app submission.

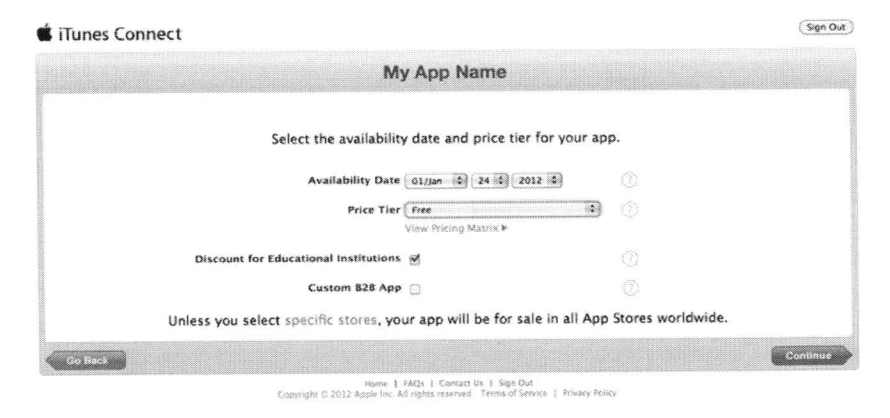

Figure 8-8: iTunes Connect – Enter Additional App Information

Chapter 8: Reserving an App Name   53

6. The following screen, shown in Figures 8-9 and 8-10, is required for entering the app's metadata. This includes information such as version number, copyright information, category selection, assignment of ratings, description, keywords, support email address, and support URL among other things. Optional fields are labeled as such. Prior to the release of iOS 6, a 512x512 pixel image was required. Now, with the release of iOS 6 and retina displays, these requirements have changed to include a 1024x1024 pixel images. In addition, a minimum of one retina screenshot is required for the viewable images seen by those browsing the App Store – which should closely resemble the app's icon in order to avoid rejection by Apple. Once this information is complete, and content uploaded, be sure to hit the "Save" button located at the bottom right corner of the screen, Figure 8-10.

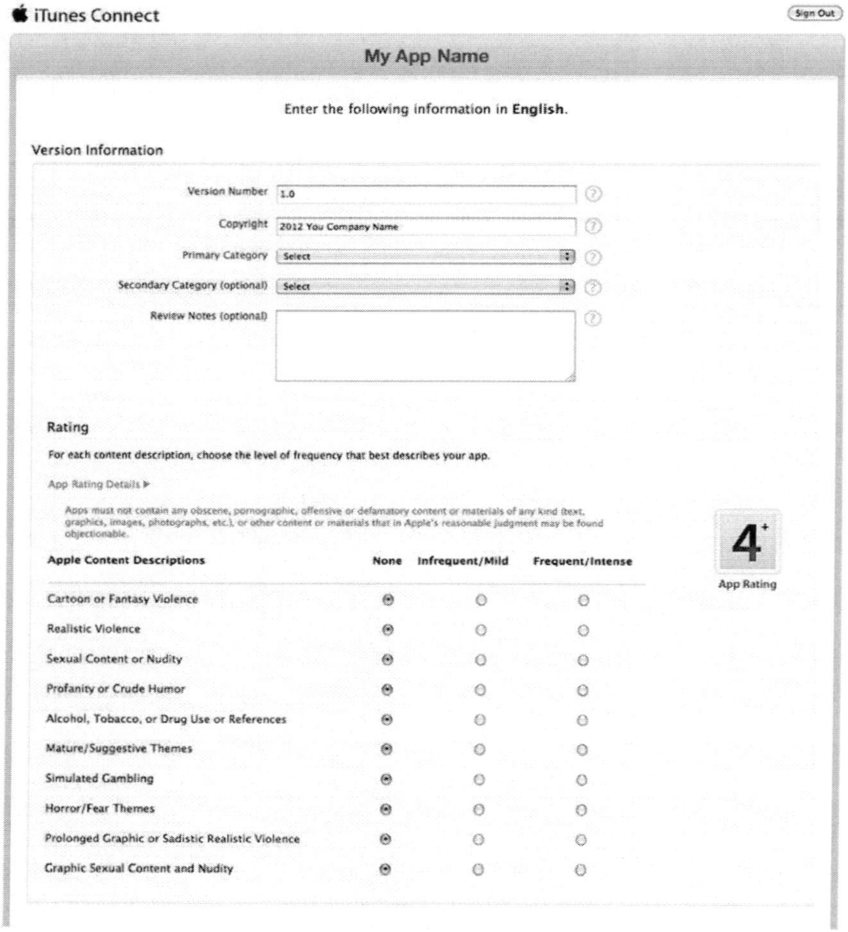

Figure 8-9: iTunes Connect – App Metadata

54   Chapter 8: Reserving an App Name

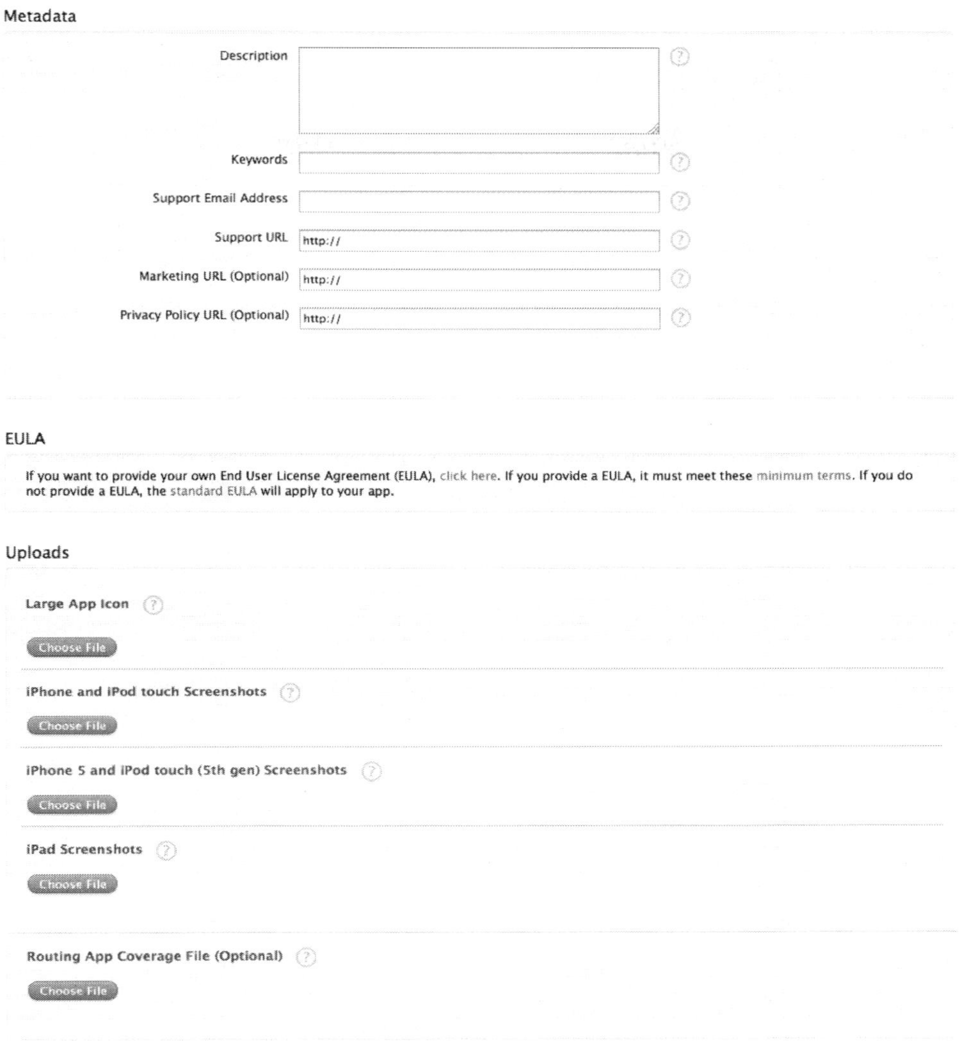

Metadata

Description	
Keywords	
Support Email Address	
Support URL	http://
Marketing URL (Optional)	http://
Privacy Policy URL (Optional)	http://

EULA

If you want to provide your own End User License Agreement (EULA), click here. If you provide a EULA, it must meet these minimum terms. If you do not provide a EULA, the standard EULA will apply to your app.

Uploads

Large App Icon
[Choose File]

iPhone and iPod touch Screenshots
[Choose File]

iPhone 5 and iPod touch (5th gen) Screenshots
[Choose File]

iPad Screenshots
[Choose File]

Routing App Coverage File (Optional)
[Choose File]

Figure 8-10: iTunes Connect – App Metadata (Continued)

Once complete, the desired app name has been successfully reserved. The metadata pertaining to the app can be modified at any time prior to submitting the app's binary to Apple. To simply reserve your desired app name, bogus information can be entered - including placeholder screenshots and a representative 1024x1024 image. Be sure to update this before actually submitting the app to Apple or the app will be rejected.

It is important to note that if your app contains prolonged graphic or sadistic realistic violence or graphic sexual content and nudity, or violates any other Apple guidelines, then it will be rejected by Apple's app review process.

This completes the walk-through on reserving app names.

8.3 Chapter Overview

Reserving an app name is critical in the development process. It ensures that the name is available before development begins. Once the app name passes trademark checks and is reserved, the developer can begin development of the app as well as start taking steps to protect their intellectual property (e.g. consider hiring your lawyer to file for your own trademark!), and begin marketing the app itself. Details regarding app metadata, keywords, screenshots and necessary steps to upload the app for distribution are covered in more detail in Chapter 38: Preparing for Submission.

Preparing for Development

In order to debug apps on a physical device, several requirements need to be met. First, a developer certificate from Apple must be created and installed so that it can be used for code signing. Next, the developer must also add devices to the developer's list of approved devices and provision them to allow for debugging on the device. Once this is complete, the developer will be able to build, run and debug projects on approved devices.

> **TIP:** If you decide to explore Xcode in this chapter, skip ahead to ***Chapter 16: Introduction to Xcode***; and then return here.

9.1 Requesting a Development Certificate

A development certificate must be requested from Apple and approved by the Team Admin of the developer account. This is the individual assigned the Apple ID of the special super-super-user that you originally used to agree to the iTunes Connect and Developer Program contracts. (Additional details about Team Admin accounts are discussed within Chapter 41.1: Team Admins: There Can Be Only One!) Requesting and configuring a development certificate are the first steps towards developing and debugging apps on physical devices. Development certificates require renewal on a yearly basis. The following steps explain how to create and install development certificates.

1. Log into the iOS Dev Center (http://developer.apple.com/ios) with your Apple ID used to sign up for the iOS Developer Program.
2. Open the "Keychain Access" application on your Mac running OS X. Keychain Access can be found by using spotlight and is located in Applications > Utilities > Keychain Access.
3. Navigate to Keychain Access > Preferences… and select the "Certificates" tab.
4. Ensure that both OCSP and CRL are set to "Off."

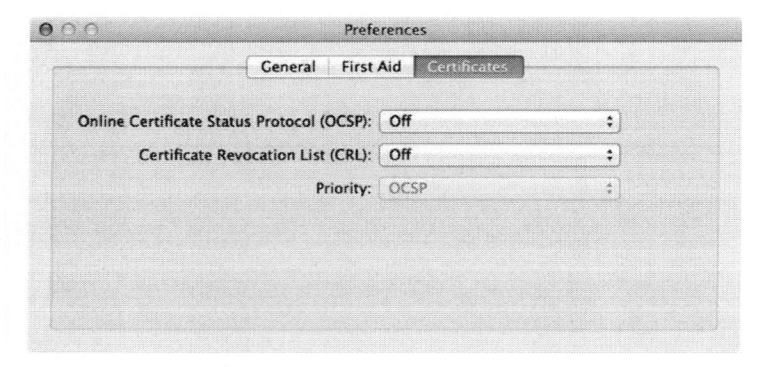

Figure 9-1: Keychain Preferences – Certificates Settings

5. Go to Keychain's Certificate Assistant by selecting Keychain Access > Certificate Assistant > Request Certificate from the Certificate Authority.

6. Enter the email address that is associated with the developer account. This should be the developer's Apple ID.

7. Enter a common name for the certificate.

8. Select "Saved to disk" and ensure that "Let me specify key pair information" is checked.

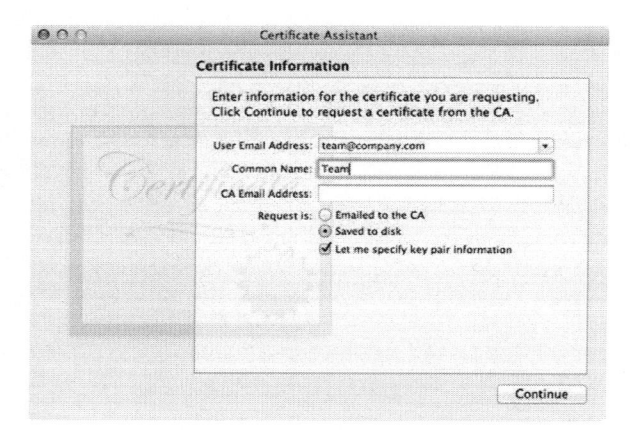

Figure 9-2: Requesting a Certificate

9. Click the button labeled "Continue" and choose the desktop as the location for the certificate to be saved.

10. Confirm that the "Key Size" is 2048 bits and the "Algorithm" is set to RSA.

Figure 9-3: Requesting a Certificate (Continued)

11. Click Continue to save it to the desktop.
12. Go to the Certificates category of the Provision Portal located in the iOS Dev Center (http://developer.apple.com/ios).
13. Click "Request Certificate" and download the WWDR intermediate certificate.
14. Install the Apple WWDR intermediate certificate by opening the downloaded file and adding it to the Keychain.

Figure 9-4: Requesting a Certificate (Continued)

15. Click on "Browse" and save the file to the desktop.
16. Click "Submit."

Chapter 9: Preparing for Development   59

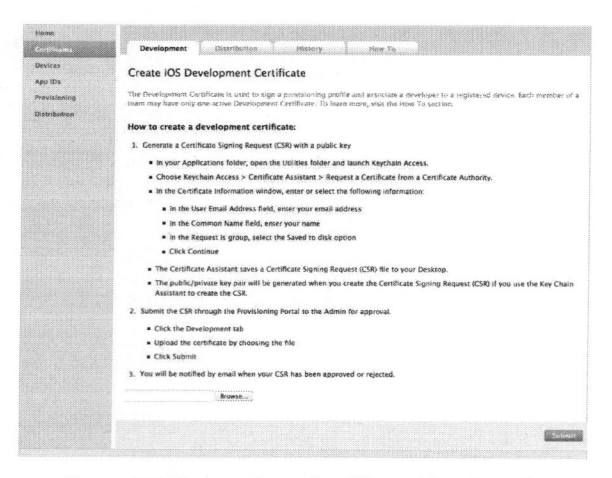

Figure 9-5: Requesting a Certificate (Continued)

17. The certificate will show as pending approval in the iOS Development Portal. A team admin must log in and approve the request. Once the request is approved, the developer can access the Developer Portal and download the certificate. Once downloaded, open the certificate file to add it to the Keychain.

9.2 Adding a Device For Development

There are two ways to add a device to the list of approved development devices. One option is to do it manually by inputting the devices Identifier Number, or UDID, into the Devices section of the iOS Provisioning Portal that is found within the Development Portal. The other option is to add the device automatically using Xcode 4.

> **TIP:** *If you decide to explore Xcode in this chapter, skip ahead to **Chapter 16: Introduction to Xcode**; and then return here.*

9.2.1 Adding a Device Manually Through the iOS Developer Portal

The manual method is useful for adding a device to the list of approved development devices if you are working with a development environment outside of Xcode, if you wish to do so from an operating system other than Mac OS X or do not have Xcode 4 installed.

1. The first step is to get the devices UDID, or Unique Device Identification number. This is done by connecting the device through USB, opening iTunes, and selecting the device on the left side as shown in Figure 9-6.

Figure 9-6: Accessing a Devices UDID through iTunes

2. Clicking on the "Serial Number" text will change it to "Identifier (UDID)" as shown in Figure 9-7. Copy the UDID to your clipboard.

Figure 9-7: Accessing a Devices UDID through iTunes (Continued)

3. Log into the iOS Developer Portal, browse to the Provisioning Portal and select Devices. From there, select "Add Device."

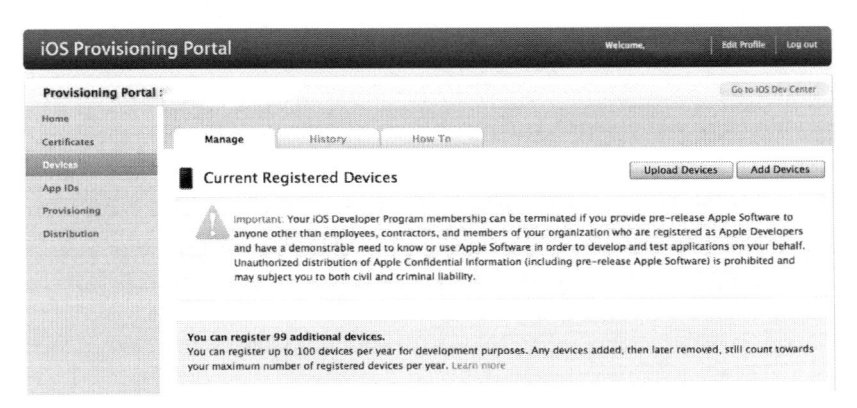

Figure 9-8: Manually Adding a Device through the iOS Developer Portal

4. On the next screen, shown in Figure 9-9, enter a name for the device as well as the UDID for that device.

Chapter 9: Preparing for Development   61

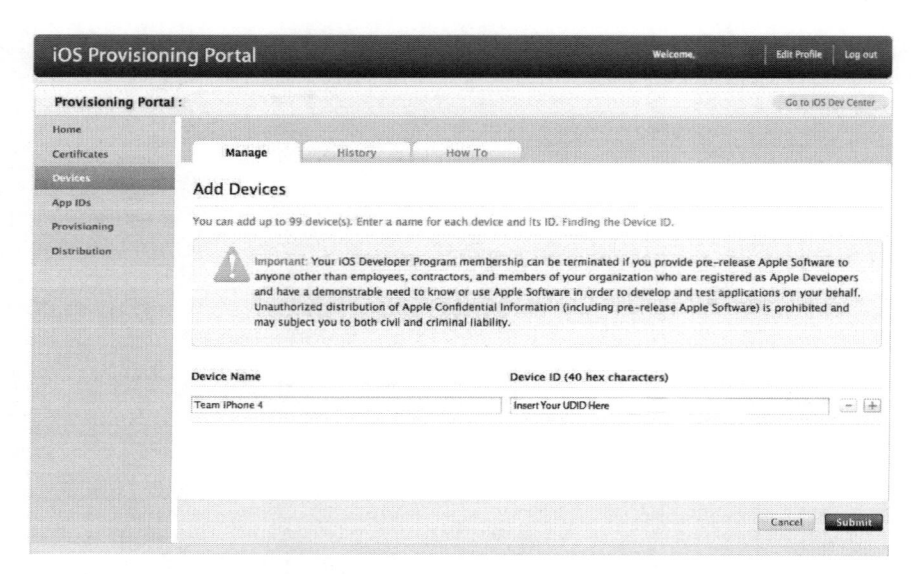

Figure 9-9: Manually Adding a Device through the iOS Developer Portal

5. Click "Submit" and your device will now be added to the list of approved development devices.

9.2.2 Adding a Development Device Using Xcode 4

Xcode 4 makes it possible to add approved development devices from within Organizer. This makes the process somewhat easier but gives you less control over the process. By adding a device through Xcode 4, the device is also added to an iOS Team Provisioning Profile, which is copied to the device itself, and necessary for developing on the device. It is important to note that developer certificates must be properly installed prior to adding devices through Xcode.

1. Plug in the device to be used for development via USB and open Xcode 4's Organizer, selecting the Devices tab.

2. Right-click on the device you wish to enable for development from the list of devices on the left panel of the Organizer and select "Add Device to Provisioning Portal."

3. Next, enter your Apple ID and password associated with the admin iOS developer account. Click login.

4. Wait as Xcode automatically adds the device to the list of approved devices as well as adds the device to the iOS Team Provisioning Profile and installs the iOS Team Provisioning Profile on the device. The device is now ready for development. The next section, "Creating Generic Provisioning Profiles" is not necessary if this method was used to add the device to the list of approved development devices.

9.3 Creating Generic Provisioning Profiles

Provisioning Profiles allow approved devices to run apps. For an app to run on an approved device, the device must be on the list of approved devices, the developer must have an active developer certificate, and the device must contain a provisioning profile that allows the app being built and the developer to build and run on the device. Provisioning profiles can also be configured in such a way that allows only specified developers and or apps to run on the provisioned device.

There are two different ways to create and update provisioning profiles. One way is to manually do so through the iOS Developer Portal and the other method can be done using Xcode 4. Manually creating provisioning profiles allows for more control over what apps can be run on the device and by what developer. The manual method also allows provisioning profile creation from any operating system that has a web browser, whereas the Xcode 4 method requires Mac OS X 10.6+ and Xcode 4. Only generic provisioning profiles will be explained for the purposes of this book but it is important to know that further configuration is available.

The device must be provisioned properly for the developer to be able to build, run and debug for the device. The following sections will explain how to create provisioning profiles manually. Provisioning devices using Xcode 4 was discussed in the previous Section 9.2.2.

9.3.1 Manually Creating Provisioning Profiles

Manually creating provisioning profiles can be done from any web browser and operating system, however it is recommended to use Apple's Safari. Creating provisioning profiles manually from the iOS Developer Portal allows developers to download provisioning profiles for use on their development devices.

1. Login to the iOS Dev Center (http://developer.apple.com/ios) and browse to the "Provisioning Profile" section.

Figure 9-10: iOS Provisioning Portal

2. Select the "Provisioning" menu item and ensure the "Development" tab is selected as shown in Figure 9-11.

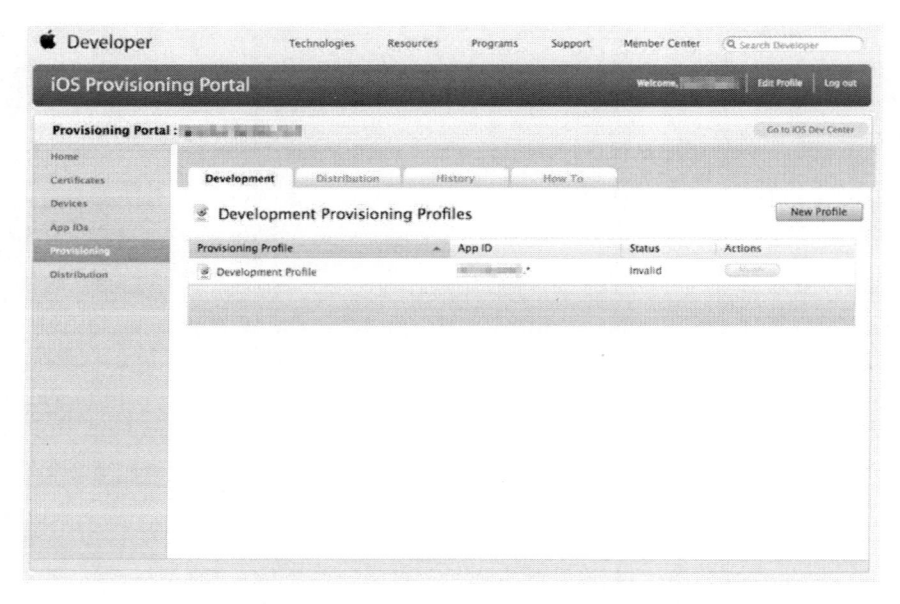

Figure 9-11: iOS Provisioning Portal – Provisioning

3. Select the "New Profile" button and fill out the fields. The "Profile Name" field is a name for the profile. The "Certificates" field is for selecting development certificates the provisioning profile will build with. The "App ID" drop down menu is for selecting what App IDs the provisioning profile will allow to run. For development, it is useful to create an App ID with a "*" as the Bundle Identifier to allow the building and running of any Bundle Identifier the developer chooses to use. Finally, from your list of approved development devices, select which devices the provisioning profile is to be applied, as in Figure 9-12.

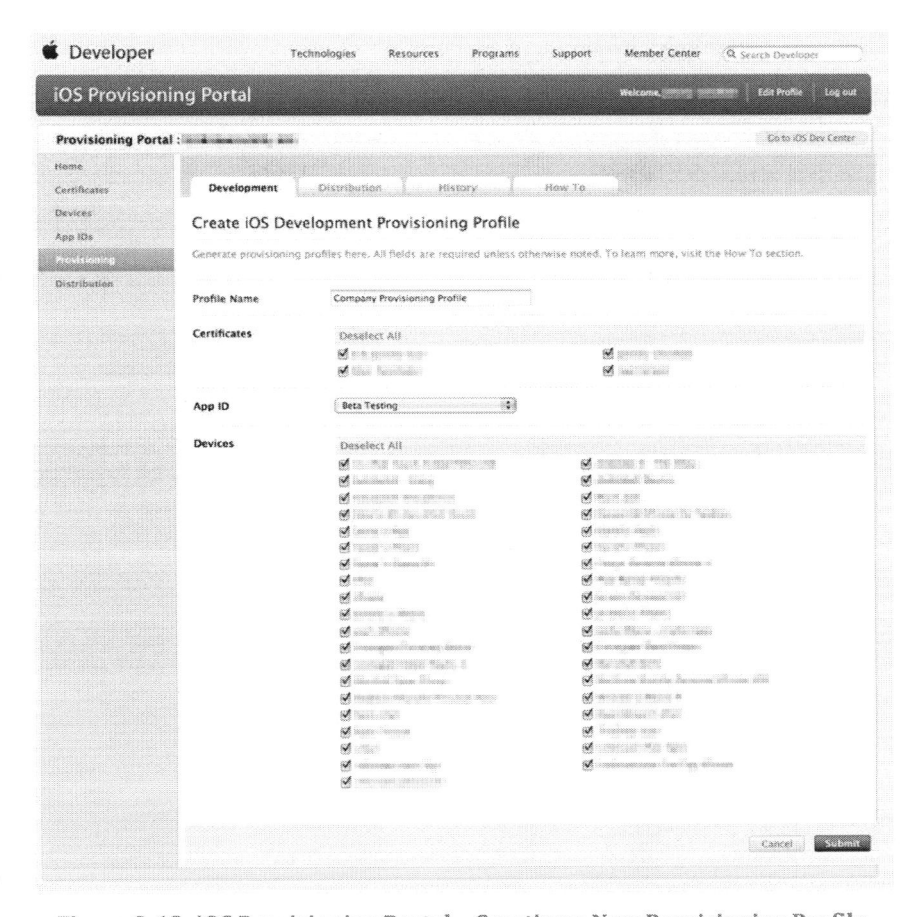

Figure 9-12: iOS Provisioning Portal – Creating a New Provisioning Profile

4. Select "Submit" and refresh the page to download the Provisioning Portal as shown in Figure 9-13.

Chapter 9: Preparing for Development   65

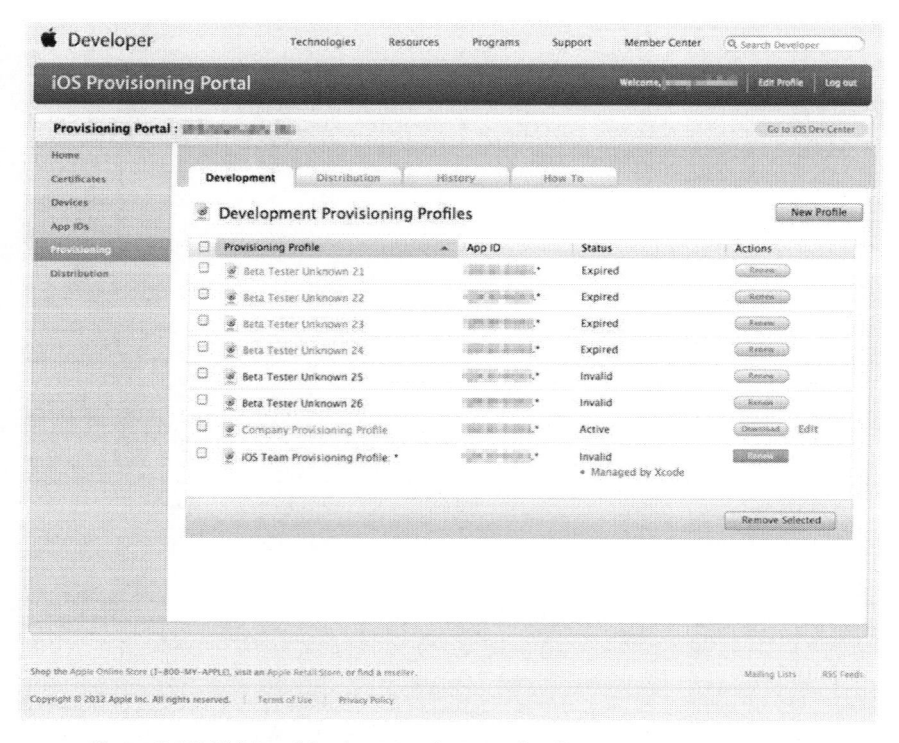

Figure 9-13: iOS Provisioning Portal – Downloading Provisioning Profiles

This concludes the section on manually creating provisioning profiles. The developer can now download the profile, distribute it, add devices and use these devices for development.

In other words, this means you've just completed the hard parts for installing your own experimental app into your own iPhone, iPad, or iPod touch – not just for beta testing, but for showing and bouncing ideas off of friends and family as well!

Checklist – Transitioning to Development

At this point in the book, several important details and necessary steps have been covered that relate to establishing an Apple Developer account (either individual or company), development and advertising accounts set up, making honest efforts towards avoiding conflict, as well as discussions on intellectual property: with the ultimate end-goal of creating a successful app!

The information covered in the previous chapters is invaluable in the sense that it provides the reader with in-depth knowledge that will facilitate their success within Apple's ever-growing App Store.

There are several different approaches to developing iOS apps that will be discussed in detail in the following section. Before moving on, it is a good idea to ensure that the following checklist has been completed:

- ✓ Form a business plan from your app idea.
- ✓ Take steps to avoid legal conflict and protect your intellectual property.
- ✓ Decide if forming a company or not.
- ✓ Create either company or individual developer accounts.
- ✓ Choose any monetization techniques – this can affect app development.
- ✓ Search for potential conflicts, and reserve an available app name.
- ✓ Configure Dev Center and iTunes Connect for app development and testing.
- ✓ Add any physical devices to the list of approved devices for app testing.

This concludes Section I. Section II will provide guidance on choosing an appropriate development method that fits the developer's skill level, as well as the requirements of the app, its content, and your goals.

SECTION II

CHOOSE A DEVELOPMENT OPTION

The following chapters will serve as a guide through the process of helping to deciding which development platform to use, based upon your existing coding experience, knowledge of scripting languages, and other factors.

From this point forward, the focus will be on authoring tools, and help you select the languages and utilities to best develop your app. The following chapters will walk you through the Apple development mentality, and the practical methods that work best based upon the collected experiences of the contributing authors.

Additionally, some chapters of this section cover options for developing your app with cross-platform support, allowing you to write an app ONE time and deploy it on MANY different devices and operating systems - to potentially reach all of the following from making a *single* project. By way of example only: Apple's iTunes App Store for iOS devices (e.g. Apple's iPhone, iPod touch, and iPad), Apple's iTunes Mac App Store for OS X computers (e.g. Apple's Mac Pro, MacBook, and MacBook Pro), as well as Google's Android Market (e.g. Google Play), and potentially others, *depending on the compiler*.

There is also a large market of non-Apple development tools. Every effort has been made to identify and detail some of the most popular third-party app-building solutions, in order to best meet your needs to develop and publish your app.

By the end of this section, you should have a good feeling for the tools you will need in order to begin development of your app!

Introduction to iOS Development Options

When developing iOS apps, there are several different approaches a developer can take. Apple recommends developers use Objective-C and Xcode due to the fact that it is supported and maintained by Apple and gives developers complete control over the operating system and device at a very low level. Objective-C is indeed a powerful programming language; however, it can have somewhat of a steep learning curve. This is especially true if the developer is not familiar with object-oriented programming and memory management principles, which are discussed in Chapter 17. While Apple recommends that developers use their method, alternative development options exist and are discussed at length in this book.

When choosing a development environment in which to create an iOS app, it is important to evaluate several things including the developer's areas of expertise, the app functionality, time, and cost. For example, if the developer's primary area of expertise is .NET and C# programming, MonoTouch might be a good solution because it allows iOS app development with the C# language and .NET libraries. In addition to evaluating the developer's area of expertise, it is important to evaluate the required app functionality. This is because some development alternatives are geared toward creating games as opposed to creating business apps. The chart in Chapter 15 is meant to help with deciding which development option is best for the developer.

The ideal way to develop an iOS app is by using Apple's Xcode IDE and the Objective-C programming language. Using this method generally produces a smaller, faster app, and can easily take advantage of all of the iOS device features such as the accelerometer and GPS capabilities. However, if you are unfamiliar with Objective-C or Xcode and learning a new programming language and environment seems daunting or too time-intensive, there are still many options available to get your app to market.

11.1   From Web Apps to iOS Apps

Many authoring tools allow you to use a wide variety of languages to write your apps, including HTML, JavaScript, CSS and others. In many cases, with a little tweaking, you can even take web apps that you've already written and convert them into iOS apps.

Two tools that work this way include PhoneGap and Titanium Appcelerator. Both require that you have the iOS SDK and Xcode installed, with PhoneGap being more integrated with Xcode than Appcelerator. In order to build a project using PhoneGap, you actually begin by starting Xcode, then selecting "PhoneGap-based Application" as the template you wish to use for your new project.

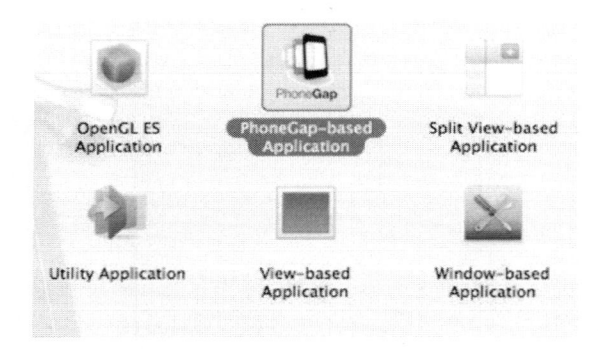

Figure 11-1: PhoneGap's template icon in Xcode

PhoneGap will build the files you need to get started, which you amend with the necessary HTML, JavaScript and CSS code using a text editor. When you are ready for testing or distribution, Xcode compiles it into a functioning app.

Appcelerator works a little differently, in that you only need to let it know where the iOS SDK and Xcode are installed when you are setting it up. You then code your app in Appcelerator using HTML, JavaScript, etc. and use Xcode to compile and test your app when you are ready to do so.

We'll go into more depth into these tools in their respective chapters later in this book.

11.2 Game-Focused Alternatives

While languages such as HTML and JavaScript are great for writing web-apps that provide information and database interaction, traditionally they are not used for developing games. Fear not, though, there are still alternatives to the Xcode/Objective-C combo that focus primarily on game-creation.

11.2.1 Independent Authoring Tools

There are a number of game-focused authoring tools that do not require Xcode for programming. These tools offer a great way for programmers and game designers, who have experience in other languages, to easily create games for iOS.

Unity3D is one of these authoring tools, which, as its name suggests, provides a full 3D engine at your disposal. Unity3D is also relatively easy to use, especially for developers who consider themselves more "artist" than "programmer." It will also handle things such as physics and collision detection, which can be significantly harder to implement using Xcode.

While 3D game development is definitely easier with Unity, developing 2D games with it tends to be more difficult, so if you are designing a 2D game, it's probably best to look at using a different authoring tool. There are also issues with implementing default iOS elements and a lack of SVN support hinders team collaboration.

11.2.2 GUI-Based Development

There are still other authoring tools that allow you to get up and running without having prior experience writing code. GameSalad, for instance, is mostly graphical and menu-driven. Designed specifically with non-programmers in mind, it provides an easy to understand, drag-and-drop interface. GameSalad remains a great introduction to game design in general, and dramatically simplifies the process needed to create and bring a game to market.

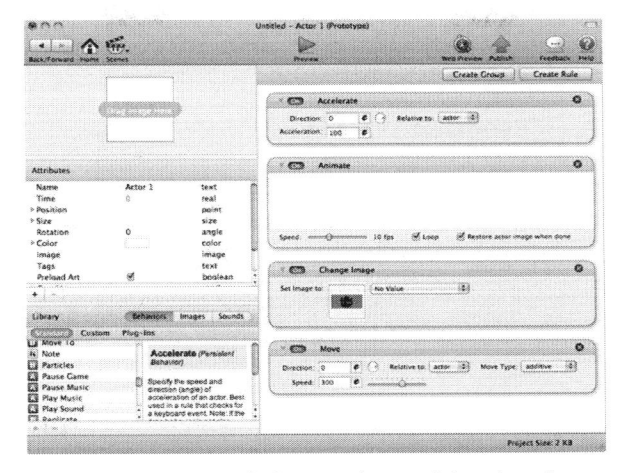

Figure 11-2: GameSalad uses a drag and drop interface

11.3 Advantages of Using Development Alternatives

The idea behind each of the alternative development tools covered in this book is to give those unfamiliar with Xcode and Objective-C a way to bring their app ideas to Apple's App Store. Learning a new programming language can be difficult and time consuming, even for experienced coders. Even if you have plenty of programming experience already, these alternatives give you a chance to use your current programming skills to experience iOS development.

Another advantage of using these authoring tools is that many of them allow you to develop for multiple platforms simultaneously. This provides the convenience of writing code once and then being able to distribute it to a variety of devices. However, this carries the disadvantage of larger, more sluggish apps due to another layer of programming for interpretation.

11.4 Disadvantages of Using Development Alternatives

The main advantage of developing in Objective-C is that it is the native language of iOS and Xcode. Developing apps with Objective-C means they will be able to take full advantage of all of the iOS device's features without using any workarounds or "hacks." Native apps also tend to be smaller, quicker and require less tweaking to make them behave as intended. All of these alternate development methods require another layer of code to interpret the non-native code to run on an iOS device. This extra layer of programming can add considerable size to the app.

This extra layer of programming can also slow down the performance of the app. Similar to the way that communication would be quicker between two people who speak the same language versus two who are communicating through an interpreter; apps that are written in Objective-C need no translation to run on iOS, and are therefore faster.

Another disadvantage with using these alternatives is cost. Most of these programs have free versions that allow you to try before you buy. The free versions are usually limited in their functionality in some way, like requiring a splash screen to be displayed upon startup, or only providing a 30-day license. This is a good option to get familiar with the environments and SDKs and their capabilities before committing to purchasing a paid version, many of which can run into the hundreds and even thousands of dollars.

Cross-Platform Development Options

Not necessarily a disadvantage, but one thing to note when using these alternate development methods is that in most cases, you will still need to have Xcode and the iOS SDK installed on your machine in order to test your game using Apples iOS Simulator. While Xcode is free, in order to make your app available on iTunes, you will need to enroll in Apple's iOS Developer Program. Membership currently costs US$99 a year, and is explained further in Chapter 6. Cross-Platform Development Options

Several of the alternative development methods mentioned in this book have cross-platform production capabilities. A cross-platform development option allows a developer to write and maintain a single code-base and deploy an app to multiple operating systems and devices. This is extremely useful in the current smart phone market, due to the fact that each of the respective major smart phone operating systems require apps to be written in different programming languages as per operating system requirements. As mentioned in this book, Apple's iOS apps are written natively in Objective-C. Android and Blackberry native apps are written in Java and require the use of an Android and Blackberry SDK respectively. A cross-platform development solution allows a developer to create an app one time and deploy it on multiple devices and operating systems at once.

Remember, there are **two** Apple stores: Their App Store, catering to iOS devices, and their Mac App Store, catering to OS X computers. Wouldn't it be fantastic to reach customers in **both** stores, from creating the ***app just once***? While it's true that Apple's Xcode is designed for building apps for either of their two stores, some third-party cross-platform environments can make that task easier and more efficient, as well as ***simultaneously*** building your app for ***completely different*** platforms – from a single project! For example: create an iPhone app, and automatically get Android, or even others - ***all at the same time! Bonus!***

Various development environments achieve cross-platform app production in a variety of different ways. Development environments, such as Unity 3D, can compile Unity scripts into native source projects for the respective platforms. Other environments, such as PhoneGap, leverage the operating systems browser and allow the developer to write one single web app, and have it run on different devices and operating systems to access native user interface elements as well as device functionality. What PhoneGap essentially does is communicate with the operating system specific browser and native functions, allowing developers to maintain one code base that will run across multiple devices and operating systems.

If your goal is to quickly create an app that runs on multiple platforms, it may be worthwhile to investigate some of the alternative development options that support cross-platform development. It is always best to ensure that the functionality requirements of your app can be achieved by using the third-party alternative development environment you choose before starting development.

Apple Recommended Development Method

Apple offers a powerful, professional group of tools for iOS and OS X development called Xcode. Note: This book, and its discussions on Xcode, focus on mobile apps running on Apple's iOS for iPhone, iPod touch, and iPad; not desktop software running on Apple's OS X for Mac.

> **TIP:** *If you decide to explore Xcode in this chapter, skip ahead to **Chapter 16: Introduction to Xcode**; and then return here.*

These powerful tools include an integrated development environment, Apple's complete API documentation for iOS and other essential tools to create apps quickly and efficiently. Among these tools are several features, which make Xcode the preferred method of iOS app development.

An integrated development environment (IDE) is a program that offers a source code editor, a compiler and a debugger. XCode's IDE offers multiple workstations for when you are working on more than one project.

Interface Builder is used to create graphic-oriented user interfaces. The source code editor with Xcode contains advanced code completion which saves you time by suggesting completed portions of code while you are programming; code folding which allows you to hide snippets of code for easier code readability; and messages that get displayed alongside your code with compiler warnings, errors and other comments.

Fix-It is a built in feature that will suggest code corrections for problematic code you have typed.

Assistant Editor is an option that, when activated, will automatically open the corresponding header file to any main file you are editing.

Snapshots offer the ability to save the current state of your code for restoration in case you make an irreversible mistake. The debugger can show data values of variables in real-time during the use of breakpoints.

Static Analysis warns you of logical fallacies in your code.

The most notable feature is the in-depth documentation that Apple has written for their developers. If you are unsure of the use of a class in a framework then you can go to Apple's iOS Developer Library and look up the Class Reference.

Quick Help is a feature in the sidebar of Xcode that is essentially Apple's documentation at a glance. When writing your program, any selected Class will activate Quick Help. It gives you the class name, declaration, iOS version availability, abstract, link to the class reference in the documentation, related documents, and even sample code that includes the class.

Xcode offers multiple compilers for C, C++ and Objective-C. The advantage of using these programming languages is the low-level access they grant you because, whether you want to access the camera or the GPS, Objective-C is the language of the classes and frameworks that talk to the device. Other programming languages have to bridge the gap between their foreign language and the native language of the frameworks in order to work properly.

For an in depth look at Xcode, please see ***Section III: Apple Supported Development Methods (via Xcode)***.

Overview of Non-Apple Development Options

The following sections will contain a brief overview of *some* of the different third-party alternative development options to Apple's Xcode, discussing what platforms each support, the cost, and the programming languages that can be used with them. It is worth noting that many of these alternatives are constantly being updated with new functionality and supported platforms added. The prices of these development options do fluctuate and are subject to change.

14.1   Unity 3D
http://www.unity3d.com/

Unity3D is a multi-platform 3D game development tool. The supported programming languages are: C#, JavaScript, and Boo. The supported workstation Operating Systems are: Windows and Mac OS X (required for iOS). Unity3D can build games for the following platforms: Android, iOS, Mac OS X, Windows, Web Player, Adobe Flash, Nintendo Wii, Sony PS3, and Microsoft Xbox 360. The software price ranges from free to $4,500 depending upon your needs. To create iOS games a $400 add on is required. The same is true for Android development. Building games for PS3, Xbox 360, or Wii requires a special license which can be obtained by contacting Unity 3D's sales department at sales@unity3d.com . The upcoming version of Unity3D may support building games for the Linux Operating System as well.

14.2   ShiVa3D
http://www.stonetrip.com/

ShiVa3D is a multi-platform 3D game development tool. ShiVa3D offers a game engine, editor, and MMO server. The supported programming languages are: Lua, C, C++, and Objective-C. The supported workstation Operating System is Windows. ShiVa3D can build games for the following platforms: Desktop (including Windows, Mac OS, and Linux), Mobile Application (including iOS, Android, BlackBerry QNX, WebOS, and Marmalade), Web Browsers, and Nintendo Wii. The price of the software ranges from free for the Web Edition, $400 for the Basic edition which allows publishing of mobile apps, and $2,000 for the advanced edition.

14.3   PhoneGap
http://www.phonegap.com/

PhoneGap is a mobile development framework. The supported programming languages are: JavaScript, HTML5, and CSS. The supported workstation Operating Systems are: Windows, Mac OS X, and Linux depending on which platform you wish to build for. PhoneGap can build apps for the following platforms: iOS, Android, Blackberry, WebOS, Windows Phone 7, Symbian, and bada. PhoneGap is free but customer support costs start at $24.95/month or $249.99 for the year.

14.4 MonoTouch
http://www.xamarin.com/monotouch/

MonoTouch is mobile software development kit. The supported programming language is C# and the .NET framework. The supported workstation Operating System is Windows. MonoTouch can build apps for iOS while its sister software Mono for Android can build apps for Android. MonoTouch ranges from $399 to $2,499.

14.5 Marmalade
http://www.madewithmarmalade.com/

Marmalade is a cross platform software development kit. The supported programming languages are: C/C++, Objective-C, CSS, JavaScript, and HTML5. The supported workstation Operating Systems are: Mac OS X and Windows. The Marmalade SDK can build apps for the following platforms: BlackBerry Playbook OS, Windows desktop (7, XP SP3), Mac OS X, LG Smart TV, Android, bada, iOS, and Symbian. Marmalade SDK licenses starts at $149 for the Community License which only supports iOS and Android.

14.6 Adobe Flash Builder
http://www.adobe.com/products/

Adobe Flash Builder is an Eclipse-based IDE for multi-platform development. The supported programming languages are: ActionScript and MXML. The supported workstation Operating Systems are: Mac OS X and Windows. Adobe Flash Builder can build apps for the following platforms: Android, Blackberry Tablet OS, and iOS. The standard version of Adobe Flash Builder costs $249 and the premium version costs $699.

14.7 Adobe Flash Professional
http://www.adobe.com/products/

Adobe Flash Professional is a multi-platform multimedia development tool. The supported programming language is ActionScript 3.0. The supported workstation Operating Systems are: Windows 7, Windows XP SP3, or Mac OS X v10.6.8, v10.7, v10.8 are supported from CS5 onward. Apps can be built for the Adobe Flash Player and Adobe AIR runtimes. These two runtimes support the following platforms: Android, Blackberry Tablet OS, iOS, Mac OS X, Windows, and browsers with the Adobe Flash Player plugin. Adobe Flash Professional CS6 costs $699.

14.8 Cocos2D
http://www.cocos2d-iphone.com/

Cocos2D is an iOS framework for building 2D games, demos, and other graphical/interactive apps. The supported languages for Cocos2D are: C/C++, and Objective-C. The supported workstation Operating System is Mac OS X with Apple's Xcode IDE. Cocos2D can build apps for iOS and Mac OS X. Cocos2d is free.

14.9 Corona SDK
http://www.anscamobile.com/corona

Corona SDK is a mobile app development framework. The programming language it uses is Lua. The supported workstation Operating Systems are: Mac OS X, and Windows. Corona SDK can build apps for the following platforms: iOS, Android, Kindle Fire, and Nook Color. The cost for Corona SDK is $199/year for the iOS or Android version; the cost for both is $349 with the inclusion of Kindle Fire and Nook Color support.

14.10 GameSalad
http://www.gamesalad.com/

GameSalad is a game-authoring tool designed for non-programmers. Instead of using a programming language it uses a graphics-based drag-and-drop game-creation interface. The supported workstation Operating Systems are: Windows and Mac OS X. GameSalad can build apps for the following platforms: Android, iOS, Mac OS X, Windows 8 and HTML5. Game salad has a free version as well as a "PRO" version that costs $299 per year. The "PRO" version supports Android Publishing as well as several important iOS features such as iAds, Game Center, and In-App Purchases.

14.11 Titanium Studio
http://www.appcelerator.com/

Titanium Studio is a multi-platform software development tool that utilizes web technology. The supported languages are: JavaScript, HTML5, and CSS. The supported workstation Operating Systems are: Mac OS X 10.6-10.7, Windows 7, and Ubuntu Linux 10.04-12.04. The Titanium SDK can build apps for the following platforms: iOS, Android, and HTML5. Appcelerator Titanium has a free version and a paid version. Contact sales for 3 tiers above free, which apply to enterprise apps, enhanced support and features such as analytics and cloud services.

14.12 MoSync
http://www.mosync.com/

MoSync is software development kit for mobile apps. The supported languages are: JavaScript, HTML5, and C/C++. The supported workstation Operating Systems are: Mac OS X, Windows PC, and Linux. MoSync can build apps for the following platforms: Android, iOS, Windows Phone 7, Blackberry, JavaME, Moblin, Symbian, and Windows Mobile. MoSync is free for private use; however if you wish to sell or distribute your app, you must publish your source code openly under the GPL2 license. If you do not want to publish your source code then you must obtain a commercial license from them, of which there is a free option or a couple paid options that come with support (annual subscription starts at €199.)

Development Options Summary

The process of making an iOS app can be done in many different ways. The ideal way is to use Objective-C and Xcode due to its support, low-level device access, and maintenance from Apple. **Alternative development options are available!** When deciding which development option to utilize, it is important to consider the requirements and possibilities of each option. The following chart is meant to help with deciding which development option is best for the developer.

Product Name	Workstation OS	Supported Languages	Destination OS	Price
Xcode	Mac OS X	C C++ Objective-C	iOS, Mac OS X	Free
Unity3D	Mac OS X Windows	UnityScript C# Boo	Android, iOS, Mac OS X, Sony PS3, Nintendo Wii, Windows, Xbox 360, Adobe Flash, Web Browser	Basic: Free iOS Basic/Pro Add-on: $400 / $1,500 Pro Version: $1,500 + Add-on For PS3, Wii, Xbox360 licenses contact: sales@unity3d.com
ShiVa3D	Windows	Lua C++	Android, iOS, WebOS, BlackBerry QNX, Mac OS X, Windows, Linux, Web Browser, Marmalade, Nintendo Wii	Basic: Free Paid Versions Starting at $400
PhoneGap	Mac OS X Windows Linux	JavaScript HTML CSS	Android, bada, BlackBerry, iOS, Symbian, WebOS, Windows Mobile, Windows Phone 7	Free Support Packages Basic: $24.95/mo Starter: $95/mo Pro: $495/mo Corporate: $1,995/mo Enterprise: Contact Support
MonoTouch	Mac OS X Windows	C# .NET	Android, iOS* *Creation on Mac OS X only	$399 - $1,499* *iOS and Android Versions Sold Separately

Product Name	Workstation OS	Supported Languages	Destination OS	Price
Marmalade	Mac OS X Windows	C C++ Objective-C HTML5 JavaScript CSS	Android, iOS, bada, BlackBerry PlayBook OS, Windows desktop (7,XP3), Mac OS X desktop, LG Smart TV	Community: $149 Indie: $499 Professional: Contact for Price
Adobe Flash Builder	Mac OS X Windows	ActionScript MXML	Android, Blackberry, iOS	Standard: $249 Premium $699
Adobe Flash Professional	Mac OS X Windows	ActionScript 3.0	Android, BlackBerry, iOS, Mac OS X, Windows	$699
Cocos2d	Mac OS X	C C++ Objective-C	iOS, Mac OS X	Free
Corona	Mac OS X Windows	Lua	iOS, Android, Kindle Fire, Nook Color	iOS or Android $199/year iOS, Android, Kindle Fire & Nook $349/year
GameSalad	Mac OS X Windows	None	Android, iOS, Mac OS X Windows 8 HTML5	Basic: Free GameSalad Pro: $299/year
Titanium Studio	Mac OS X Windows Linux	JavaScript HTML5 CSS	iOS Android HTML5	Basic: Free Paid: Contact sales for enterprise apps and enhanced support and features such as analytics and cloud services
MoSync	Mac OS X Windows Linux	HTML5 JavaScript C C++	Android, Blackberry, iOS, JavaME, Moblin, Symbian, Windows Mobile, Windows Phone 7	Free (no support) Basic 199 EUR Gold 2,999 EUR

SECTION III

APPLE SUPPORTED DEVELOPMENT METHODS (VIA XCODE)

This Section focuses initially on introducing and then driving deep through Apple's Xcode development environment, using the Objective-C programming language. If you are not a programmer, then do not fret! **There are alternatives.** The Chapters build on each other, guiding you through this process in specific steps, with clear example code, and easy to follow explanations.

Additionally, many consider the deeper parts of this Section which dive into Objective-C for existing programmers, as a "book within a book" bonus guide.

It's important to emphasize that the Chapters in this Section will slowly build on each other, starting with an introduction to Xcode itself. This gradual Chapter-by-Chapter approach was intentionally structured in sequential order, with new concepts thrown in briefly and then expanded on much later, to help avoid overwhelming newcomers to Xcode's otherwise seemingly massive environment.

After this entire Section, you will have a solid grasp of working with Xcode and Objective-C, which you can use as the foundation for developing very robust and profitable iOS apps.

Or, after installing Xcode in Chapter 16 (required for digitally signing and uploading apps to Apple): If you wish to avoid Xcode as much as possible and Objective-C altogether, then you CAN simply jump right to the next Section for alternatives: Starting in Chapter 25!

16

Introduction to Xcode

Xcode is the one-stop-shop for Apple platform development. It provides a complete set of editing, building, debugging, profiling, testing, and source control features. While third-party development and cross-platform tools do exist, as you have read in Chapters 11-15, Xcode is generally the center of development workflow.

> **TIP:** *Xcode is a massive environment. This chapter starts with merely naming some of Xcode's terms and components, which will first be defined in an introductory style. The chapter will then walk you through a few details while showing each component, starting with 16.2 Xcode Installation, 16.3 Xcode's Windows, and then leads into using Xcode's feature-set for the traditional "Hello World" (written in Objective-C merely to showcase Xcode's features: Objective-C is introduced later in Chapter 17).*

Gradually, each of the following Chapters in this Section will build on Xcode, covering more and more of Xcode's extensive feature-set. This was intentionally done to help avoid completely overwhelming newcomers to Xcode's immense scope and capabilities, and to help become productive with it as fast as possible.

In September 2012, Apple released the Xcode 4.5 Gold Master to developers and shortly thereafter, the world. Due to an update in the Xcode 4.5 GM release, all external libraries must be updated to be compiled with the GM build. All apps that use libraries not compiled with Xcode 4.5 will experience build errors. This is an industry wide issue driven by Apple's updates and will impact your SDKs from all third-party vendors. It is recommended to look to those vendors for updates to their apps for Xcode 4.5 and iOS 6 compatibility. This is common when Apple releases a new iOS.

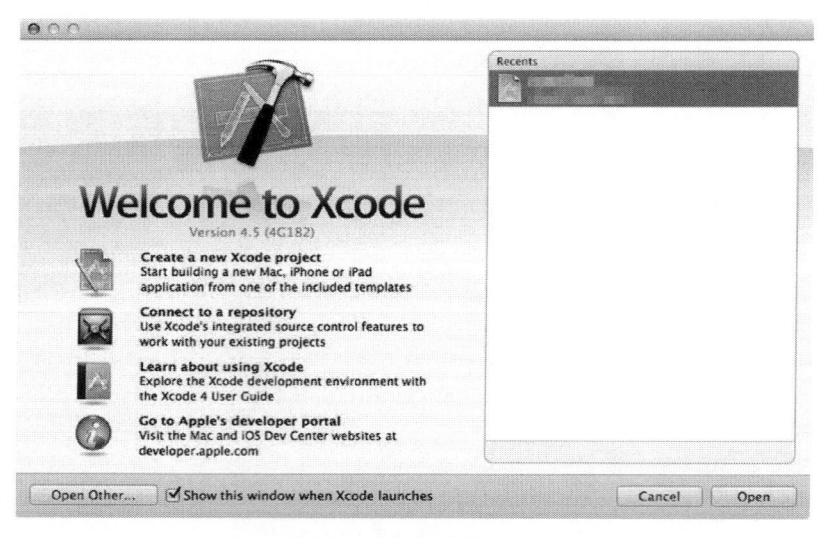

Figure 16-1: Xcode 4 Welcome Screen

16.1 Main Feature Overview of Xcode

Xcode comes equipped with development tools for all Apple platforms: Mac OS X, Safari, and iOS. Though mainly interested in iOS development tools within this book, it's a good idea to be aware of everything this powerful utility has to offer; this is especially important for seasoned programmers that are new to the Apple Mac environment. This chapter will start with merely naming and showing some of the major components, leading gradually into downloading, installing, and getting started with Xcode.

16.1.1 Interface Builder

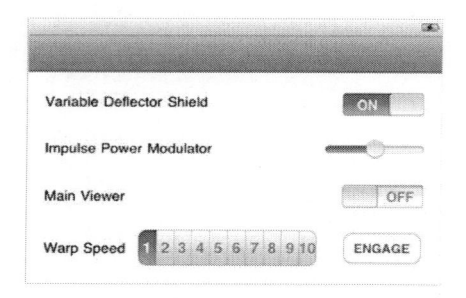

Figure 16-2: UI made with Interface Builder

The interface builder allows you to quickly throw together a UI for all platforms. Visual elements, like buttons and lists, are simply dragged and dropped into place. The interface builder also connects these elements to objects and methods in your code. It's used to create interfaces for Mac OS X native apps, as well as iOS apps. It will be covered in detail later.

88   Chapter 16: Introduction to Xcode

16.1.2 Simulator

Notice the careful choice of words: iPhone and iPad *simulator*, not *emulator*. The simulator allows testing of iOS apps on a Mac. It provides enough functionality for basic testing of layouts and simple code changes. It runs reasonably fast and it's easy to use. Xcode will build, install, and run your app on the simulator in just one click.

Figure 16-3: iOS Simulator

Unfortunately, the simulator is just that: a simulator. Unlike the Google Android platform, the environment is not an emulator. A simulator mimics the behavior of a device while an emulator mimics the hardware of the device. This isn't necessarily a bad thing, but it can become a double-edged sword if the developer isn't aware of potential pitfalls.

16.1.3 Instruments

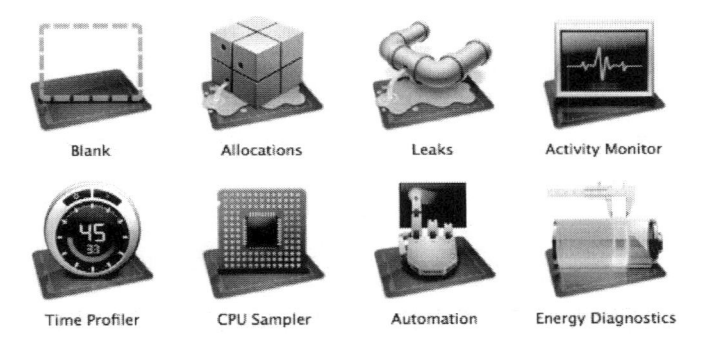

Figure 16-4: Instrument Templates

Xcode includes integrated software profilers, known as instruments. They provide various performance metrics for your software such as memory usage, memory leaks, function/method CPU time, resource usage, and much more. Instrument data can be used to identify issues in app performance and reliability, as well as a benchmark for comparison between code revisions.

16.1.4 Other Notable Features

- **Assistant**
    - A Side-by-side, two-pane view where Xcode displays the relevant file in the pane opposite to the active editor. So while you are editing your User Interface, Xcode will automatically show you the corresponding controller.
- **Fix-it**
    - Similar to grammar checkers in word processors, Xcode can now detect common programming mistakes and provide possible solutions.
- **Version Editor**
    - Xcode integrates version control (Subversion and Git) with the editor interface. This allows for intuitive browsing through code revisions via a timeline.

16.2 Xcode Installation

Now that we've learned a bit about Xcode, download and install it along with the iOS SDK.

1. Go to http://developer.apple.com and click on the iOS Dev Center link.

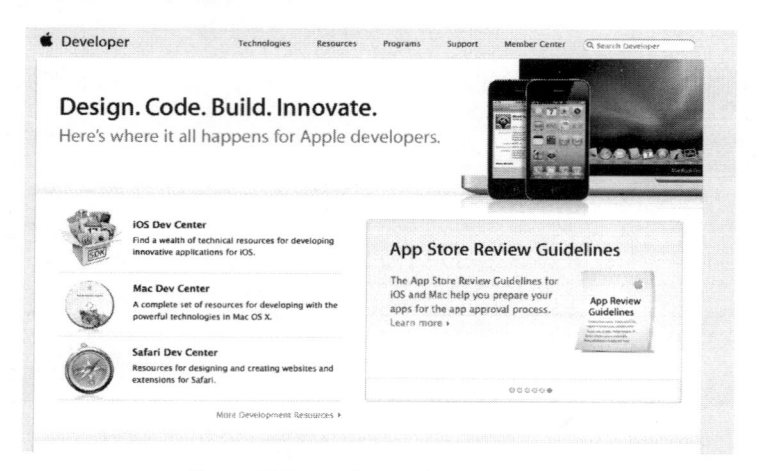

Figure 16-5: Apple Developer Website

2. Once the page loads, click on the link to download Xcode 4. Doing so will prompt you to login.

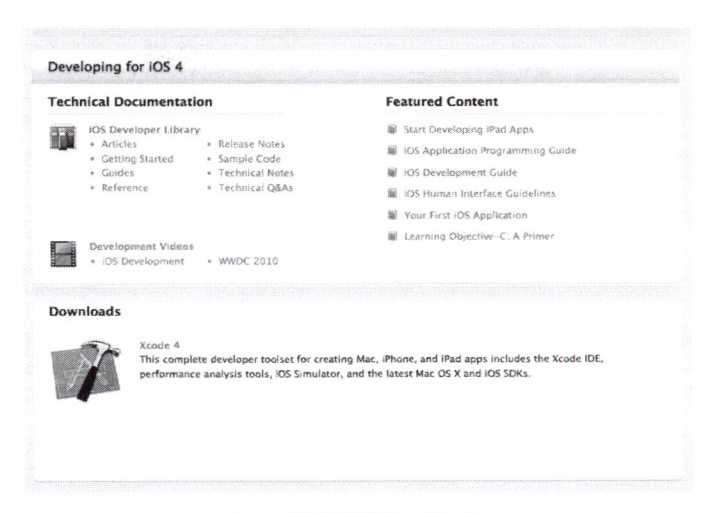

Figure 16-6: iOS Dev Center

3. Once logged in you can download Xcode 4 and the iOS SDK by clicking the provided link, **"Looking for additional developer tools? View Downloads"** (see Figure 16-7). *At the time of writing this book the current version of Xcode is 4.5 and iOS SDK 6.*

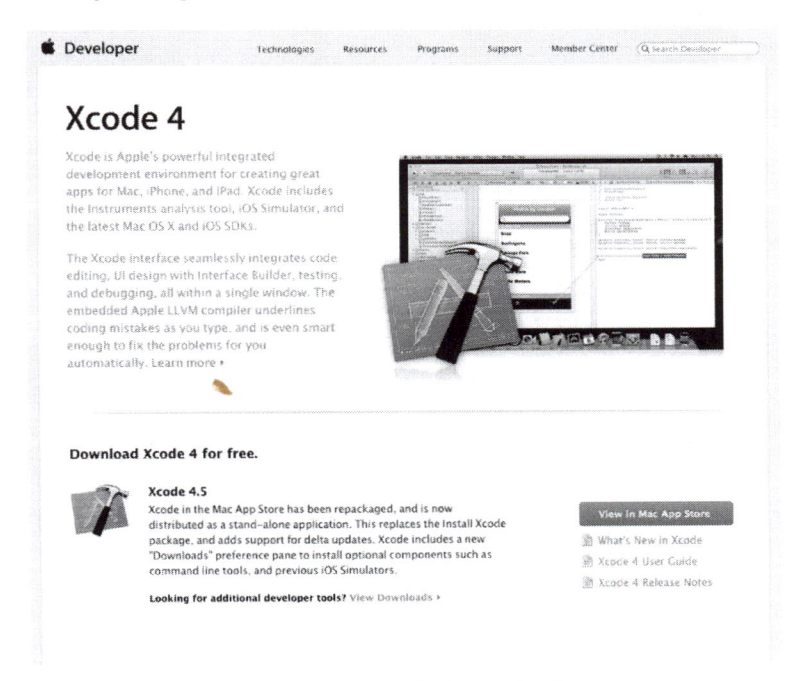

Figure 16-7: Xcode 4 Download

4. Run the installer, which will walk you through the installation process.

Figure 16-8: Install Xcode

5. Once the installation is complete - you can now launch Xcode and start learning its features, which will be reviewed in the next section.

16.3 Xcode's Windows

When you launch Xcode for the first time you will be asked to create a new project. For the purpose of seeing the IDE, create a new View-Based Application by selecting its icon from the set of templates shown in Figure 16-9 and name it "Introduction."

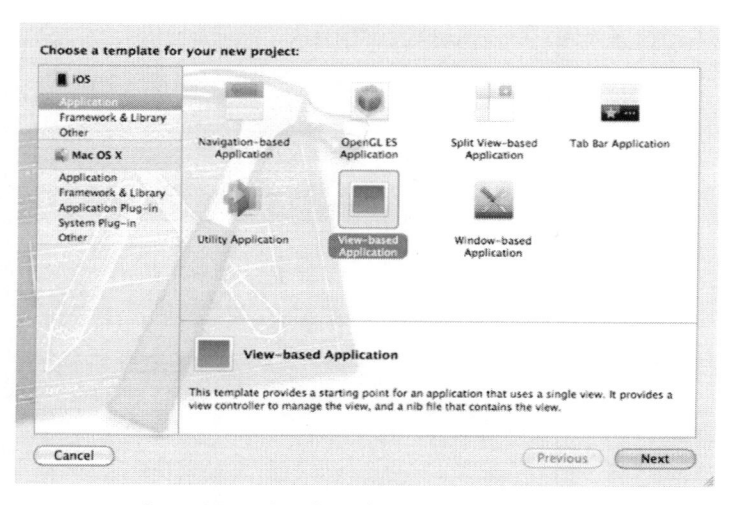

Figure 16-9: View-based Application template

After creating a project, the main Xcode window will appear and should look similar to Figure 16-10.

92   Chapter 16: Introduction to Xcode

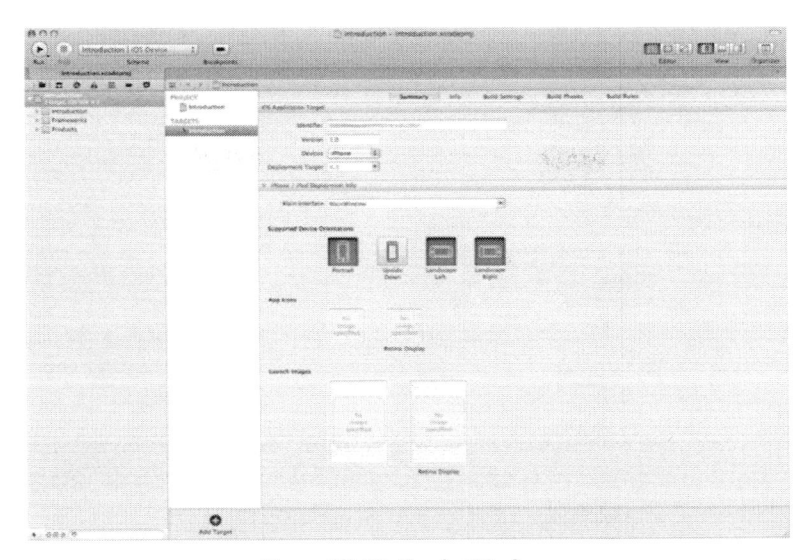

Figure 16-10: Xcode Window

The top right corner of Xcode contains buttons that allow you to select which areas you want displayed, which kind of editor you want to use, and launch the organizer. This can be seen in Figure 16-11.

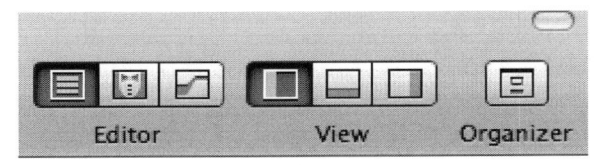

Figure 16-11: Editor, View, and Organizer controls

16.3.1 Editors

There are three different options for the Editor pane. They are the Standard editor, the Assistant editor, and the Version editor.

16.3.1.1 Standard Editor

The Standard editor is displayed by default when you open Xcode. It displays one file at a time and changes depending on the type of file. For example, clicking on a xib file will change it to Interface Builder editor.

16.3.1.2 Assistant Editor

The Assistant editor is useful when you want to have more than one file displayed at a time. For example, it can be used to view and edit two separate parts of the same file at the same time. Other than configuring the Assistant editor manually you can set it up to one of several other options, one of the more useful ones is the counterpart option. This sets up the Assistant editor to display the counterpart to the current open file; meaning you could have the .m and .h files displayed simultaneously, which is useful because they are counterparts to each other.

16.3.1.3   Version Editor

The Version editor is used with version control to compare revisions of files. There are several options with the Version editor that include showing a file comparison and timeline, changing the logs for a file, and individual change annotations for each line of a file.

16.3.2   Navigator View

The left hand area is called the Navigator and is currently showing the Project Navigator. Other than the Project navigator there are six other navigators that can be used to navigate through various things. These navigators are as follows: Project Navigator, Symbol navigator, Search navigator, Issue navigator, Debug navigator, Breakpoint navigator, and the Log navigator.

16.3.2.1   Project Navigator

The Project Navigator shows all the files in the open project, as seen on the left side of Figure 16-12. If you expand the folders you can see the files in the Introduction project just created. If you select the IntroductionViewController.h file you will see the file's contents in the Editor area, as seen on the right side of Figure 16-12.

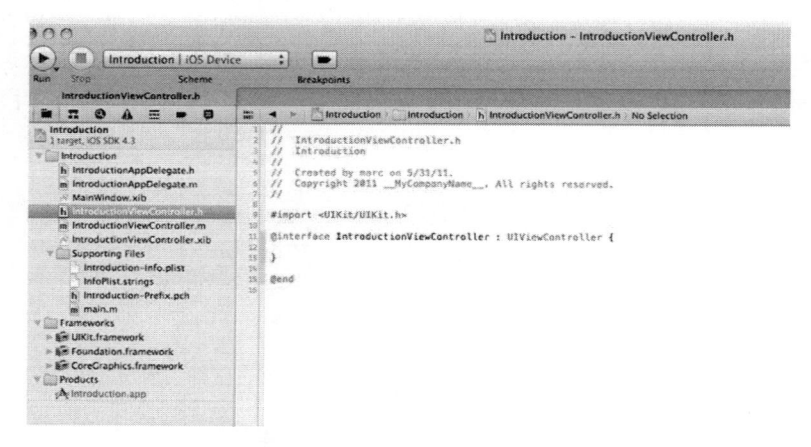

Figure 16-12: IntroductionViewController.h

16.3.2.2   Symbol Navigator

The Symbol Navigator, shown in Figure 16-13, lets you browse through the symbols in your project. Symbols can be displayed in a hierarchical or flat list, and the three buttons on the bottom can refine the displayed symbol list. The first button hides symbols other than classes and their members like protocols and enums. The second button hides symbols that are defined in external frameworks, resulting in displaying only symbols defined in the project or workspace. The third button hides member symbols that are defined within the classes.

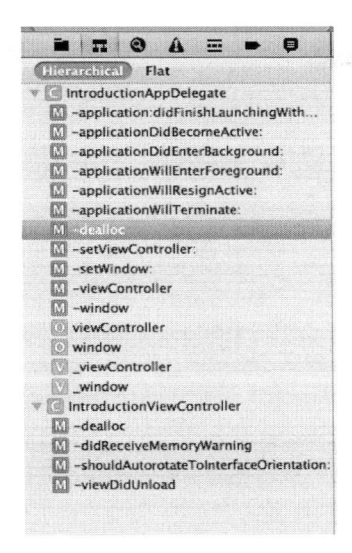

Figure 16-13: Symbol Navigator

16.3.2.3   Search Navigator

The Search Navigator can be used to find text strings in your project or workspace. Figure 16-14 shows the Search Navigator.

Figure 16-14: Search Navigator

16.3.2.4   Issues Navigator

The Issues Navigator lists messages, warnings, and errors in your code and project. The Issues Navigator is shown in Figure 16-15.

If, there are errors when building the project, the Issues Navigator will be opened automatically. Clicking on an issue will display it in the source editor to the right.

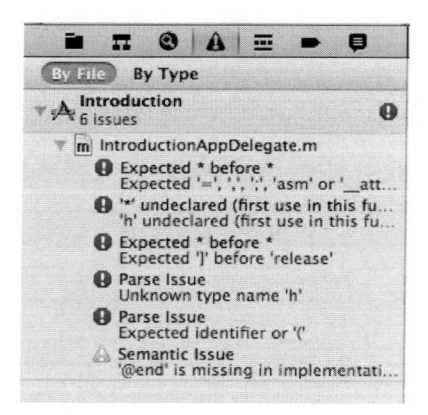

Figure 16-15: Issues Navigator

16.3.2.5 Debug Navigator

The Debug Navigator shows the current state of the stack in each running thread while you are in the process of debugging. If you click an item in the Debug Navigator, it will show you the values of variables in the debug area. This navigator is shown in Figure 16-16.

Figure 16-16: Debug Navigator

16.3.2.6 Breakpoint Navigator

The Breakpoint Navigator allows you to see all your breakpoints in one place. It is also possible to use this navigator to add conditions and options to breakpoints. The Breakpoint Navigator is shown in Figure 16-17.

Figure 16-17: Breakpoint Navigator

16.3.2.7 Log Navigator

The Log Navigator logs operations such as building or analyzing your code, debug sessions, and other similar operations that are normally recorded. Selecting an item in the Log Navigator will show the log in the editor pane. This navigator is shown in Figure 16-18.

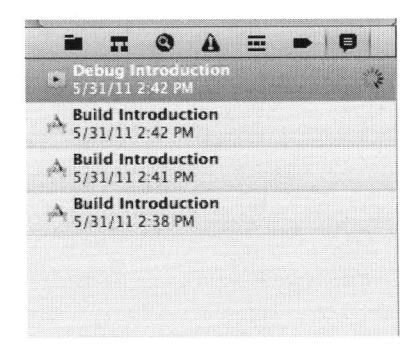

Figure 16-18: Log Navigator

16.3.3 Debug view

The bottom view is the Debug area, shown in Figure 16-19. It can be configured to display the console pane and/or the variables pane.

Figure 16-19: Debug View

16.3.4 Utilities view

The right-most view is the Utilities area, which holds the different inspectors and libraries. The Utilities area can be seen in Figure 16-20.

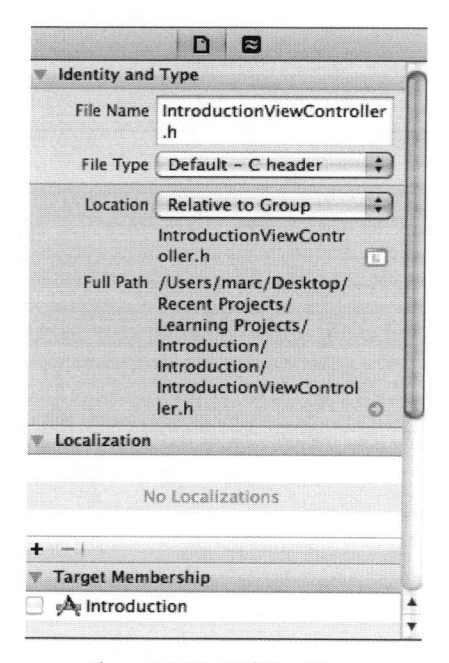

Figure 16-20: Utilities View

If a file other than a *xib* is selected, the File Inspector and Quick Help are available. However if a *xib* file is selected, four extra inspectors are available. They are the Identity Inspector, Attributes Inspector, Size Inspector, and Connections Inspector, which are shown in Figures 16-21, 16-22, 16-23 and 16-24.

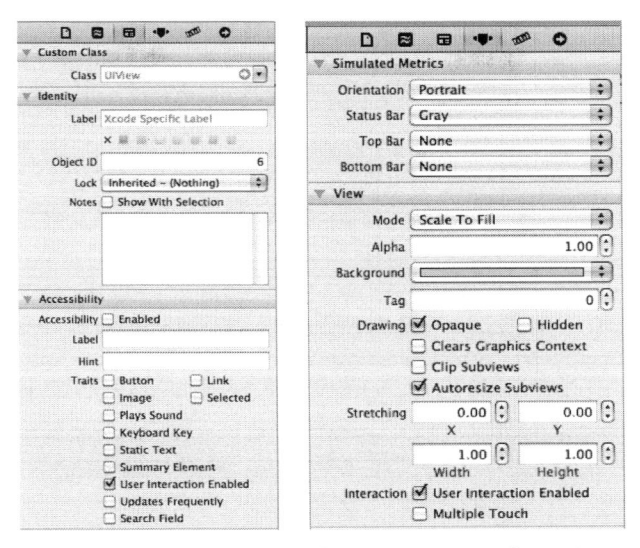

Figure 16-21: Identity Inspector and Figure 16-22: Attributes Inspector

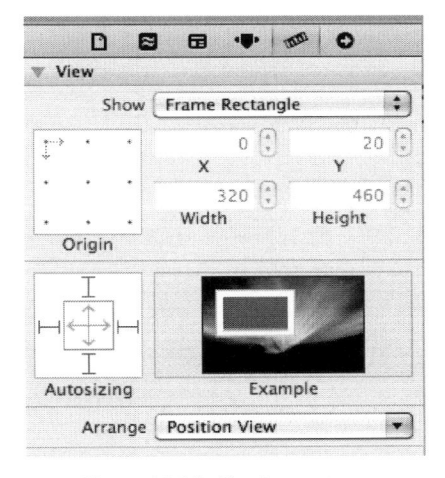

Figure 16-23: Size Inspector

Chapter 16: Introduction to Xcode   99

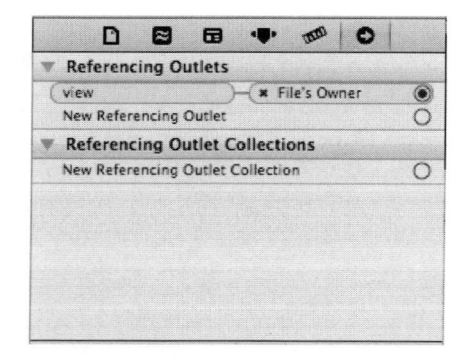

Figure 16-24: Connections Inspector

More information will be given on the various Interface Builder Inspectors in Chapter 19.

16.3.5 The Organizer Window

In the top right hand corner of the workspace window there is a button to launch the Organizer window. The Organizer window displays documentation, project organization, source control, and access to mobile devices.

16.3.5.1 Documentation

In the Organizer window there is a button that, when selected, allows you to browse through iOS, Mac OS X, and Xcode Developer libraries. These libraries can easily be browsed through, searched, and bookmarked for later reference. The Documentation window is displayed in Figure 16-25.

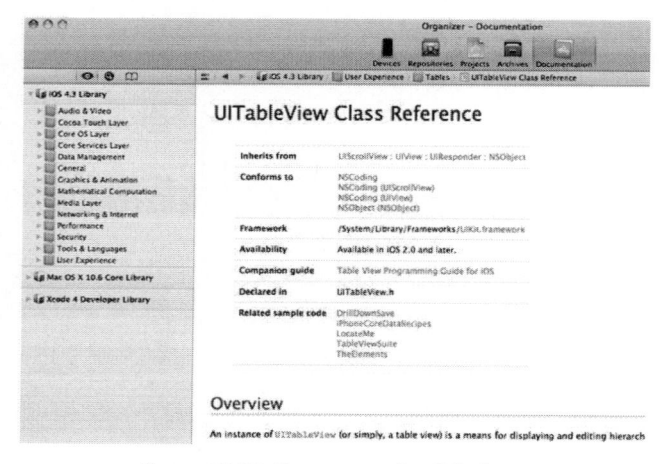

Figure 16-25: Documentation Window

16.3.5.2 Devices

In the Organizer there is a button that will display a list of your devices. There are also a lot of other useful options in this view. Developer and provisioning profiles can be managed through here as well and uploaded onto devices. Screenshots and Device logs can also be viewed through this section. The Devices window is shown in Figure 16-26.

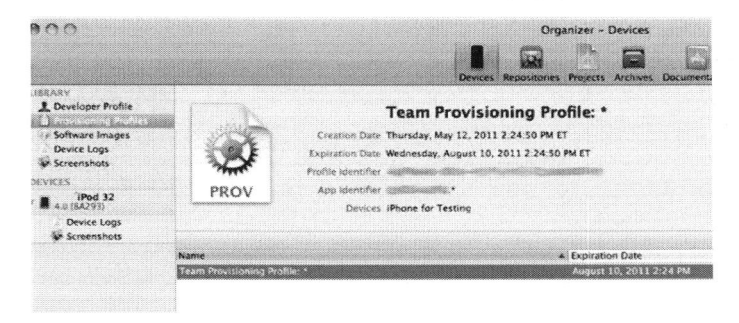

Figure 16-26: Devices Window

16.3.5.3 Repositories

In this section you can manage Source Control Repositories to help keep track of changes and coordinate amongst multiple programmers. Xcode 4 provides support for both Subversion and Git. The Repositories window is shown in Figure 16-27.

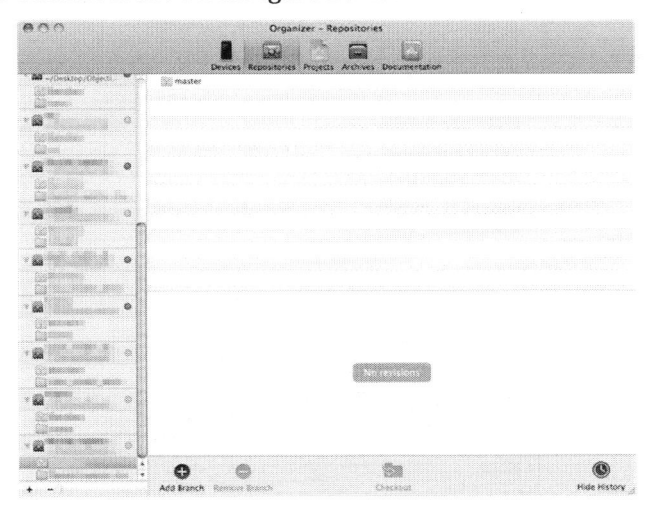

Figure 16-27: Repositories Window

16.3.5.4 Projects

In this section you can manage your projects and the Snapshots that are saved for each of them. A Snapshot saves a copy of the current version of your workspace. This is useful, for instance, if you are going to refactor your code or change something throughout your code. The Projects window is shown in Figure 16-28.

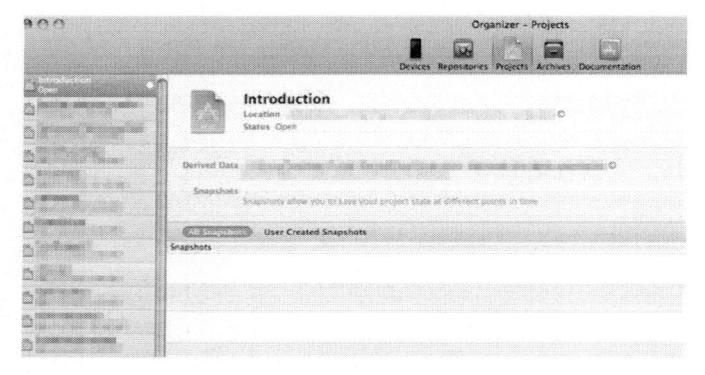

Figure 16-28: Projects Window

16.3.5.5 Archives

This section contains the archives of your projects, which is a save of the project and build configuration to be used later to submit the app to the App Store, to validate that it is ready for submittal, or to share with others. The Archives window is shown in Figure 16-29.

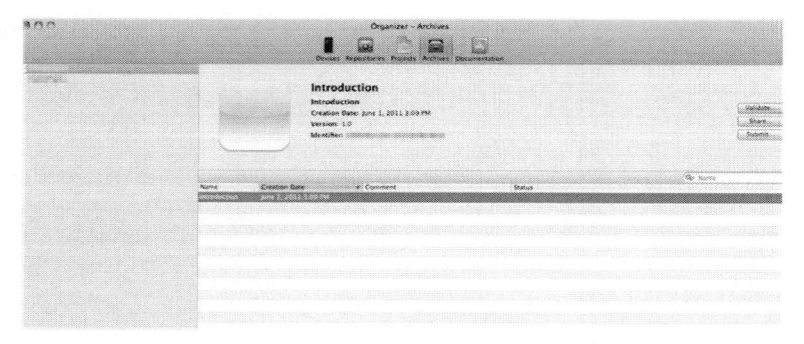

Figure 16-29: Archives Window

16.3.6 Hello World!

Start with a very basic "Hello World!" style program. We'll go back and explain how it works later. For now, the book will keep a focus on Objective-C itself, as opposed to the iOS SDK specific details.

To get started, open Xcode and click the "Create a new Xcode project" button in the welcome screen, as shown in Figure 16-30.

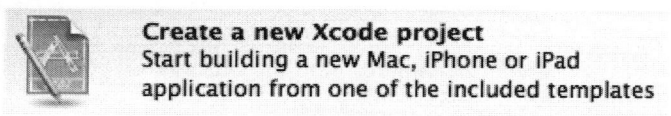

Figure 16-30: Create New Project

If the welcome screen isn't visible, you can access it through the "Window" menu under "Welcome to Xcode." Then select "Application" under the Mac OS X category and Command Line Tool, then click Next, as shown in Figure 16-31.

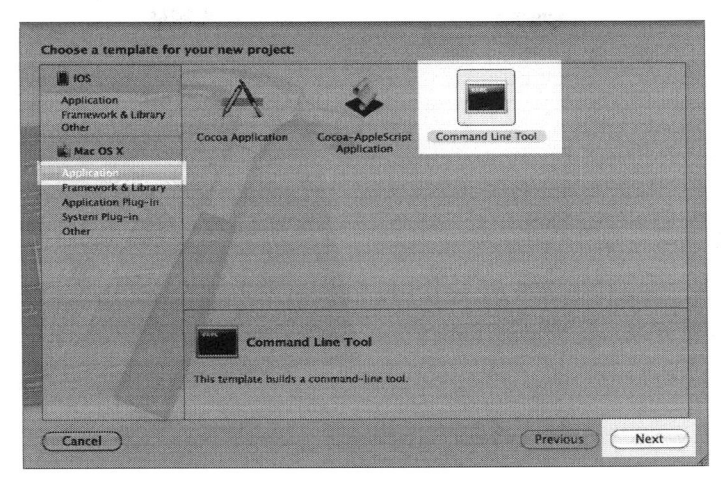

Figure 16-31: Command Line Tool Project

At the next screen, you will need to enter a product name. You'll be using "HelloWorld!" Make sure you've selected Foundation in the type selector.

Click the "Next" button. A popup window will appear where you will indicate where to save the project. Pick your location and continue.

Xcode will continue to load, and at this point, you now have a working project.

Figure 16-32: Xcode Four Shortcuts

Go ahead and run the project and see what happens. To do this, click the run button that is shown in Figure 16-32. The program will run and seemingly nothing will happen. Make sure the debug information view is open by toggling it; this is also shown in Figure 16-32.

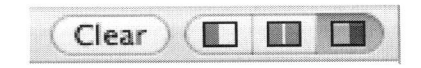

Figure 16-33: Debug Area Options

To display the output in the debug area, toggle the console by selecting the button shown in Figure 16-33. Look inside the console and you will see the "Hello, World!" text.

Chapter 16: Introduction to Xcode   103

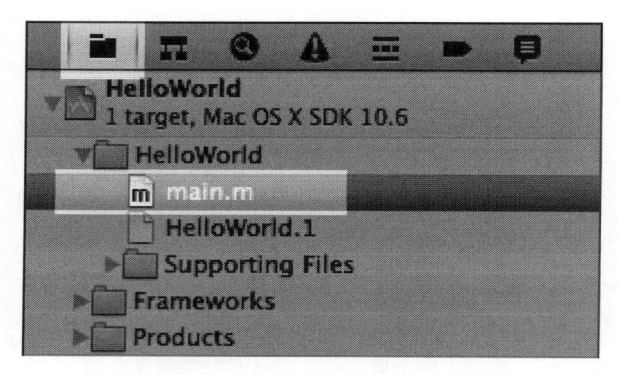

Figure 16-34: Project Navigator

On the left column in the Xcode window, click the project navigator icon, which looks like a folder. Now you will notice a tree structure with files and folders.

> **TIP:** This Chapter will now briefly tangent, providing a lead-in for the following Objective-C code chapters. Don't worry about what exactly is happening with the code so much at the moment, as this is merely an introduction for building your first "Hello World" app in Xcode using its native Objective-C (covered in detail, starting with the next Chapter).

All files with the .m extension are Objective-C source code files, similar to the .cpp extension C++ uses. Similarly, the .h files are header files that behave almost identically to those found in C and C++ code. Go ahead select the main.m file depicted in Figure 16-33. The contents of the file will then show up in the editor to the right. Take a look at the code, line by line:

```
#import <Foundation/Foundation.h>
int main (int argc, const char * argv[])
{
  NSAutoreleasePool * pool = [[NSAutoreleasePool alloc] init];
  // insert code here...
  NSLog(@"Hello, World!");
  [pool drain];
    return 0;
}
```

The first line of code shown below is a preprocessor directive:

```
#import <Foundation/Foundation.h>
```

Like C and C++, preprocessor directives begin with the # symbol. In Objective-C, the #import directive works similarly to the #include directive found in classic C, but has one additional feature: it guarantees no duplicate inclusions. This functionality is nothing new; C/C++ programmers have been doing it for years using #ifndef directives in header files. The #import directive is just a shortcut convention used in Objective-C.

In this case, you are importing the Foundation Framework. (Side-note: Again, NEXTStep is the basic framework in which Apple's OS X, and later iOS were developed, which is why so many things start with "NS").

The next line is the header for the main function:

```
int main (int argc, const char * argv[])
```

This is a standard C function header. Remember; since Objective-C is a superset of C, C code is valid. Just like in C, this is also the program entry point.

The first line in the main function is creation of an autorelease pool:

```
NSAutoreleasePool * pool = [[NSAutoreleasePool alloc] init];
```

Autorelease pools are an interesting topic unique to Objective-C. They are collections of objects that must be released (deleted or freed from memory) later. This is one of Objective-C's memory management strategies. The important thing to know is that Cocoa, both on Mac OS X and iOS, expects there to be an instance of `NSAutoreleasePool` in order to operate. A detailed exposition on memory management and autorelease pools is found in Chapter 17.

This is also your first exposure to Objective-C's messaging syntax. Don't let the brackets scare you; the guide explains how it all works in the next Chapter.

The next line displays the text "Hello, World!" to the output.

```
NSLog(@"Hello, World!");
```

The `NSLog` function is analogous to C's `printf` function. They are almost identical, differing only in syntax. The first obvious difference is the @ symbol in front of the string literal. In Objective-C, this is a syntax shortcut for a string literal object, as opposed to a character array literal, as in C. So the main difference between `printf` and `NSLog` is that `printf` expects a C-string as its first parameter, whereas `NSLog` expects an `NSString` object as its first parameter. The rest of the C format specifiers also work in Objective-C. It's worth noting that the specifier for a C-string, `%s`, does not work for `NSString` objects. The proper specifier for an `NSString` object is `%@`.

The next line sends the drain message to the pool created earlier, shown below:

```
[pool drain];
```

Basically, this line drains the autorelease pool, thus freeing all the memory that's expected to be autoreleased.

The last line a simple return statement:

```
return 0;
```

This is the end of the function and it returns an integer value, exactly as expected in standard C/C++.

Congratulations on your first Objective-C app! The next Chapters will build upon, and cover in detail, Objective-C's classes and syntax.

Introduction to Objective-C

What is Objective-C? Objective-C is the proprietary object-oriented programming language used by Apple with a history that's actually interesting!

But first, it's important to reiterate that this book does not try to teach you how to program from ground-zero. Rather, this section of the book guides the existing programmer though becoming proficient in Object-C as quickly and efficiently as possible.

When the late Steve Jobs left Apple in 1985, he founded NeXT Computer, Inc. The object-oriented NeXTSTEP operating system was so innovative, that Apple purchased NeXT in 1996 - which returned Mr. Jobs to Apple in 1997. Apple's use of NeXTSTEP as a basis for OSX is apparent through Objective-C. Much of the Objective-C foundation framework starts with the abbreviation "NS" stemming from its NeXTSTEP roots.

Cocoa is the environment for Mac OS X and iOS. In the architecture of iOS, the application layer is called Cocoa Touch, which is a framework that provides functionality for audio, video, graphics, animation, data management, networking, Internet, and integration with user applications. These functionalities can only be delivered by object-oriented techniques. Because of this, Objective-C is the programming language used to create iOS apps with the Cocoa Touch frameworks.

Objective-C is easy to learn because its syntax is small and unambiguous. It supports the same basic syntax as C, while the syntax for object-oriented features was adopted from Smalltalk.

The following are some features related to the data types and syntax of C and Objective-C that are similar:

- Comments
- Case-sensitivity
- Identifier declarations
- Mixing up code and declarations of variables
- Arithmetic, comparison, logical, bitwise, and ternary operators
- Basic data types: char, short, int, long, long long, float, double and long double as well as their unsigned counterparts
- The void type
- Arrays
- Constants
- Enumerations
- Structs

- Pointers
- Loops including for, while, and do...while loops
- The switch...case selections
- The break and continue statements

This chapter covers the features of Objective-C that were added from the C language, since Objective-C is derived from the C language, as well as some differences between the Objective-C and C++ languages.

17.1 Objective-C Basic Types

As covered within the Introduction section, all basic data types from the C language can be used in any Objective-C program. In addition, Objective-C has introduced other basic types, which will be described in this section.

17.1.1 Boolean Types

The Boolean type represents logical values (usually denoted *true* and *false*). In Objective-C, it is known by the BOOL keyword. The standard values of BOOL are YES and NO rather than true or false, or 1 or 0. However, BOOL variables can be set with YES, NO, 1, 0, true or false. It is good practice to use BOOL YES and NO when programming in Objective-C for clarity and correspondence with the language.

17.1.2 Foundation Framework Data Types

Basic Objective-C classes are defined in the Foundation Framework. The Foundation Framework contains classes that represent the basic data types, collections, system information, communication ports, as well as the root object class. Section 17.5 describes the root class (NSObject), while in Chapter 18 you will see several other important types of classes. Because the goal is to introduce you to Objective-C programming, only the basic data types are reverence here.

The NSInteger and NSUInteger data types defined by the Foundation Framework are dynamic type definitions. They are synonyms for an integer and unsigned integer respectively. You usually want to use NSInteger or NSUInteger when you do not know what kind of processor architecture your code might run on, so you may want the largest possible int or unsigned int type.

String management is another important point to know when starting to learn a language. In Objective-C, to construct and manage a string that can be changed after it has been created, you must use the NSMutableString class; otherwise, use the NSString class. For more information, Chapter 18 discusses strings in depth.

17.2 Objective-C Classes

There are two main parts to an Objective-C class: the interface and the implementation. The interface and the implementation are both in separate files: .h and .m files respectively. In Objective-C, methods and instance data are separate, as opposed to C++ where methods and attributes are declared together inside of the class braces. Because of this, private-like methods can be implemented and left out of the header file, whereas with C++ private methods are declared alongside public methods, which tends to clutter the header file with methods that do not need to be exposed to the interface. Note that these methods are not truly private in that they are not fully hidden from access.

17.2.1 Class Interface

Objective-C only supports single inheritance, meaning that a class can only inherit from a single class. In place of multiple inheritance, Objective-C has protocols and categories. Following is an example of the syntax for a class interface declaration:

```
@interface MyClass: MySuperClass {
/*
    Instance variable declarations
*/
}
/*
Method declarations
*/
@end
```

The `@interface` and `@end` directives indicate the beginning and end of the interface declaration. The first line of the declaration tells you that the new class (`MyClass`) inherits from its superclass (`MySuperClass`). If `MySuperClass` and the colon were to be removed then `MyClass` is declared as a root class. It is recommended that all classes have a superclass, where the root class of the hierarchy is the `NSObject` class. This means that all classes are subclasses of the `NSObject` class, which in turn means they can take advantage of all functionalities that it provides. For more information about the `NSObject` class, see section 17.5.

17.2.2 Class Implementation

In the implementation file, you first need to import the interface file:

```
#import "ClassName.h"
```

Then, the implementation occurs between the `@implementation` and `@end` directives. The syntax for class implementation is:

```
#import "MyClass.h"
@implementation MyClass
/*
class methods implementation
*/
@end
```

17.2.3 Instance Variables

An instance variable is a variable that is unique in each instance of the class it is in. The syntax for an instance variable is:

```
visibility SomeVarType varName;
```

The visibility of instance data can be `@public`, `@protected`, `@private` or `@package` with `@protected` being the default. The three first terms have the same meaning that they do in C++. The `@package` keyword means the instance variable is accessible only from the executable image that implements the class. This access level is similar to `private_extern` for C variables and functions, `internal` for C#, or the default visibility in Java.

In Objective-C, all variables are created as instance variables. Objective-C does not support class or static variables as in C++, but you can simulate its behavior using class constructors.

17.2.4 Methods

Methods are always public in Objective-C. However, it's possible to have a method with similar functionality to a private method in Objective-C. As mentioned earlier, this is done by implementing the method in `@implementation` and leaving it out of `@interface`, or by using a category class.

The syntax for method declaration follows this format:

```
prefix (RType)methodName;
prefix (RType)methodName:(PType)parameter;
prefix (RType)methodName:(PType)paramOne paramLabel: (PType)paramTwo;
```

Where `RType` is the method return type, and `PType` is the `parameter` type.

In Objective-C a class can declare either an instance method or a class method. A `prefix` token with a minus (-) sign, indicates an instance method and one with a plus (+) sign indicates a class method. The difference between the two is that instance methods require an instance of the class to be created while class methods do not. If you have several instances of a class, you can run an instance method on each of them separately without affecting the other.

The parameters in a method can have labels associated with them, they are the names just before the ":". These labels are parts of the method name and thus modify it when changed. Following is an example of some methods and their selectors as indicated within the comments.

```
@interface MyClass : NSObject {
}
- (void)method;                        // Method name="method"
- (void)method:(int)x;                 // Method name="method:"
- (void)method:(int)x:(int)y;          // Method name="method::"
// Method name="method:paramTwo:"
- (void)method:(int)x paramTwo:(int)y;
// Method name: "method:newParam:"
```

```
- (void)method:(int)x newParam:(int)y;
// Method name: "method:paramTwo:paramThree: "
- (void)method:(int)x paramTwo:(int)y paramThree:(int)z;
@end
```

In other programming languages methods with the same name but different input and output types, also known as method overloading, can exist. However, in Objective-C, methods are differentiated between each other by their parameter labels. Therefore, Objective-C methods and C functions cannot be overloaded. The following example shows a class with possible methods and how some of them have the same name, which is not allowed by the Objective-C compiler.

```
@interface MyClassWithOverloading : NSObject {}
- (void)method:(int)x;
- (void)method:(float)y;              // BAD: Same method name
- (void)method:(int)x:(int)y;         // GOOD
- (void)method:(int)x:(float)y;       // BAD: Same method name
- (void)method:(int)x param:(int)y;   // GOOD
- (void)method:(int)x param:(float)y; // BAD: Same method name
- (void)method:(int)x newParam:(int)y;// GOOD
@end
```

Another particular feature is that Objective-C, like C, allows the use of methods with a variable number of parameters by using "..." as the last argument.

In addition, Objective-C differs from other languages like C++ and C# because it does not allow you to set default values for parameters in the method declarations.

17.3 Objects

An object is the instantiation of a class that consists of an identity (name), state (variables) and methods (operations). For example, if you have a program that performs mathematical operations on shapes, you will probably have a different class for each type of shape. There might be a class to represent a Circle object with instance variables for the center and radius. The Circle class would then need several methods; a method to set the center of the circle, one to change the radius, one to compute its area, one to compute its circumference, and finally one to create it.

The `id` type is the basic type for objects regardless of class. When you call a method on an object, it does not need to know the object's type. It is sufficient that the method exists. The code below defines a variable of `id` type:

```
id anObject;
```

Suppose you have two classes of shapes: Circle and Rectangle, both with the draw method implemented. The following code uses the id type:

```
Circle *myCircle =              // myCircle creation
```

```
Rectangle *myRectangle =// myRectangle creation
id myShape;
/* Draw a circle */
myShape = myCircle;
[myShape draw];
/* Draw a rectangle */
myShape = myRectangle;
[myShape draw];
/* release memory of myCircle and myRectangle*/
```

17.3.1 Messaging

Messaging an object tells that object to apply a method in order to do something. The syntax of a message is composed of the method signature along with the parameter information the method needs. In Objective-C, brackets are used to enclose messages as shown below.

```
[receiver message];
[receiver message:parameter];
[receiver message:parameter1 andParam:parameter2];
```

When using the Class object, `MyClass`, from section 17.2.4 you can send it the following messages as shown below.

```
MyClass *anObject= [[MyClass alloc] init]; // anObject creation
[anObject method];
[anObject method:1];
[anObject method:1 :2];
[anObject method:3 paramTwo:4];
[anObject method:5 newParam:6];
[anObject method:1 paramTwo:2 paramThree:3];
```

To avoid declaring numerous local variables to store temporary results, Objective-C allows you to nest messages. For instance,

```
/* Get the director's name of a movie */
NSString *name = [[movie director] name];
/* Replace the first element of myArray with the last one */
[myArray replaceObjectAtIndex:0 withObject:[myArray objectAtIndex:[myArray count]-1]];
```

Sending a message to `nil` is perfectly valid in Objective-C. If you send a `nil` message that would normally return an object, it will instead return `nil`. Because of this, you do not need to check to see if an object is `nil` before messaging the object.

If you want to call a method from the superclass, you can use the `super` keyword, similar to Java.

In Objective-C, if a subclass defines the same method of its superclass, the method is overridden by default. Thus all methods in Objective-C are virtual.

Another important thing to know when using Objective-C is that there is no need to use downcasting since you can send a message to an object even if it appears that it will not recognize that message. For example, there will be times where you have a pointer to a parent class but want to send a message to the subclass object. In other programming languages you would need to cast it to the subclass's type before calling the method. However, you can still cast the object to the desired type in order to fix any compiler warnings, as shown in the following Director example.

In this example, `Director` is a subclass of `Person`. The `Director` class has a `printMovies` method that prints all the movies directed by `aDirector`.

```
Director *aDirector = [[Director alloc] init];
Person *aPerson = aDirector;
[(Director*) aPerson printMovies];
```

17.3.2 Selectors

In Objective-C, *selector* is the term used to refer to the method name in a message. However, *selector* has another meaning in Objective-C. It can be thought of as a sort-of "method pointer" that is used when code is compiled to refer to a method instead of having to use the method name. These selectors are of type `SEL`.

At compile time, you use the compiler directive `@selector` to create a selector as seen in the following line of code.

```
SEL aSelector = @selector(methodName);
```

Using the defined methods of `MyClass` from section 17.2.4, the following statements are valid.

```
SEL aSelector = @selector(method);

SEL aSelector = @selector(method:);

SEL aSelector = @selector(method::);

SEL aSelector = @selector(method: paramTwo:);

SEL aSelector = @selector(method: newParam:);

SEL aSelector = @selector(method: paramTwo: paramThree:);
```

At runtime, you use the `NSSelectorFromString` function, where the string passed as an argument is the name of the method. For instance:

```
SEL aSelector = NSSelectorFromString(@"method: paramTwo:");
```

You can invoke a method using a selector with the following code.

`performSelector:`,

`performSelector:withObject:`, or

`performSelector:withObject:withObject:`

The syntax to call this method is:

```
[anObject performSelector:aSelector];
```

Calling `respondsToSelector:` will check if an object can trigger that method. For example, the following code checks if method "f" exists, and calls it.

```
SEL aSelector = @selector(method);
if ([anObject respondsToSelector:aSelector])
    [anObject performSelector:aSelector];
else NSLog(@"The class does not respond");
```

17.3.3 Properties

Accessor methods allow "getting" and "setting" the private information of an object, in a way that its structure is protected from the outside. Objective-C uses object properties that handle the internal attributes of an object. When defining properties, dot syntax can be used for invoking get and set accessor methods. The typical syntax for these methods is as follows.

```
- (propertyType) propertyName;           // this is a get method
- (void) setPropertyName: (PropertyType) newValue;  // this is a set method
```

Below is an example of using dot syntax to invoke these accessor methods.

```
itemsCount = [myArray count];   /* To get the quantity of elements of an array:*/
directorName = aDirector.name;  /* To get the name of a director*/
aDirector.name = aName;         /* To set a name to a director.
                                   This is equivalent to [aDirector setName: aName] */
aPoint.x = 10;     /* To set a name to the x coordinate to a point.
                      This is equivalent to [aPoint setName:10]*/
```

When you want to refer to the current object, use the keyword `self,` which is similar to `this` in C++. If you want to directly access an instance variable, you must use its own name. Otherwise, use `self.propertyName` for using the related property.

If an instance variable is specified in the interface declaration with the keyword `@property` (and some attributes), the Objective-C compiler automatically generates the setter and getter methods. The syntax for declaring a property is shown below.

```
@property PropertyType propertyName;

@property(attributes) PropertyType propertyName;  // for defined attributes
```

Here are the different possible values for attributes:

- `assign`, `retain`, or `copy`: used to specify how the value is stored inside of its setter. The default is `assign`.

- `readwrite` or `readonly`: specify if the property should have both the getter and setter accessor methods (`readwrite`) or only the getter method (`readonly`). The default is `readwrite`.

- `getter=getterName`, `setter=setterName`: used to specify the get and set accessor method names.

- `nonatomic`: prevents thread-safety guards from being created. Default is atomic, however there is no keyword for atomic.

The attributes `assign`, `retain`, and `copy` affect the way the data member is modified inside its setter. The `assign` attribute makes a simple assignment. The `retain` attribute also makes a simple assignment, but it indicates that this variable is referenced one more time. Finally, the `copy` attribute assigns a new object that is a copy of the original. Section 17.4.1 explains how these types of assignation help with memory management.

The implementation of properties can be done by using the `@synthesize` or `@dynamic` directives.

- `@dynamic` means the developer needs to provide both the getter and setter accessor methods. If the property is declared as `readonly` then only the getter needs to be provided.

- `@synthesize` means that the compiler will generate the accessor methods, unless the developer has already created them.

This example shows a `Movie` class with attributes for storing the title and the director of a movie.

```
@interface Movie : NSObject {
NSString *title;
Person *director;
}
@property (readonly, copy) NSString *title;
@property (assign) Person *director;
@end
@implementation
@synthesize title;     // The compiler generates the title and setTitle methods
@dynamic director;     // The get/set accessors must be implemented
  /* The getter for director */
 - (Person *) director {
//…
}
  /* The setter for director*/
```

Chapter 17: Introduction to Objective-C   115

```
- (void) setDirector:(Person *)value {
//…
}
@end
```

Note that if a getter or setter method for the `title` property is implemented, the compiler will use it instead of generating a new one.

When using `@synthesize`, a property does not necessarily need to have the same name as the data member:

```
@interface MyClass : NSObject {
    OneClass* _myObject;
}
@property OneClass* myObject
@end
@implementation MyClass
        /* Use "_myObject" as the instance variable for the property */
@synthetize myObject =_myObject
@end
```

17.3.4 Instantiation

Creating instances of objects in Objective-C consists of first allocating memory to the instance and then initializing it. To first allocate memory to an object the `alloc` message is sent to the object's Class. The `alloc` method will allocate memory for the object and its instance variable dynamically. After the memory is allocated it then needs to be initialized, so the `init` method is called.

Usually these two steps are completed on the same line as follows:

```
MyClass *anObject = [[MyClass alloc] init];
```

Determining whether and object is created or not is done by testing against `nil`.

Generally, an initializer is implemented as follows:

```
- (id) init {
 if (!(self = [super init]))
return nil;
 return self;
}
@end
```

The following example shows an initializer method for a two-dimensional point class, named

Point2D.

```
@interface Point2D : NSObject {
int x;
int y;
}
-(id) initWithPoint:(int)newX yCoordinate:(int)newY;
@end
implementation Point2D
-(id) initWithPoint:(int)newX yCoordinate:(int)newY {
  if(!(self = [super init]))    // call initializer of the superclass
        return nil;             // return nil in case of failure
        /* In case of success initialize variables */
  x = newX;
  y = newY;
  return self;              // return the object
}
@end
```

Using the class created above, this is how a `Point2D` object can be created:

```
Point2D *myPoint2D = [[Point2D alloc] initWithX:1 andY:2];
```

17.3.4.1 Default Constructor: Designated Initializer

Objective-C does not have a default constructor, but you can simulate this behavior by having a designated initializer. This method contains some code that must be executed in all initializers, and the other initializers will call it.

The following code shows the initializers for a three-dimensional point class.

```
@interface Point3D : NSObject {
    int x;
    int y;
    int z;
}
-(id) init;
-(id) initWithPoint:(int)newX;
-(id) initWithPoint:(int)newX yCoordinate:(int)newY;
// Designated initializer
```

Chapter 17: Introduction to Objective-C   117

```
-(id) initWithPoint:(int)newX yCoordinate:(int)newY zCoordinate:(int)newZ;
@end
@implementation Point3D
-(id) init{
    return [self initWithPoint:0 yCoordinate:0 zCoordinate:0];
}
-(id) initWithPoint:(int)newX{
    return [self initWithPoint:newX yCoordinate:0 zCoordinate:0];
}
-(id) initWithPoint:(int)newX yCoordinate:(int)newY{
    return [self initWithPoint:newX yCoordinate:newY zCoordinate:0];
}
-(id) initWithPoint:(int)newX yCoordinate:(int)newY zCoordinate:(int)newZ{
 if (!(self = [super init]))
       return nil;
 x = newX;
 y = newY;
 z = newZ;
 return self;
}
@end
```

17.3.4.2 Class Constructors

In Objective-C, classes do not have a "constructor" as they do in C++. They do however have initializers which are essentially the same thing. The `NSObject` class declares the class method:

```
+(void) initialize;
```

This method is inherited from `NSObject` by all of its subclasses and it is called automatically for each class before it is used in the program for the first time.

In addition, the `initialize` method can be used to do the initialization of global and static variables declared inside of the implementation file. In this way, you can simulate the same behavior of static instance variables. This static variable can be used within any class or instance method of the class. The following code shows how to do this.

```
static int numberOfInstance;
@interface MyClass : NSObject {
```

```
    }
+(void) initialize;
-(id) init;
-(void) displayNumberOfInstances;
@end

@implementation MyClass
+(void)initialize {
 numberOfInstance = 0;
}
-(id)init {
 if (!(self = [super init]))
        return nil;
 numberOfInstance++;
 return self;
}
-(void)displayNumberOfInstances {
NSLog(@"Number of instances of this class: %d", numberOfInstance);
}
@end
```

17.3.4.3 Copy Constructor

Objective-C does not offer operator overloading. Copying is done through copy methods. The `-(id)copyWithZone:(NSZone*)zone` method returns a cloned object with `zone` as a parameter. The argument is an area of memory (zone) to be used for allocation of the new object. If the argument is the default zone, use the `-(id) copy` method that encapsulates a call to `copyWithZone`. To use this method, conform to the `NSCopying` protocol.

17.3.5 Destructors

The `dealloc` method is the destructor for Objective-C. The `dealloc` method should never be called explicitly; if you want to release an object from memory, the simple way is using either the `release` or `autorelease` methods. The `release` method decreases the object's reference count. If this count reaches 0 then the `dealloc` method is called. Section 17.4.1 describes the `autorelease` method in detail.

The following example shows how to implement a `dealloc` method:

```
@interface MyClass : NSObject {
UsedClass * myUsedClass;
```

```
}
-(void) dealloc;

@end

@implementation MyClass

-(void) dealloc {

[myUsedClass release]; /* all objects from instance variables must be released first */

[super dealloc];

}
@end
```

17.4   Memory Management

When managing memory in Objective-C it is important to try and keep memory freed up as much as possible. Ensuring that objects are deallocated when they are no longer required, will accomplish this. However it is important to be careful not to deallocate objects that are still being used.

Objective-C supports three methods for memory management, which use two separate models: automatic garbage collection and reference counting. Two of the three methods for memory management, MRR (Manual Retain-Release) and ARC (Automatic Reference Counting) use the reference counting model, the third method being garbage collection. With garbage collection, you do not have to worry about keeping track of memory because the system automatically takes care of it for you. However this is only supported in Mac OS X and not iOS. So this leaves you with only the option of reference counting if you wish to develop for iOS. Automatic reference counting or ARC uses the same model as MRR however the system performs the necessary operations at compile-time. MRR will be covered in the next section and used throughout this book as the memory management method of choice.

17.4.1   Reference Counting

When using reference counting to manage memory each object has a reference counter that keeps track of the number of times that it is referenced. This reference counter is initially set to 1 when an object is newly created. If an object is returned by reference there is no ownership of the object, thus the developer is not in control of when it is deallocated. To take ownership of the object it is necessary to increment its reference counter by sending it a `retain` message. When you are done with the object you can send it a `release` message, which will decrease the counter by 1. After sending a `release` message, if the reference counter reaches 0, the `dealloc` message is sent. If the release message is sent to an object that has already been destroyed, it will result in an error. The methods `alloc`, `mutableCopyWithZone:` and `retain` increase the reference counter. Thus, when using any of them it is necessary call either the `autorelease` or `release` methods.

Sometimes methods will need to create objects using the `alloc` method, and thus need to relinquish ownership of the object by releasing it. However, if a `release` message is sent it will be deallocated immediately. In situations like these, the `autorelease` message is useful. This is because it adds the object to an autorelease pool. When the pool is destroyed, all of the objects it contains will be sent a `release` message.

The code below implements a method for adding two rational numbers.

```
-(Rational*) add:(Rational*)num1 and:(Rational*)num2 {
    Rational *result = [[Rational alloc]init];
    result.numerator = num1.numerator * num2.denominator +
    num1.denominator * num2.numerator;
    result.denominator = num1.denominator * num2.denominator;
    /* add to autorelease pool for deletion later */
    [result autorelease];
    return result;
}
```

When creating an app using the UIKit framework it will create an autorelease pool automatically at the beginning of an event loop and destroy it at the end.

If you are not using the UIKit framework then you should create an autorelease pool and nest the code inside the autorelease pool. If you have multiple autorelease pools nested inside of each other, the object that sent the `autorelease` message would be added to the innermost pool that it is nested in. The autorelease pool can be drained when it is no longer needed, by sending it the `drain` message as shown below.

```
int main(int argc, char* argv[]) {
    NSAutoreleasePool *mainPool = [[NSAutoreleasePool alloc] init];
    // Objects created and added to pool here
    [mainPool drain];
    return 0;
}
```

17.4.2 Garbage Collection

With garbage collection it isn't necessary to worry about `retain` and `release` because the app's memory is managed for you.

Garbage collection is not available on iOS, it is only available on Mac OS X. In a garbage-collected app, the equivalent of the `dealloc` method is the `finalize` method:

```
- (void)finalize {
    // supposing that a method for closing a file is implemented
    [myFile close];
    [super finalize];
```

```
}
```

17.5 Root Class

The `NSObject` class is the root class of every Objective-C class. This means that every class will inherit from this class, or be a subclass of `NSObject`. These subclasses can override the `NSObject` class's methods.

The following code tests the equality of the pointer values (identity):

```
if (object1 == object2) {
    //Same exact object instance
}
```

An `isEqual` method tests object attributes. An example is shown below.

```
if ([object1 isEqual: object2]) {
    // Logically equivalent, but may be different object instances
}
```

`NSObject` implements a description to return an object represented in format strings using the code %@ as seen below.

```
[NSString stringWithFormat:@"The object's description is: %@", myObject];
```

Or an object's description can be printed with:

```
NSLog([anObject description]);
```

17.5.1 Introspection

Introspection is the ability of an object to provide information about itself at runtime. There are several methods in the `NSObject` class for manipulating introspection. These methods are described in detail below.

- `isMemberOfClass`: tells if one object is an instance of the given class.
- `isKindOfClass`: tells if one object is an instance of the given class or any of its subclasses `isSubclassOfClass`: checks to see if object is a subclass of the given class.
- `class`: returns the `Class` object of the receiver's class.
- `superclass`: the same as `class` but it returns the receiver's superclass.
- `respondToSelector` and `performSelector` are described in section 17.3.2.

Here is an example.

```
BOOL isCircle = [Circle isKindOfClass: [myCircle class]];
if (isCircle)
  NSLog(@"I am an instance of the Circle class");
```

17.6 Protocols

In Objective-C, a protocol is similar to an interface in Java. One of its purposes is to declare methods that will be implemented by the adopting class. The two types of protocols, formal and informal will be discussed in the next sections. A formal protocol is a list of methods that must be implemented by the adopting class. Grouping methods in a category declaration creates an informal protocol, where the methods are optionally implemented.

17.6.1 Formal protocols

A class can adopt a protocol by either declaration or inheriting it from a superclass. The syntax for adopting a protocol is as follows:

```
@interface Class : Superclass <ProtocolOne, ProtocolTwo>
```

As shown above, multiple protocols can be adopted.

The following example shows the declaration of the `Circle` class. The `Sizeable` protocol has methods to inflate and deflate `Circle` objects. The `Shape` protocol has a method to draw shapes.

```
#import "Point2D.h"

@protocol Sizeable

- (void)inflate:(int)amount;

- (void)deflate:(int)amount;

@end

@protocol Shape

- (void) draw;

@end

@interface Circle: Point2D <Sizeable, Shape> {
    float radius;
}

@end
```

Using the `@optional` directive you can specify methods that are optional. The `@required` directive means the methods are required to be implemented, this is the behavior by default. Consider the following declarations:

```
@protocol ProtocolOne

// Implementation required by default

-(void)requiredMethodOne;

@optional

-(void)optionalMethodOne;

-(void)optionalMethodTwo;

@required
```

Chapter 17: Introduction to Objective-C   123

```
-(void)requiredMethodTwo;

@end
```

The protocol itself does not implement any of the methods that it declares; this is left to the class that adopts the protocol. In order for a class to conform to a protocol it must adopt the protocol (or inherit it from a class that does), then it must implement all required methods.

The `conformsToProtocol:` message can be used to check whether or not a class conforms to a protocol. The following is an example:

```
if ([aCircle conformsToProtocol: @protocol(Shape)])
```

In any declaration of a method, function, or instance variable, it is possible to specify that it will conform to a protocol. It is most common to use the dynamic object type, `id`, for declarations of instance variables and parameters, for instance:

```
id <Sizeable, Shape>  mySizeableShape;

Circle *myCircle;

mySizeableShape = myCircle;

// The parameter must conform to the Shape protocol
- (void)anMethod:(id <Shape>) sender;
```

Cocoa provides several examples of protocols. An interesting one is the `NSObject` protocol.

17.6.2 Categories

Categories are handy when you want to extend the functionality of a class by adding methods, without subclassing it. When creating a category it is important to know that the methods that are added will become part of the class type. This means that all subclasses of that class will also inherit these methods.

The implementation and declaration of a category is very similar to that of a subclass with the difference being that the name goes in parentheses next to the @interface and @implementation directives. An example of this is shown below. The `move:` method is added to the `Point2D` class (example from section 17.3.4) that moves the origin (x,y) of the point. In the header and implementation files for the category, the code would be written as shown below

```
#import "Point2D.h"

@interface Point2D (Moveable)

- (void)move:(Position*)newPosition;

@end


#import "Point2D.h"

@implementation Point2D (Moveable)

-(void)move:(Position*)newPosition {
```

```
    // implementation code...
}
@end
```

17.7 Exception Handling

Exception handling in Java programming is similar in Objective-C. This is because the syntax is nearly identical. The @try, @catch, and @finally compiler directives are used to define blocks of code that deal with exceptions. The @throw directive is used to throw an exception.

When dealing with code that might throw an exception, it should be placed in the @try block. When an exception is thrown in this block, the @catch block is used to handle the exception. The @finally block is used when it is necessary to execute code regardless of whether or not an exception is caught.

In Objective-C, the NSException object is used to throw an exception. Other objects besides the NSException objects can be thrown, however there are certain situations where the Cocoa frameworks will not catch these objects as shown in the example below.

```
#import <Foundation/NSException.h>
@interface ArithmeticException : NSException {
}
@end


@implementation ArithmeticException
@end


#import <Foundation/NSException.h>
@interface FinallyCheck : NSObject {
}
-(void)f;
@end


@implementation FinallyCheck
-(void)f {
@try
{
int a = 1, b = 0;
if (b == 0) {
```

Chapter 17: Introduction to Objective-C   125

```objectivec
        NSLog(@"Throwing  an arithmetic exception");
        NSException *e = [ArithmeticException
                    exceptionWithName:@"ArithmeticException"
                    reason:@"Division by 0"
                    userInfo:nil];
        @throw e;
    }
    else {
        float c = a/b;
      NSLog(@"Division: OK");
  }
}
@finally {
    NSLog(@"Cleaning up");
}
}
@end


#import <Foundation/NSException.h>
@interface MyClass : NSObject {
}
@end

@implementation MyClass
FinallyCheck *testFinally = [[FinallyCheck alloc] init];
@try {
    [testFinally f];
}
@catch (ArithmeticException *e) {
    NSLog(@"Division is not possible");
}
@finally {
    NSLog(@"Latest clean up…");
```

```
}
@end
```

17.8 Compiling Objective-C with C++

C++ and Objective-C can be compiled together to run harmoniously. In order for the compiler to know that C++ is included some file extensions must be modified. The main file's extension that contains the C++ code must be `.mm` instead of `.m` and the same goes for any main file that calls any C++ method or C++ code. For example, if a class is written with C++ that class's main file must have a `.mm` extension and any main file that instantiates that class must also have a `.mm` extension.

18

Foundation Framework

The Foundation framework contains interfaces, methods, and data types that are useful for any project. These classes consist of strings, values and collections to name a few. The most import class is NSObject. It is the root class of all Objective-C classes.

Every class in the Foundation Framework has the prefix NS. As mentioned in the previous chapter, it stands for NeXTSTEP, which is an operating system that Apple acquired in 1996 and used as the basis for Mac OS X. The most important class is the `NSObject` class. As you will recall from Chapter 17, it is the root class of all the other classes. Some classes have two variations, static and mutable. Static classes can only be defined at the time of their creation. Mutable classes can be modified, or mutated, anytime during the app's runtime.

Figure 18-1 below, is an overview flow chart of the hierarchy. Note that the chart does not show all classes in the Foundation framework.

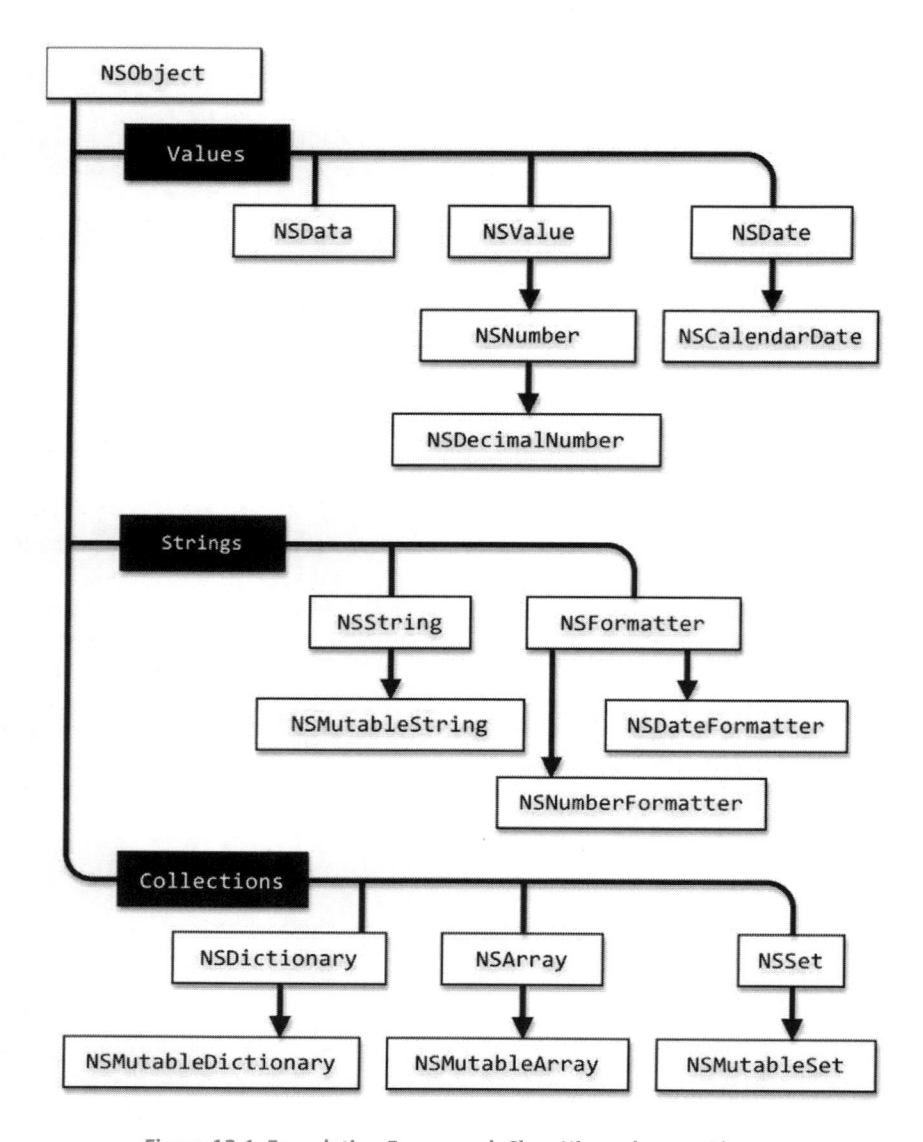

Figure 18-1: Foundation Framework Class Hierarchy at a Glance

In this chapter, the most notable classes will be discussed briefly with an example of how to use an object of the class in an app. For the entire list of classes included in the Foundation framework please refer to the Foundation Framework Reference document in Apple's Developer Library.

18.1 Data Types and Storage

18.1.1 NSData

The `NSData` class creates a static object for data storage. Below you will see an example of how to use `NSData` to get the contents of a text file:

```
NSString *pathOfFile = [[NSBundle mainBundle]
                        pathForResource:@"file" ofType:@"txt"];
NSData *myData = [NSData dataWithContentsOfFile:pathOfFile];
NSString *fileContents = [NSString stringWithUTF8String:
                          [myData bytes]];
```

First the path of the text file is stored into an `NSString` object, and then the contents of the file are loaded into an `NSData` object. The `NSData` object stores the data in a binary format that is unreadable to humans. To convert the binary data into readable text you need to convert the binary data into an `NSString` object.

18.1.2 NSValue

The `NSValue` class creates a container for a single data item. The purpose of this container is for storing data types into a collection that otherwise would not be able to store these data types. This is because collections only store objects. Structures, pointers, and integers, to name a few, aren't objects. `CGPoint` is an example of one of these structures. A `CGPoint` structure defines two dimensions, an X and a Y. In order to store a `CGPoint` in a collection it must be stored in an `NSValue` object first. Below is an example of how to convert a `CGPoint` into an `NSValue` object and then convert it back.

```
CGPoint point = CGPointMake(5, 10);
NSValue *myValue = [NSValue valueWithCGPoint:point];
CGPoint newPoint = [myValue CGPointValue];
```

18.1.3 NSNumber

`NSNumber` is a subclass of `NSValue` designed to store numerical values. However, the data type of the number stored is not preserved. This means you can retrieve the value of an `NSNumber` as any data regardless of the data type that was stored. Below are some examples of storing numbers as well as a `Boolean` into `NSNumber` objects and then retrieving them.

```
NSNumber *number1 = [NSNumber numberWithInt:100];
NSNumber *number2 = [NSNumber numberWithFloat:3.14f];
NSNumber *number3 = [NSNumber numberWithBool:YES];
int intVariable =   [number1 intValue];
float floatVariable = [number2 floatValue];
bool booleanVariable = [number3 boolValue];
```

18.1.4 NSString

A string is a series of characters or symbols or both. The `NSString` class declares an interface for managing strings. The quickest way to define an `NSString` is show below.

```
NSString *myString = @"String Literal";
```

As shown above, Objective-C string literals start with the @ symbol. The longer way to construct a string is:

```
NSString *myString = [[NSString alloc] initWithString:@"String Text"];
```

18.1.5 NSMutableString

The `NSMutableString` type is similar to `NSString` except its contents can be changed. If you were to append an `NSString` to another `NSString` then a new `NSString` object is generated. Appending to an `NSMutableString` does not create a new object; instead it just modifies the existing object.

```
NSMutableString *myString = [[NSMutableString alloc]
        initWithString:@"String Text"];

[myString appendString:@"More Text"];
```

18.1.6 NSRange

`NSRange` is not a class but a structure type. An `NSRange` structure is used to reference a portion of a series. It consists of two integers: location and length. The location integer specifies the start of the range. The length integer defines how long the range is. An example is show below.

```
NSRange range = NSMakeRange(2, 5);
```

18.1.7 NSDate

The `NSDate` class creates an object that represents a point in time. All date objects include time and therefore date objects cannot be created to represent a day without time, e.g. a whole day.

```
NSDate* now = [NSDate date];
```

18.2 Utilities

18.2.1 NSTimer

The `NSTimer` class is used to create an object that calls a method either once or repeatedly. The timer must be given a time interval, which determines how often the given method is fired. In the example, an interval of 1/60th of a second is given, making the timer fire the method "methodToRun" 60 times every second.

```
NSTimer *timer = [NSTimer scheduledTimerWithTimeInterval:1.0/60.0f

target:self selector:@selector(methodToRun)

userInfo:nil repeats:YES];
```

18.2.2 NSCalendar

The `NSCalendar` object provides information about the year, such as the number of days in a certain month.

```
NSDateComponents *components = [[NSCalendar currentCalendar]
components:NSDayCalendarUnit | NSMonthCalendarUnit | NSYearCalendarUnit
fromDate:[NSDate date]];

NSInteger month = [components month];

NSInteger day = [components day];

NSInteger year = [components year];
```

18.2.3 NSFormatter

The `NSFormatter` class is an abstract class that is extended by subclasses to produce a textual representation of other classes. Each formatter requires a format that is used to specify the way each class is formatted. For more information on data formatting see the Data Formatting Guide in Apple's Developer Library.

18.2.4 NSDateFormatter

The `NSDateFormatter` class formats `NSDate` objects into a string for output.

```
NSDateFormatter *dateFormatter = [[NSDateFormatter alloc] init];

[dateFormatter setDateFormat:@"MMM dd, yyyy HH:mm:ss"];

NSDate *now = [[NSDate alloc] init];

NSString *dateAsString = [dateFormatter stringFromDate:now];
```

With this code the `dateAsString` object will be "Sep 12, 2012 10:01:37".

18.2.5 NSNumberFormatter

The `NSNumberFormatter` class formats any number into a string for readability or functionality.

```
NSNumberFormatter *numberFormatter = [[NSNumberFormatter alloc]init];

[numberFormatter setPositiveFormat:@"#,###,##0.0#"];

[numberFormatter setNegativeFormat:@"#,###,##0.0#"];

float floatNumber = 1.2345;

NSNumber *myNumber = [NSNumber numberWithFloat:floatNumber];

NSString *output = [numberFormatter stringFromNumber:myNumber];
```

The contents of the string `output` would be "1.23".

18.2.6 NSUserDefaults

The `NSUserDefaults` class is used to store and retrieve information for customizing an app based on the user's preferences. Information that is stored can be retrieved even after the app restarts. Basic data types like Booleans, doubles, floats, and integers can be stored. A pointer value can also be stored. Each stored value is assigned a key that is used for retrieving the value.

```
[[NSUserDefaults standardUserDefaults] setBool:true
                            forKey:@"myBoolean"];
[[NSUserDefaults standardUserDefaults] setDouble:1.234
                            forKey:@"myDouble"];
[[NSUserDefaults standardUserDefaults] setFloat:3.14f
                            forKey:@"myFloat"];
[[NSUserDefaults standardUserDefaults] setInteger:10
                            forKey:@"myInteger"];
[[NSUserDefaults standardUserDefaults] setValue:[[NSObject
                      alloc]init] forKey:@"myObject"];


bool a = [[NSUserDefaults standardUserDefaults]
                            boolForKey:@"myBoolean"];
double b = [[NSUserDefaults standardUserDefaults]
                            doubleForKey:@"myDouble"];
float c = [[NSUserDefaults standardUserDefaults]
                            floatForKey:@"myFloat"];
int d = [[NSUserDefaults standardUserDefaults]
                            integerForKey:@"myInteger"];
id e = [[NSUserDefaults standardUserDefaults]
                            valueForKey:@"myObject"];
```

18.3 Collections

18.3.1 NSArray

The `NSArray` class is a type of collection. Collections are containers for multiple objects. The `NSArray` class stores objects in a series. The placement of an object in an array is called the index. The size of an `NSArray` is static meaning that it cannot be changed. The class provides you with utility methods for searching, sorting, querying, and more. Start by taking a look at the common syntax for creating and initializing an `NSArray`:

```
NSArray *myArray;
[myArray initWithObjects:object1, object2, object3, nil];
```

The most important detail to take notice is the termination of the list with `nil`. Here's how to access an element:

```
[myArray objectAtIndex:index];
```

That's pretty much it for basic array functionality. Now it's time for a couple more nice features. To find the size of the array:

```
[myArray count];
```

To get a sub-array from the array:

```
NSArray *mySubArray;
NSRange myRange;
myRange.location = 1;
myRange.length = 2;
mySubArray = [myArray subarrayWithRange:myRange];
```

When an `NSArray` object is released, the `release` message is sent to all the objects it contains. This saves you from the responsibility of explicitly releasing each individual object.

18.3.2 NSMutableArray

The major difference between `NSArray` and `NSMutableArray` is that the latter allows you to add and remove objects on the fly. So essentially, `NSMutableArray` is a non-static variation of `NSArray`.

Create a mutable array and add some objects to it:

```
NSMutableArray *myMutableArray = [[NSMutableArray alloc] init];
[myMutableArray addObject:object1];
[myMutableArray addObject:object2];
[myMutableArray addObject:object4];
```

So it appears that these objects were accidentally inserted in the wrong order. No problem; as you can insert objects at any index, like so:

```
[myMutableArray insertObject:object3 atIndex:2];
```

To remove any arbitrary object by its index:

```
[myMutableArray removeObjectAtIndex:3];
```

Or it might be convenient to remove the last object, which makes the `NSMutableArray` act like a queue or stack (depending on where the objects are being inserted into):

```
[myMutableArray removeLastObject];
```

And finally, if you need to swap out an object, here's an easy way to do it:

```
[myMutableArray replaceObjectAtIndex:2 withObject:myObject];
```

So that sums up the important details about the `NSArray` and `NSMutableArray` classes. They have much more functionality than covered here, so be sure to check out the class reference pages on Apple's website.

18.3.3 NSSet

The `NSSet` class represents another type of collection, a set. Sets are unordered collections of elements. Sets can replace arrays when the order of elements is not important. Below is an example of initializing an `NSSet` with objects:

```
NSSet *mySet = [[NSSet alloc] initWithObjects:
                object1, object2, object3, object4, nil];
```

To find the count of objects in the set:

```
[mySet count];
```

To check for membership, there are two main methods. The first simply just checks for membership and returns `Boolean` value YES if the object is a member of the set, or NO if otherwise:

```
[mySet containsObject:myObject];
```

Most of the time you will be interested in retrieving an object that is a member of the set. There is a shortcut for this that returns the object if, and only if, it is a member of the set:

```
[mySet member:myObject];
```

To check if a set is a subset of, intersects with, or is equal to another set (respectively):

```
[mySet isSubsetOfSet:myOtherSet];

[mySet intersectsSet:myOtherSet];

[mySet isEqualToSet:myOtherSet];
```

To enumerate through all objects in the set:

```
NSEnumerator *myEnumerator = [mySet objectEnumerator];
id currentObject;
while ((currentObject = [myEnumerator nextObject])) {
    // at this point, currentObject is a pointer to
    // an object in the set.
}
```

18.3.4 NSMutableSet

The main functionally gained with the mutable variant is that you can add and remove objects from the set. Because of this, you are also given the ability to perform the following set arithmetic operations: union, intersection and subtraction.

To allocate and initialize an `NSMutableSet`:

```
NSMutableSet *mySet = [[NSMutableSet alloc] init];
```

To add objects:

```
[mySet addObject:myObject];
```

To remove objects:

```
[mySet removeObject:myObject];
```

When performing set arithmetic, the result is stored in the set object that is performing the arithmetic. For example, in the line

```
[mySet unionSet:myOtherSet];
```

The `myOtherSet` object remains unchanged, but now `mySet` is the union of both sets.

To perform set subtraction:

```
[mySet minusSet:myOtherSet];
```

To perform set intersection:

```
[mySet intersectSet:myOtherSet];
```

18.3.5 NSDictionary

Dictionaries behave like an associative array or a map. For every object in the dictionary, there is a string, known as a key, associated with it. Querying the dictionary with these keys will grant access to the objects associated with them. Keep in mind that a key is only associated with one object, but an object may have more than one key associated with it. Following the pattern with all collection classes, there is a static and mutable version.

Typically, you create dictionaries out of parallel arrays. For example:

```
NSArray *objects = [[NSArray alloc] initWithObjects:object1,
        object2, object3, nil];
NSArray *keys = [[NSArray alloc] initWithObjects:key1, key2,
        key3, nil];
NSDictionary *myDictionary = [[NSDictionary alloc]
        initWithObjects:objects forKeys:keys];
```

With the above example, `object1` is associated with `key1` and so on. In order to retrieve an object for a given key you would use the following code:

```
[myDictionary objectForKey:myKey];
```

But since one object may have several keys associated to it, it's sometimes also useful to get all keys for a given object. This is how to get an array of all keys associated with an object:

```
[myDictionary allKeysForObject:myObject];
```

The nice thing about the mutable variant is that you can add or remove associations as needed. To make a dictionary:

```
NSMutableDictionary *myDictionary = [[NSMutableDictionary alloc] init];
```

To add associations:

```
[myDictionary setObject:myObject forKey:myKey];
```

And to remove them:

```
[myDictionary removeObjectForKey:myKey];
```

18.4 Summary

The previously covered classes are just some of the classes offered in the Foundation Framework. For the entire list of classes included in the Foundation framework please refer to the Foundation Framework Reference document in Apple's Developer Library.

19

App Features, Structure, and Lifecycle

The iOS development style uses the Model View Controller (MVC) design pattern. This design pattern defines the communication between three types of objects: model, view, and controller objects.

The *model* represents the backend code, usually dealing directly with the data upon which the app operates. When a model changes its state, it sends a notification to the controller. The model's state change may be triggered either internally or by an update message from its controller.

The *view* is the user interface front-end. It is literally what is viewable from the user's perspective: buttons, lists, images, and so on. Its role is to present the model to the user in a form suitable for interaction. When the view changes its state due to user interface interaction, the controller is notified.

The *controller* acts as a mediator between the model and the view. It receives input in the form of model notifications or user actions and sends messages to its attached model and views. The controller may also cause a state change by updating the view with new model data.

The whole point of the MVC programming paradigm is to isolate application logic from the input and presentation layer; thus, permitting independent development, testing, and maintenance of each individual part.

In this chapter you will learn how to build apps using the MVC design pattern as well as the role that each object plays in app development.

19.1 Views

User Interface elements in their most basic form are called views, and are represented by the `UIView` class. In other words, all views are subclasses of `UIView`. Typical apps will contain many views, but only one window. Views can be thought of as form elements, while windows can be thought of as the entire screen, including the status bar at the top. All user interface classes are in the UIKit framework and will be discussed further in Chapter 20.

There are two ways of creating and configuring your app's main window and views: manually or using Interface Builder.

19.1.1 Interface Builder

XIB (Xcode Interface Builder) files store the app's user interfaces. This file contains information about the design of visual aspects of an app and the configuration of controller objects. XIB files allow you to visually create a CocoaTouch interface, making it easy to adjust UI properties such as the text in a label, colors, fonts, and so on. During compile-time, XIB's are compiled down to *NIB* (NeXTStep Interface Builder) resources. These *NIB* resources are loaded at run-time to construct the interface. Even though the interface builder files have the extension XIB, most documentation calls them NIB files because apps use the compiled NIB files and not the XIB files.

A good way to explain each one of the interface builder elements is through the classic "Hello World" example. Throughout this section, you are going to build this example, step-by-step, by following the guid.

To get started, create a new Xcode project and choose the View-based Application template (Fig 19-1).

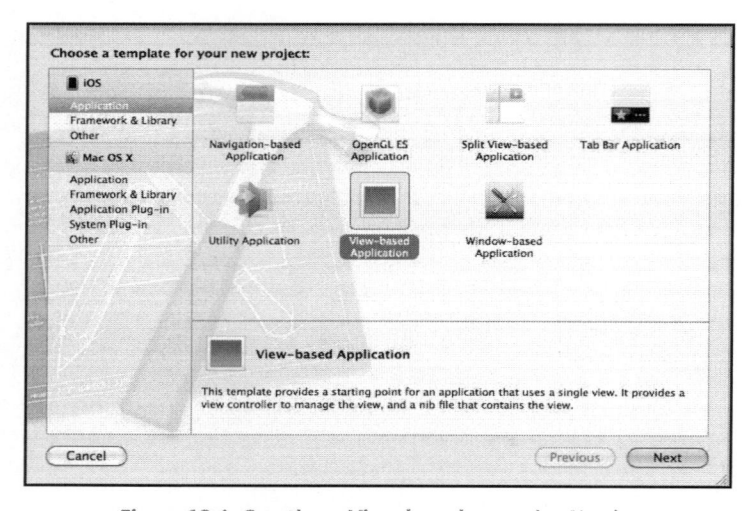

Figure 19-1: Creating a View-based app using Xcode

Use "HelloWorld" for the project's Product Name and make sure that "Include Unit Tests" is unchecked, as shown in Figure 19-2.

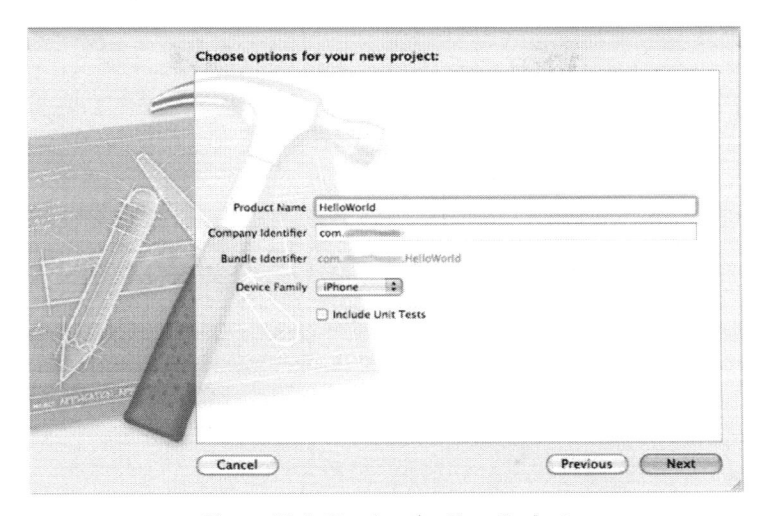

Figure 19-2: Naming the New Project

The next screen will show some basic information and options regarding the app's properties, including the app's version number, supported devices, iOS deployment target, supported orientations, as well as app icons and launch images as in Figure 19-3.

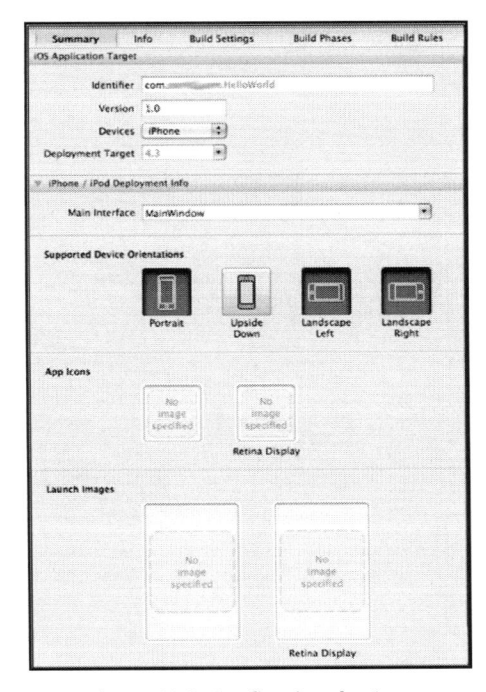

Figure 19-3: Configuring the App

Chapter 19: App Features, Structure, and Lifecycle   141

To access Interface Builder, select an XIB file from the project navigator on the left-hand panel of Xcode. When you create a View-based Application, `Xcode` automatically generates an initial XIB as shown Figure 19-4. Once selected, Interface Builder will display in the main editor window. For this example, select the "HelloWorldViewController.xib" in the project navigator. You will see an empty view in the editor window.

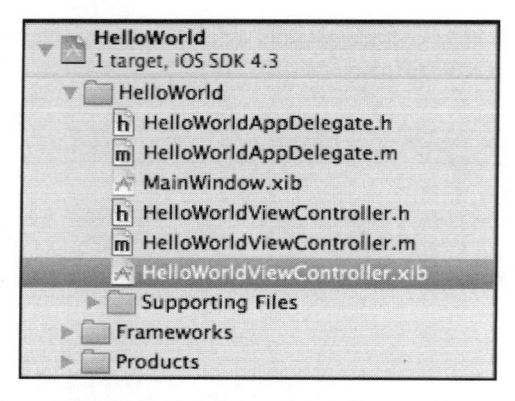

Figure 19-4: Project Navigator for HelloWorld Example

Another important part of Interface Builder is the Library pane. It has four libraries to select: the File Template Library, Code Snippet Library, Object Library, and Media Library. This section uses the Objects Library that is shown in Figure 19-5. This library contains useful objects used to design user interfaces and is used with Interface Builder to build XIB files.

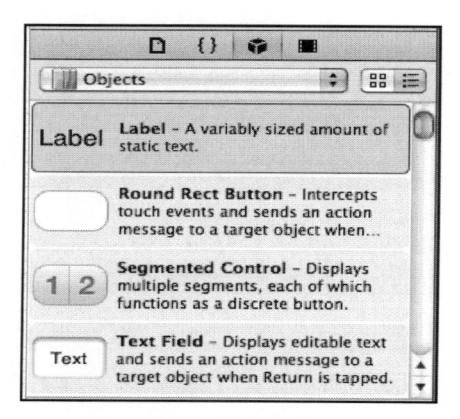

Figure 19-5: Objects Library

To add an object to a view, drag the desired object onto the view itself. For the "HelloWorld" example, add a `UILabel` and a `UIButton` to your main view.

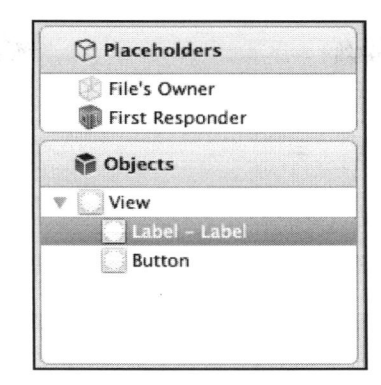

Figure 19-6: Placeholders and Objects of the HelloWorld example

A vertical column located to the right of the Project Navigator has a window that displays objects added from Interface Builder as seen in Figure 19-6. Also, it contains placeholders that store references to objects that live outside of the NIB file. It is important to understand that File's Owner is the main link between your app code and the contents of XIB file.

Some properties of each visual object of an app, as well as File's Owner, can be modified using the Inspector Pane (Figure 19-7) located above the Library on the right-hand side of Xcode. This column has six inspectors available: File, Quick Help, Identity, Attributes, Size, and Connections Inspectors.

Figure 19-7: Inspector Pane

The Identity Inspector is used to change the class type of an object. This is useful when sub-classing objects.

The Attribute Inspector (Figure 19-8) allows you to change object attributes. For example, for a `UILabel`, you can modify attributes such as alignment, text color, background color, alpha, font type, and so on. Attributes are control specific, however many controls have similar attributes. These attributes can also be modified programmatically.

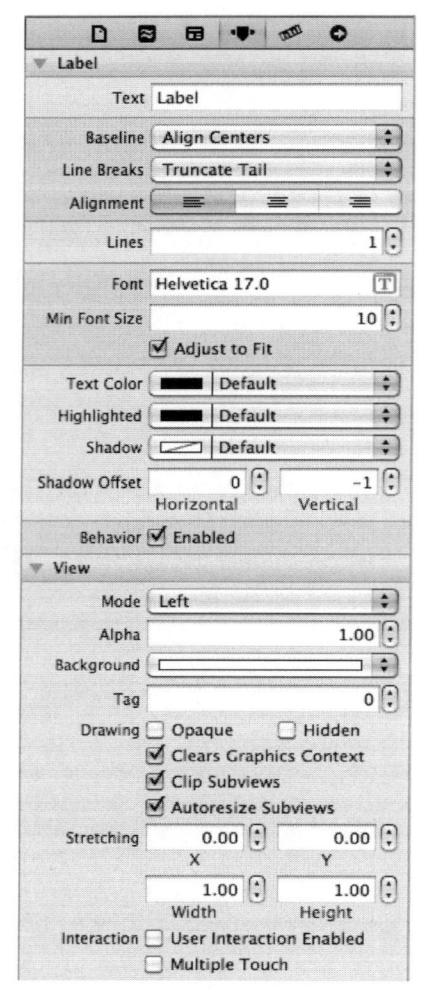

Figure 19-8 Attribute Inspector

The Size Inspector is used for changing the position and size of objects.

The Connection Inspector is a tool for managing events as well as outlet connections on the selected object. The Outlets section contains pointers to associations with other objects in code. The Sent Events section contains methods in the class file that are set to receive events from controls; such as Touch Up Inside, Value Changed, Touch Down, and so on.

Modify the attributes values of the added items. Click on the Label and set its Text attribute to an empty string so that it is blank. Modify the title of the button to the string "Click Me."

The next goal is to make the label display "Hello, World!" when you click on the button. To accomplish that, you will need to add some lines to your code, and then wire up your connections using Interface Builder.

Open Hello World project's view controller header file, "HelloWorldViewController.h" and add the following lines of code:

```
@interface HelloWorldViewController : UIViewController {
    IBOutlet UIButton *helloButton;
    IBOutlet UILabel  *helloLabel;
}
```

The `IBOutlet` modifier for either variable or property allows the view controller to store references to objects using an outlet. This way, Interface Builder will synchronize the display and the connection of outlets with Xcode.

Next, define a function, which later will be connected to the button created in the Hello World example. The following line of code should go just before the `@end` directive:

```
- (IBAction) buttonClick;
```

The `IBAction` modifier is equivalent to the void type, however it is good practice to use IBAction instead of void for use within Interface Builder.

Now, in the Hello World project's implementation file named `HelloWorldViewController.m`, implement the `buttonClick` method by adding the following lines of code:

```
- (IBAction) buttonClick {
    helloLabel.text = @"Hello, World!";
}
```

Next, wire up the view objects to the code by going back into the interface builder. In the left column, right click (or CTRL + click) on File's Owner. Interface builder determines what elements are available to be connected by reading the class's header file.

In the popup window, drag the circle next to `helloButton` down to your `UIButton` in the interface. Do the same for `helloLabel` and your `UILabel`. Finally, connect `buttonClick` under "Received Actions" to the `UIButton` as shown in Figure 19-9. When a second popup shows a selection of different events, select "Touch Up Inside."

Chapter 19: App Features, Structure, and Lifecycle   145

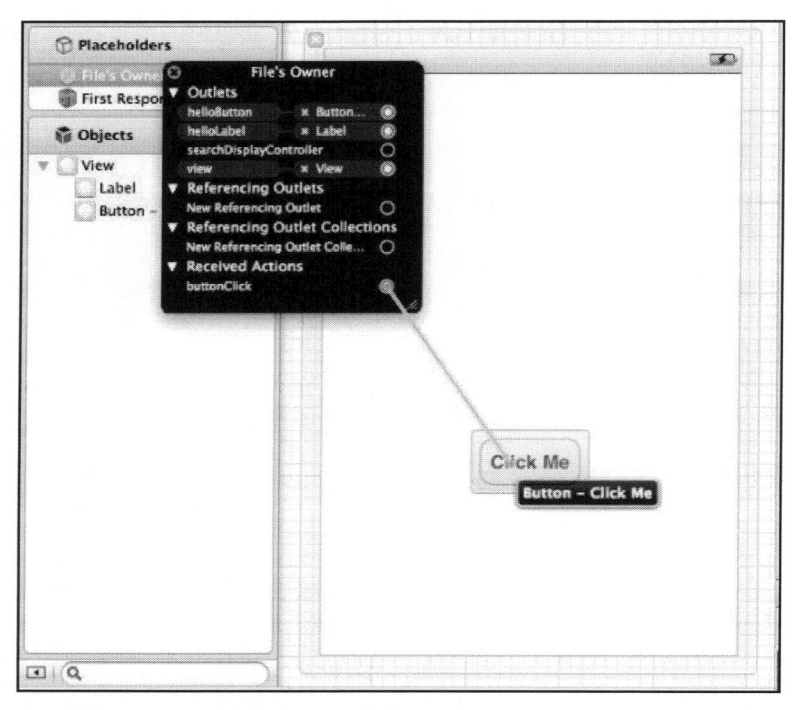

Figure 19-9: Connecting an Action with an Event on the Button

Run the app to see the results. Touching the button now sets the label text to "Hello World."

You are going to expand on this example with a few changes to display "Hello", "Hey", or "Good Morning" greetings randomly when the button is touched.

Back in the implementation file, modify the `buttonClick` method as follows:

```
- (IBAction) buttonClick {
    /* Create an array consisting of three strings */
    NSArray *introStrings = [[NSArray alloc]
        initWithObjects: @"Hello", @"Hey", @"Good Morning", nil];
    /* Randomly pick one element from the array */
    int myRand = rand() % 3;
    NSString *pickedString = [introStrings objectAtIndex: myRand];
    /* Change the text of the label */
    helloLabel.text = pickedString;
}
```

When the user clicks the button, a greeting is randomly chosen from the `introStrings` array and is displayed in the label.

Interface Builder can save time and requires less code to be written to achieve the same behavior when implementing everything manually. However, not all operations are possible through Interface Builder, such as setting certain attributes on objects.

19.1.2 Creating interfaces manually

It is important to know how to create view interfaces manually. This approach to creating view interfaces is useful when inserting elements dynamically, and making sure the order and the position in which they are going to be placed is correct. This section describes some of the view's properties and how to create views programmatically.

19.1.2.1 Views' properties

A view has several properties to configure. They include appearance and behavior in relation to the coordinate system, such as:

- *Frame* specifies the size and origin of the view in its superview's coordinate system.

- *Bounds* specify the size of the view in its local coordinate system.

- *Center* specifies the position of a view in the superview's coordinate system.

- *Transform* contains a `CGAffineTransform` structure with the transformations to apply.

The `UIView` coordinate system, which has its origin on the top-left corner, allows you to keep control of the visual space that an app can use. It extends down and to the right from the origin point.

Next, some of the structures related to view's properties are explained, including `CGPoint`, `CGSize`, `CGRect` and `CGAffineTransform`.

`CGPoint`: represents a point in a two-dimensional coordinate system. To create a point, you can write:

```
CGPoint aPoint = CGPoint(0, 0);
```

Where the arguments, which in the above example are "0,0", indicate the x and y coordinates respectively of the point.

`CGSize`: represents the dimensions of a rectangle. To create a `CGSize`, you can do the following:

```
CGSize aSize = CGSize(100, 100);
```

The two parameters, which in the above example are 100,100, have values for width and height respectively that define a dimension.

`CGRect`: represents the origin and size of a rectangle. The syntax to create a rectangle with four equal sides is:

```
CGRect aRect = CGRectMake(0, 0, 100, 100);
```

The two first parameters indicate the x and y coordinates of the origin point, while the other two define the width and height of a rectangle. Here is how to access the values of a rectangle with four equal sides:

```
aRect.origin.x = 10;
aRect.origin.y = 10;
aRect.size.width = 50;
aRect.size.height = 50;
```

The `CGAffineTransform` structure is used to define transforms, such as translating position, scaling, or rotating. Each one of these transformations has defined specific functions. For example, to define a rotation you can use the following function:

```
CGAffineTransform xform = CGAffineTransformMakeRotation(M_PI/4.0);
```

`M_PI` is the constant pi defined in the math library. The passed argument indicates an angle of 45 degrees.

19.1.2.2 Creating views

Views include defined methods that allow its creation and managing.

To create views, use the `initWithFrame:` method, which sets the initial size and position of the view relative to its parent view. The following line of code is the syntax for creating a new generic `UIView` object:

```
UIView *myView = [[UIView alloc] initWithFrame: aRect];
```

In the same way, you can create any object that its type is a subclass of `UIView`, for example:

```
CGRect aframe = CGRectMake(10, 10, 100, 100);
UILabel *aLabel = [[UILabel alloc] initWithFrame:aframe];
```

The `addSubview`, `insertSubview` and `removeFromSuperview` methods can be used to arrange the views in a hierarchy. For example, adding a label to a view looks like:

```
[aView addSubview:aLabel];
```

To define a custom view, it must inherit from `UIView` or any subclass of it. Also, you have to implement `dealloc` as well as initialization methods. If the creation is done manually, override the `initWithFrame:` method. If you want to draw in the view, you must override the `drawRect:` method. For example:

```
- (void)drawRect:(CGRect)rect{
    CGRect frame = CGRectMake(100, 100, 100, 30);
    UILabel *aLabel = [[[UILabel alloc] initWithFrame:frame] autorelease];
    aLabel.text = @"Hello!";
    [self addSubview:aLabel];
}
```

Touch events can be handled and responded to by overriding the `touchesBegan:withEvent:`, `touchesMoved:withEvent:`, `touchesEnded:withEvent:`, and `touchesCancelled:withEvent:` methods. For more information about this, please refer to Chapter 23.

19.1.3 View's Life Cycle

A view controller manages the view's life cycle, which is an instance of `UIViewController`. You can call any method of a view inside methods of your view controller. This section describes the most important methods of the `UIViewController`.

The `initWithNibName:bundle:` method creates a new `UIViewController` and returns it. The first parameter indicates the name of the associated NIB file and the second specifies the bundle in which to search for the NIB file. This method is called when you use Interface Builder and the specified NIB file exists. Take the following the statement:

```
MyViewController *aViewController = [[MyViewController alloc]
initWithNibName:@"MyViewController" bundle:nil];
```

It assigns a new object of type `MyViewController` in the main bundle to the object, `aViewController`.

If there is no NIB file created, the NIB name parameter is `nil`. In this case, you must override the `loadView` method to create your view manually. For example:

```
- (void)loadView {
    self.view = [[UIView alloc] initWithFrame: CGRectMake(0, 0, 320, 480)];
    self.view.backgroundColor = [UIColor blueColor];
}
```

This code creates a view for the view controller with blue as its background color.

The `viewDidLoad` method is called after the NIB file is loaded. Thus, any additional initialization should be placed here. For example, the following code snippet creates a label and inserts it into the view:

```
- (void) viewDidLoad {
    [super viewDidLoad];
    CGRect frame = CGRectMake(0.0, 0.0, 100, 100);
    UILabel *aLabel = [[[UILabel alloc] initWithFrame:frame]  autorelease];
    aLabel.text = @"Hello";
    [self.view addSubview: aLabel];
}
```

Note that superviews automatically retain their subviews. Thus, after using these methods, you should `release` that subview, or use `autorelease` when it's created.

The `viewWillAppear` and `viewDidAppear` methods are respectively called each time, before and after, a view is visible on the screen. Other methods, `viewWillDisappear` and `viewDidDisappear` are called when the view will disappears and did disappear respectively.

The `viewDidUnload` method is called when a view controller releases a view. It is good practice to override this method to perform any necessary cleanup.

19.1.4 Delegates and Data Sources

Delegates and data sources are helper objects that add functionality to a class without having to subclass. A delegate controls the user interface, while a data source controls the data. Both must adopt a protocol and implement the required methods of that protocol. There are several classes in the UIKit framework that use these kind of objects, such as `UITextField`, `UITableView`, `UIPickerView`, and so on.

For example, if you want to do some operations when editing inside of a text field, you must implement the `UITextFieldDelegate` protocol. Then, if you want to dismiss the keyboard when the user presses the Return button, you can implement the `textFieldShouldReturn:` method. Going back to the "HelloWorld" example, see how this works with only three steps:

1. Make the class conform to the `UITextFieldDelegate` protocol in the "HelloWorldViewController.h" file by adding the following line of code:

```
@interface HelloWorldViewController : UIViewController <UITextFieldDelegate>
```

Now, open "HelloWorldViewController.m" and do the following two steps.

2. In the `viewDidLoad` method, add the following lines:

```
// Sets the return button type to display "Done"
nameField.returnKeyType = UIReturnKeyDone;
// The delegate functions are implemented on self
nameField.delegate = self;
```

3. Add the following function:

```
- (BOOL) textFieldShouldReturn: (UITextField *)theTextField {
    [theTextField resignFirstResponder];
    return YES;
}
```

Run the example, enter text into the text field and observe the behavior of the "Done" button.

19.2 The Application Life Cycle

All Cocoa apps start with its main function, placed in the main.m file. This function calls the following `UIApplicationMain` function:

```
int retVal = UIApplicationMain(argc, argv, nil, nil);
```

This line of code creates an instance of `UIApplication` and then scans the app's Info.plist file. This property list file contains information about the app such as its name, icon, orientation, status bar style, and system requirements. The `NSMainNibFile` field is used to specify the NIB file that will be loaded when the app launches.

The application delegate object, an instance of a class that conforms to the `UIApplicationDelegate` protocol, participates in the app lifecycle by monitoring its behavior. The `UIApplication` object sends messages to its delegate object when events occur. The app moves through several states.

The following are delegate methods accessible through the `UIApplicationDelegate` class:

```
- (void)applicationDidFinishLaunchingWithOptions:(UIApplication *)application
(NSDictionary *)launchOptions;
```

The `application:didFinishLaunchingWithOptions:` message is sent to the application object's delegate once it has completed its setup. If `launchOptions` dictionary is empty, it indicates that the user launched the app directly. Otherwise, it gives information on the reason for the launch. By putting code in this method, you can configure the user interface and perform other tasks before the app is displayed.

The following method is called before the app terminates. It is useful to save state information before an app terminates.

```
- (void)applicationWillTerminate:(UIApplication *)application;
```

The following method is called before the app moves from an active state to an inactive state.

```
- (void)applicationWillResignActive:(UIApplication *)application;
```

The following method is called immediately after the app enters the background.

```
- (void)applicationDidEnterBackground:(UIApplication *)application;
```

19.2.1 Local and Remote Push Notifications

iOS has the ability to inform the user regarding some event outside of the app through local and remote notifications. These notifications are displayed as an alert message or badge visible on an app's icon, and sometimes can be accompanied with a sound.

Local notifications are scheduled and displayed using the same device; while a remote server sends remote notifications, also called push notifications. This section describes how to configure and display local notifications.

19.2.1.1 Local Notifications

To implement a local notification, you need to inform the operating system that you want to schedule a new notification. In this way, iOS will display the notification at a specified time. A local notification object is an instance of the `UILocalNotification` class. Here are some of its properties:

- *fireDate*: specifies the date and time for when the system will deliver the notification. This is the most important property and should never be left out.

- *repeatInterval*: specifies a regular interval, such as: daily, weekly, yearly, in which the notification should be delivered.

- *alertBody*: the alert message.

- *alertAction*: the title of the action button

- *soundName*: the sound to play.

- *applicationIconBadgeNumber*: the app icon badge number.

- *userInfo*: a dictionary that includes custom data.

The following example shows how to create and schedule a local notification in the simplest way using a `buttonClick` action.

```
- (IBAction) buttonclick {
    NSDate *now = [NSDate date];
    NSLog(@"Now is %@",now);
    NSDate *myNewDate = [now dateByAddingTimeInterval: 60];
    // Scheduling a notification that will deliver 1 minute from now
    [self scheduleNotificationForDate: myNewDate];
    NSLog(@"The event will occur on %@", myNewDate);
}
- (void) scheduleNotificationForDate: (NSDate*)date {
    UILocalNotification *localNotification = [[UILocalNotification
                                                alloc] init];
    localNotification.fireDate = date;
    NSLog(@"Notification will be shown on:
                        %@",localNotification.fireDate);
    localNotification.timeZone = [NSTimeZone defaultTimeZone];
    localNotification.alertBody = [NSString stringWithFormat:
                                                @"Event occurred"];
    localNotification.alertAction = NSLocalizedString(@"Details",
                                                nil);
    localNotification.soundName =
                        UILocalNotificationDefaultSoundName;
    [[UIApplication sharedApplication] scheduleLocalNotification:
                                    localNotification];
}
```

As you can see, in order to ensure that the notification is delivered, you must use the `scheduleLocalNotification:` method. If you want to deliver the notification immediately, which will ignore the fire date property, use the `presentLocalNotificationNow:` method instead.

When a notification is delivered, the `application:didReceiveLocalNotification:` delegate method is called if the application delegate implements it. This method is useful to check the `UILocalNotification` object properties, including access to any custom data from the `userInfo` dictionary.

In addition, when a notification occurs and the app is not visible or active, the alert message is displayed, the app icon badge is modified, and the sound is played. If the app is active, then none of the aforementioned will occur.

You can call `cancelLocalNotification:` or `cancelAllLocalNotifications` to cancel specific or all local notifications.

To handle the actions when notifications are delivered, while the app is in the background, you can use the `applicationDidEnterBackground` method.

19.3 Basic Debugging

To show how to debug, you first need to introduce a bug into the project as an example. Go back to the "Hello World" example and in the `buttonClick` method, modify the line of code used to generate a random number to the following:

```
int myRand = rand() % 4;
```

Previously, the modulus 3 was yielding numbers from 0 to 2, inclusive. You only have 3 objects in the array, so this makes sense. However, now that you are using modulus 4, there is the possibility of running the array out of bounds and causing an error. The compiler will not catch this, since it is a logical error and not a symbolic or syntax error. Try it out. Continue to click the button until it crashes.

The simplest form of debugging is to insert temporary `NSLog` calls throughout your code. For example:

```
NSLog(@"Count:%i Picked Index:%i", [introStrings count], myRand);
```

But this method is cumbersome. Once done finding the error, you need to go back and remove all the `NSLog` calls and it just gets ugly. Additionally, where you place the call can make a difference in results. For example:

```
- (IBAction) buttonClick {
    /* Create an array of three strings */
    NSArray *introStrings = [[NSArray alloc] initWithObjects:
                    @"Hello", @"Hey", @"Good Morning", nil];
    /* Randomly pick one element of the array */
    int myRand = rand() % 4;
    NSString *pickedString = [introStrings objectAtIndex: myRand];
```

```
        NSLog(@"Count: %i    Picked Index: %i", [introStrings count],
                                                    myRand);
    /* Change the text of the label */
        NSString *helloText = [NSString stringWithFormat: @"%@, %@!",
                                        pickedString, nameField.text];
        nameField.text = helloText;
    }
```

In this example, the app will crash before it reaches the `NSLog` statement. There is nothing wrong with this form of debugging but it is not always the best option. `NSLog` is useful when the program is not crashing, but is producing intermittently incorrect results. `NSLog` provides you with a very convenient way to check the history of values and figure out what conditions cause an error. Xcode also provides several tools to debug your code that you do not need to write `NSLog` calls all the time. Here is an overview of these tools:

The *Issues Navigator* shows all problems found while building. If you select any error or warning, it will show the line where it occurs.

The *Log Navigator* shows the history of your console run and debug. To open the source editor to an error or warning, double-click on it. You can also click the list icon at the most right of a line to see the build command and results.

The *Breakpoint Navigator* lists all breakpoints of the project. Breakpoints are a flexible way to inspect code. They allow you to verify logical flow, as well as to inspect values of variables. Additionally, you do not need to manually remove them for release builds. Xcode makes it easy to turn them on and off with the click of a button. There are conditional breakpoints, which are triggered when a specific conditional happens. Another kind of breakpoint is triggered when a specific exception is thrown or caught. Symbolic breakpoints are triggered when a specific method or function executes. You can insert these kinds of breakpoints clicking the Add (+) button at the bottom of the breakpoint.

Before continuing the example, clean up the code by removing all remaining `NSLog` calls.

To add a breakpoint, click to the left of the line on which you would like to create the breakpoint. A blue tab will appear; this means that the breakpoint has been added. If you click it again, it'll turn semi-transparent. This means that the breakpoint still exists, but it has been disabled. To delete a breakpoint, right click on it and select "delete breakpoint" or click and drag it out of the window.

Make a breakpoint as shown in Figure 19-10 and run the app. Upon clicking the button, the app will pause execution and wait for you to decide what to do.

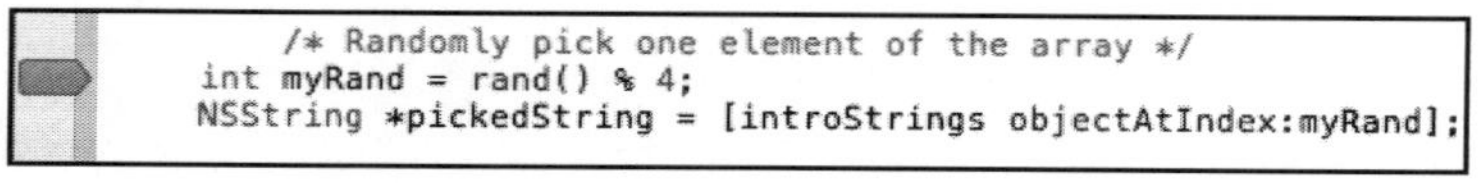

```
        /* Randomly pick one element of the array */
    int myRand = rand() % 4;
    NSString *pickedString = [introStrings objectAtIndex:myRand];
```

Figure 19-10: Putting a breakpoint in code

At this point, you can inspect the values of variables in the current scope at the bottom of the window in the debug pane, as in Figure 19-11. Now, if you look at myRand's value, you will notice it has not been initialized because you broke on the line. In other words, a line break happens before the execution of the line of code it's breaking on. You can also read the value of a variable and its fields by hovering your mouse cursor over it or by clicking on the disclosure triangle.

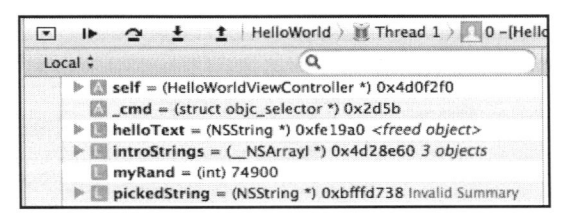

Figure 19-11: Application variables in the debug area

To continue execution, you have the four options: to continue, step over, step into, and step out, shown here from left to right:

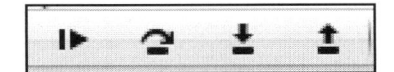

Figure 19-12: Buttons for debugging

- *Continue*: keeps executing normally until another break point is hit
- *Step Over*: executes the next line of code
- *Step Into*: evaluates the current line of code, detects the first function being called, then enters its source code and pauses on the first line
- *Step Out*: resumes normal code execution until the current function returns, then pauses as it returns to its calling function

For now, you can go ahead and continue. If the random number had a modulus of 3 the program will have crashed, if not repeat until it does actually crash. Now, the debugger will have detected an error and halted the program on the line triggering the error.

19.4 Summary

This concludes this chapter. Topics learned in this chapter include: Views and their lifecycle, How to create interfaces manually as well as with Interface Builder, Application lifecycle, and some basic debugging. The next chapter will cover Designing iOS Apps.

20

iOS Design Basics

This chapter will teach you the basics of how to design iOS apps. It describes types of important views and controls that can be used for various purposes, and how to manage these views through their view controllers. Thus, this chapter is divided in 3 sections that will cover views, view controllers, and an example that shows how all these objects work together.

20.1 Designing Views

As you saw in Chapter 19, all visual components are subclasses of the `UIView` class. This section describes several views, which you can have in your app according to your needs, and their principal features.

20.1.1 Displaying images

The `UIImageView` class can be used to display one or several images animated together. There are methods to create instances of the class, such as the `initWithImage:` method and the `initWithImage:highlightedImage:` method. You can also use the `initWithFrame:` method, since it inherits from the `UIView` class. For example:

Using `initWithImage:`

```
UIImageView *anImage = [[UIImageView alloc] initWithImage:
                            [UIImage imageNamed: @"Image1.png"]];
```

Using `initWithFrame` method from the `UIView` class:

```
CGRect anImageRect = CGRectMake(0, 0, 320, 480);

UIImageView *anImage = [[UIImageView alloc]
                            initWithFrame: anImageRect];

[anImage setImage: [UIImage imageNamed: @"Image1.png"]];
```

To display a list of images with a certain interval between them, which is called animation; use the `animationImages` property, like this:

```
NSArray *imageArray = [NSArray arrayWithObjects:
                            [UIImage imageNamed: @"Image1.png"],
                            [UIImage imageNamed: @"Image2.png"],
                            [UIImage imageNamed: @"Image3.png"], nil];
```

```
UIImageView *anAnimatedView = [UIImageView alloc];
[anAnimatedView initWithFrame:[self bounds]];
anAnimatedView.animationImages = myImages;
anAnimatedView.animationDuration = 0.60; // time in seconds
[anAnimatedView startAnimating];
```

As you can see, here another class named `UIImage` is used. This object represents the data of an image, which can be adjusted by its properties (imageOrientation, size, scale, etc.). In addition to these properties, the `UIImage` class provides methods for drawing the image. The `imageNamed` method returns the image object associated with the specified filename (the argument).

20.1.2 Selecting elements from a list

For selecting either single or multiple objects, such as months, numbers or options, you can use `UIPickerView`. You can see this object in the library, as shown in Figure 20-1.

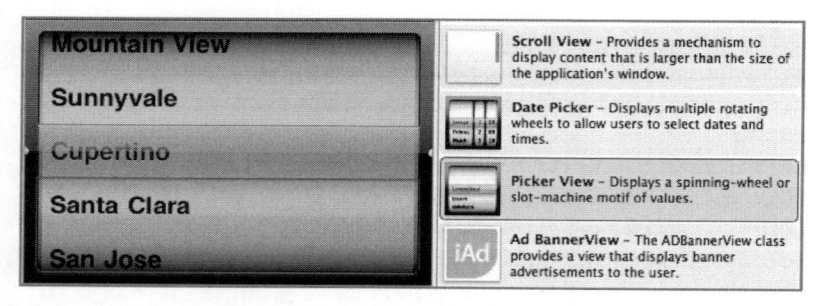

Figure 20-1: A PickerView object when dragged from the Library and dropped into the Editor

Once you have a `UIPickerView`, you need to connect the `dataSource` and `delegate` to the File's Owner. This allows you to populate the `UIPickerView` with data and implement the delegate methods to react to events, like selecting a row. To do this Ctrl + Click on the `UIPickerView` and then click on the `dataSource` and `delegate` and drag them to the File's Owner.

After making connections, you have to set up an `IBOutlet` for the `UIPickerView` in your view controller, like this:

```
@interface PickerExampleViewController: UIViewController {
  IBOutlet UIPickerView *picker
}
@end
```

158   Chapter 20: iOS Design Basics

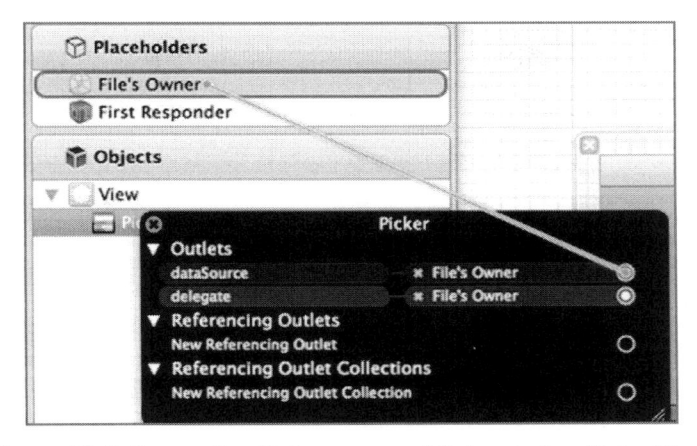

Figure 20-2: Connecting Datasource and Delegate to a PickerView

In addition, you must connect the IBOutlet (picker variable) to the UIPickerView. Finally, in order to configure and provide the picker with some data, the view controller should conform to the UIPickerViewDelegate and UIPickerViewDataSource protocols, as follows:

```
@interface PickerExampleViewController : UIViewController
                                    <UIPickerViewDelegate,
                                     UIPickerViewDataSource> {

    IBOutlet UIPickerView *picker;
}
@end
```

To set the number of components and rows in each component, you should implement the numberOfComponentsInPickerView: and pickerView: numberOfRowsInComponent: methods of the UIPickerViewDataSource protocol. For instance, to set up the picker with two components and twenty rows:

```
- (NSInteger) pickerView: (UIPickerView *)pickerView
  numberOfRowsInComponent: (NSInteger)component {
      return 20;
}
- (NSInteger) numberOfComponentsInPickerView: (UIPickerView *)pickerView {
      return 2;
}
```

If you want to configure what is displayed on each row and also what happens when a row is selected, the UIPickerViewDelegate protocol provides methods to set the appearance and behavior of the rows. You must implement either the titleForRow method or the viewForRow method. The viewForRow method is used to display a custom view, for example a UIImageView containing an image. The `titleForRow` function is used to display a simple string in the rows, like this:

```
-(NSString *)pickerView:(UIPickerView *)pickerView
          titleForRow:(NSInteger)row
          forComponent:(NSInteger)component {
    return [NSString stringWithFormat: @"%i:%i", row, component];
}
```

To perform an operation when selecting an element in a component you need to implement the `pickerView:didSelectRow:inComponent:` method:

```
- (void)pickerView:(UIPickerView *)pickerView
       didSelectRow:(NSInteger)row
       inComponent:(NSInteger)component {
    NSLog(@"Selected row:%i and component:%i", row, component);
}
```

In this case, if you change the selected element in the `UIPickerView` it prints out the row and component indices to the console.

20.1.3 Displaying web content

The `UIWebView` class is used to display content on screen, like documents or URLs. You can display content loaded dynamically or content loaded from a file, using the `loadHTMLString:baseURL:` method:

```
NSString* html = @"<html><center><font size=20 color='red'>
                    An error occurred</font></center></html>";
[(UIWebView *) aView loadHTMLString: html baseURL: nil];
```

Alternatively, you can use the `loadData:MIMEType:textEncodingName:baseURL:` method, for example:

```
NSData *pdfData = [NSData dataWithContentsOfFile:@"/Users/aUser/aDoc.pdf"];
[(UIWebView *)aView loadData:pdfData MIMEType:@"application/pdf"
                 textEncodingName:@"utf-8"
                 baseURL:nil];
```

Also, the `UIWebView` class allows the content to be loaded from a network using the `loadRequest:` method, like this:

```
[aWebView loadRequest:[NSURLRequest requestWithURL:[NSURL URLWithString:
@"http://www.apple.com/"]]];
```

Other methods are defined in the `UIWebView` class, such as, the `stopLoading` method, to stop the `UIWebView` from loading. Also, there are the `goBack` and `goForward` methods that move through the webpage history.

You can implement methods that are called when the web content is loaded. These methods are defined by the `UIWebViewDelegate` protocol, and they are:

- webView:shouldStartLoadWithRequest:navigationType:
- webViewDidStartLoad:
- webViewDidFinishLoad:
- webView:didFailLoadWithError:

The following code snippet shows how to implement the last one:

```
- (void) webView: (UIWebView *)webView didFailLoadWithError: (NSError *)error
{
    NSString* errorHTML = @"<html><center><font size=20 color='red'>
                                        An error occurred
                                            </font>
                                        </center>
                                    </html>";
    [(UIWebView *)aView loadHTMLString: errorHTML baseURL: nil];
}
```

20.1.4 Displaying alerts

Alerts are the semi-transparent blue popup windows seen in iOS apps, which are mainly used for notifications. For example, when the user receives a text message, iOS displays a basic alert. When an alert message is displayed, its appearance will vary as it depends on whether it's of the UIAlertView or UIActionSheet class.

20.1.4.1   UIAlertView

An alert view includes either zero or more button actions. The alert is called "Timed Alert View" if it has no button; this case is only to indicate the user that something is happening in the background. If the alert has just one button, usually it's the "OK" button. If you want to customize an alert, use more than one button.

Figure 20-3: Displaying a Simple Alert View

To create an alert you can use the
`initWithTitle:message:delegate:cancelButtonTitle:otherButtonTitles:`
method. The alert of Figure 20-3 is displayed using this code:

```
UIAlertView *myAlert = [UIAlertView alloc]
                        initWithTitle:@"Warning"
                        message:@"Do you want to continue?"
```

```
                    delegate:self
                    cancelButtonTitle:@"Cancel"
                    otherButtonTitles:@"OK", nil];
```

```
        [myAlert show];

        [myAlert release];
```

To customize the operations used by an alert, its delegate must conform to the **UIAlertViewDelegate** protocol. It contains the **alertView:clickedButtonAtIndex:** method, used to intercept and handle the user's button clicks in order to perform actions.

```
        -(void) alertView:(UIAlertView *)theAlertView
        clickedButtonAtIndex:(NSInteger)index {

            if (buttonIndex == 1) {

                    UIAlertView *alert = [[UIAlertView alloc]
                                    initWithTitle:nil
                                    message: @"Completed"
                                    delegate: self
                                    cancelButtonTitle: @"OK"
                                    otherButtonTitles: nil, nil];

                [alert show];

                [alert release];
            }
        }
```

Since **UIAlertView** is a subclass of **UIView**, you can add any type of view to the alert view that you want. For example, the following code specifies the creation of an alert view containing a text field, shown in figure 20-4.

```
        UIAlertView *myAlert = [[UIAlertView alloc]
                                initWithTitle: @"Enter Name"
                                message: @" "
                                delegate: self
                                cancelButtonTitle: @"Cancel"
                                otherButtonTitles: @"Ok", nil];

        UITextField *txtField = [[UITextField alloc]
                    initWithFrame: CGRectMake(20.0, 45.0, 245.0, 25.0)];

        txtField.borderStyle = UITextBorderStyleRoundedRect;

        txtField.placeholder = @"Enter Name";

        txtField.returnKeyType = UIReturnKeyDone;

        txtField.delegate = self;

        [myAlert addSubview: txtField];

        [myAlert show];

        [txtField becomeFirstResponder];   // to get focus to txtField
```

```
[myAlert release];
```

Figure 20-4: Displaying a Text Field Alert View

20.1.4.2 *UIActionSheet*

Action sheets are another way of presenting an alert. They are perfect for forcing the user to make a decision, but they may also be used as status indicators.

Figure 20-5: UIActionSheet example

The figure labeled 20-5 is the simplest representation of a `UIActionSheet` object. This is the ideal candidate for situations where the user must make a decision before the app can continue execution. Like other alert methods, it is another form of a modal. In other words, the app will cease to execute until the user chooses a decision.

The following code shows how to create the object represented by figure 20-5.

```
UIActionSheet *mySheet = [[UIActionSheet alloc]
            initWithTitle: @"Perform Action"
      delegate: self                              cancelButtonTitle:
@"Cancel"                                destructiveButtonTitle:
@"Destroy"
         otherButtonTitles:@"Don't Destroy", @"Never mind", nil];

[mySheet showInView: self.view];

[mySheet release];
```

Just like `UIAlertView`, `UIActionSheet` has delegate methods that allow you to customize its behavior. These methods are part of the `UIActionSheetDelegate` protocol.

20.1.5 Controls

Controls are instantiated by subclasses of the `UIControl` class. They are a particular case of a view that allows user input and interaction. This means the `UIControl` class inherits from the `UIView` class, but it also defines action messages that are called when certain events occur, such as clicking a button or editing text. There are many types of controls, such as buttons, text fields, sliders, etc. Buttons are instances of the `UIButton` class. Using sliders (`UISlider` class) is a convenient way to pick a single value out of a range of values. Its current value is represented graphically by the position of the indicator along the track. Text fields (`UITextField` class) are single-line fields. Switch controls are instances of the `UISwitch` class. They allow the user to interact with the `Boolean` state of a value. Another important control is the `UIDatePicker` control, which provides an easy way to enter a date and time.

All of these controls use the target-action mechanism for handling communication between objects in a program. In this case, a target is the receiving object, and the message sent to the target (by the control) is called the action. The action selectors can be defined in the following ways:

- (void)action
- (void)action:(id)sender
- (void)action:(id)sender forEvent:(UIEvent *)event

The parameter `sender` is used to identify the control that sent the action message.

You can define the targets and actions programmatically or by using Interface Builder. Using Interface Builder would look like figure 20-6, where the `buttonClick` action is connected to an event that occurs when the button is clicked.

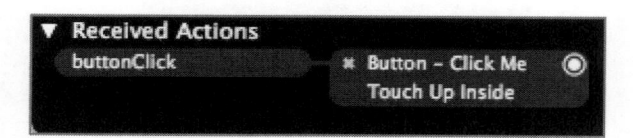

Figure 20-6: Action connected to an event

If you choose to set the target-action programmatically, you can use the `addTarget:action:forControlEvents:` method, as in the following code:

```
[helloButton addTarget: self
            action: @selector(buttonClick)
            forEvents: UIControlEventTouchUpInside];
```

The next section will show you a complete example using a type of Control called UISegmentedControl.

20.1.5.1 A Control example: UISegmentedControl

`Segmented Controls` allow the user to quickly choose between a few options. They can be used to select different settings, answer a multiple-choice question, and a multitude of other things.

When you add a new `UISegmentedControl` to the view, it defaults to two segments. This can be easily changed through the attribute inspector by modifying the number of segments, as indicated in figure 20-7.

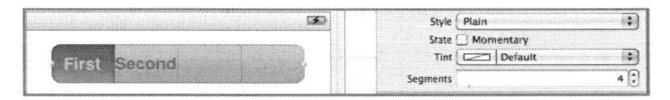

Figure 20-7: UISegmentedControl with 4 segments

To change the segment names, change the field titled "Segment" in the attribute inspector, then modify its Title field with its segment number. It will look like figure 20-8.

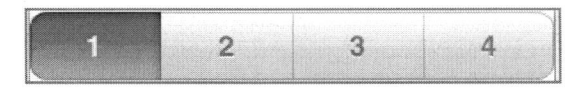

Figure 20-8: A UISegmentedControl with its segment names changed

To connect actions to events, first you have to define the actions in the header file, like this:

```
@interface SegmentExampleViewController : UIViewController{

}
- (IBAction) segmentSelected: (UISegmentedControl *)sender;
@end
```

Now in the NIB file you have to connect this action just created with the "Value Changed" event of the `UISegmentedControl`, like Figure 20-9 shows.

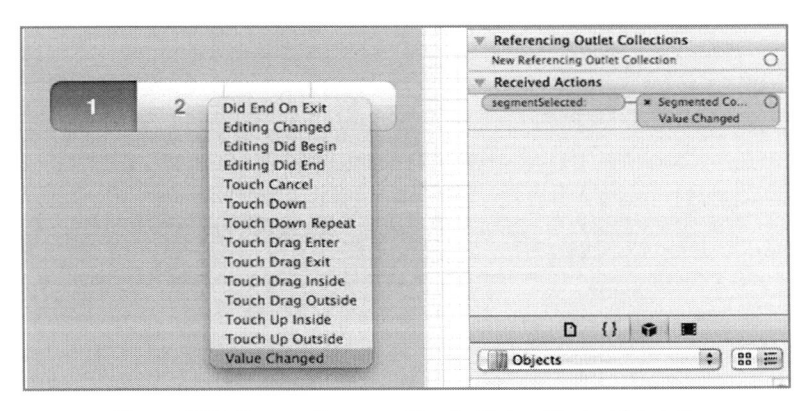

Figure 20-9: Connecting an action to Value Changed event

Finally, you should define the action in the implementation file, for instance:

```
@implementation SegmentExampleViewController
```

Chapter 20: iOS Design Basics   165

```
- (void) segmentSelected: (UISegmentedControl *)sender{
    NSLog(@"Segment selected index: %i",
                    sender.selectedSegmentIndex);
}
```

In this method, the sender object is a pointer to the `UISegmentedControl` that is calling this method. Therefore, depending on which segment is selected; you can perform different tasks by making conditional statements.

20.1.6 Table Views

Table views, represented by the `UITableView` class, display large sets of data to the user. They consist of a single column, but can have any number of rows grouped by sections. The class supports vertical scrolling, since it inherits from the `UIScrollView` class.

With some creativity, almost anything is possible with table views, but there are two basic styles of table views offered by the SDK: plain and grouped. The style of a table is a `readonly` property that can only be set when you initialize the table with the `initWithFrame:style:` method. If the table is initialized using the `initWithFrame:` method of the `UIView` class, the plain style is used by default. If you use interface builder, the style property can be set using the Attribute inspector, as shown in the Figure 20-10. This way, the value for the style property is set to `UITableViewStylePlain` or `UITableViewStyleGrouped`.

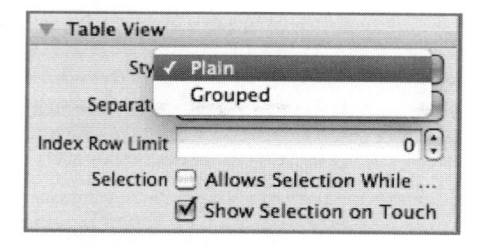

Figure 20-10: Selecting the table view's style

Both styles of table view contain a single table header (situated at the top of a table), single table footer (situated at the bottom), and several sections (in the middle). Each section has a header, footer, and many cells. The difference between the two styles is that the plain style shows all sections combined just separated by a header, and the grouped style shows the sections separated as group of rows.

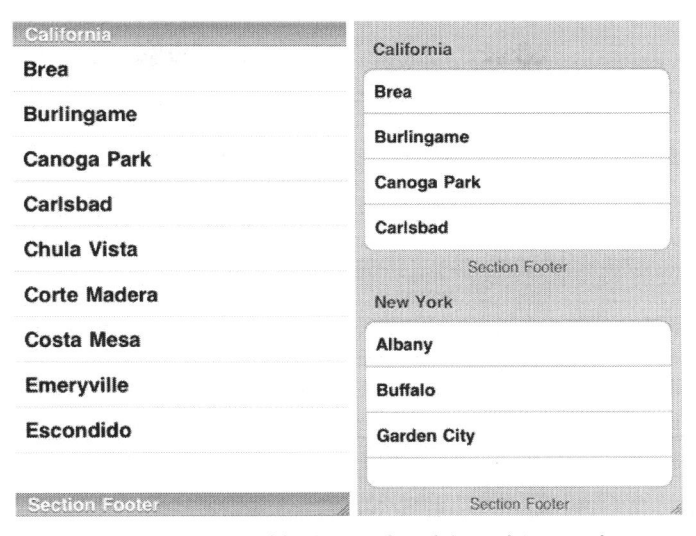

Figure 20-11: UITableView Styles Plain and Grouped

20.1.6.1 Table View's Cell

Cells of a `UITableView` are of type `UITableViewCell`. Custom cells can be created by subclassing the `UITableViewCell` class and implementing a custom control layout. Another way to change the design of cells is by creating the `UITableViewCell`, using one of four different style constants as the first parameter in the `initWithStyle:reuseIdentifier:` method. Figure 20-11 shows a table view with the four different styles, UITableViewCellStyleDefault, UITableViewCellStyleSubtitle, UITableViewCellStyleValue1, and UITableViewCellStyleValue2, in that order:

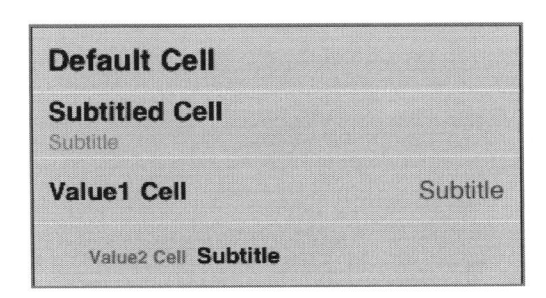

Figure 20-12: Table View Cell's Styles

Inside of a `UITableViewCell` there is certain content that can be accessed through several properties. The `textLabel` property of a `UITableViewCell` is a `UILabel` that can be used to set the main text of the cell. The second property is the `detailTextLabel` property, which is a `UILabel` that can be used to set the text of the secondary label of the cell, like the "Subtitle" text in Figure 20-11. The third property is the `imageView` property, which is a `UIImageView` that can be used to display an image inside of the cell.

Chapter 20: iOS Design Basics   167

Cells have a property named `accessoryType`, which indicates the element that appears in the right side of the cell. This accessory is typically used to indicate that the user can click the accessory for more information. Figure 20-12 shows the four values for `UITableViewCellAccessoryType;` `UITableViewCellAccessoryNone,` `UITableViewCellAccessoryDisclosureIndicator,` `UITableViewCellAccessoryDetailDisclosureButton,` and `UITableViewCellAccessoryCheckmark,` in that order.

Figure 20-13: Accessory types of Table View Cells

There are also several other properties and methods of the `UITableViewCell` class that are used for purposes such as accessing views of the cell, managing accessory views, managing cell selection and highlighting, editing the cell, and managing content indentation.

20.1.6.2 Table View Data Delegation

The `UITableView` class uses `delegates` to dynamically update the table view as needed. This is ideal for memory management because it is not required to load all cells into memory at once. When working with a `UITableView`, the `ViewController` either needs to be a subclass of `UITableViewController` or it needs to implement the `UITableViewDataSource` and `UITableViewDelegate` protocols.

The `UITableViewDataSource` protocol has various methods that are used to set up the `UITableView` and configure it with data. However, only two of them are required methods, because of their role in the configuration of a table view, which are `tableView:cellForRowAtIndexPath:` and `tableView:numberOfRowsInSection:` methods. The following code is an example of how to implement these methods:

```
- (NSInteger) tableView: (UITableView *)tableView
  numberOfRowsInSection: (NSInteger)section {
    return [contentsArray count];
}
- (UITableViewCell *) tableView: (UITableView *)tableView
        cellForRowAtIndexPath: (NSIndexPath *)indexPath {
    static NSString *CellIdentifier = @"Cell";
    UITableViewCell *cell = [tableView
            dequeueReusableCellWithIdentifier: CellIdentifier];
```

```
        if (cell == nil) {
            cell = [[[UITableViewCell alloc]
                    initWithStyle: UITableViewCellStyleDefault
                    reuseIdentifier: CellIdentifier] autorelease];
        }
        cell.textLabel.text = [contentsArray objectAtIndex: indexPath.row];
        return cell;
    }
```

As you can see, to represent a position of a cell in a table, the `NSIndexPath` class is used. This class identifies the row and section indices that can be used to locate a particular cell. In this example, the number of rows is set to the size of an array (`contentsArray`) and the text of the individual `UITableViewCell` is set to the object in the array indexed by the row number (`indexPath.row`).

The `UITableViewDelegate` protocol contains several methods to manipulate the cell properties, to handle user interaction with the cells, as well as, handle specific events like beginning or ending editing.

An important thing to know when working with `UITableView` is that cells rarely stay selected, so that's where the `UITableViewDelegate` protocol comes into play. The method `tableView:didSelectRowAtIndexPath:` is called when a user selects a row. In this method is where you want to place the actions that should occur when a user selects the cell, which is usually presenting a new view to the user.

Other than performing an action when a row is selected, it is also possible to only allow certain rows to be selected. This is useful if a cell can only be selected under certain conditions. Using the `tableView:willSelectRowAtIndexPath:` method you can prevent the cell from being selected by returning `nil`.

20.2 Controllers

You have worked with simple view controllers so far. Now, it is time to learn that there are some kinds of controllers such as navigation, tab bar, and table view controllers that can make app building easy. You will learn how to use interface builder or configure your controllers manually. With a large example that uses all three of these controllers, the process is explained step-by-step.

20.2.1 Navigation Controller

A navigation-based application uses a navigation controller object to manage a stack of views. It has a `UINavigationController` object and one or more `UIViewController` objects that show and manage the different views. Each added view controller is pushed on the top of a navigation stack and it is shown on the screen. This operation is done by the `pushViewController:animated:` method. For example, calling this method would look like this:

```
[self.navigationController pushViewController: viewController
                                     animated: YES];
```

To remove a view controller from a navigation stack you can use the back button provided by the navigation controller or call the `popViewControllerAnimated:` method.

You can customize the navigation bar using the `navigationItem` property, which is of `UINavigationItem` type. This class describes the appearance of the navigation bar, so you can customize the left, middle, and right side it.

In the middle of the navigation bar, there can be a title string (`title`) or custom title view (`titleView`), for example:

```
self.navigationItem.title = @"Various Fruits";
```

or

```
self.navigationItem.titleView = [[UIImageView alloc] initWithImage:
                                                        anImage];
```

The left button (`leftBarButtonItem`) and right bar button (`rightBarButtonItem`) are instances of the `UIBarButtonItem` class, which is a button designed for items in navigation bars and toolbars with special appearance and behavior. These buttons have an associated target and action, just like a regular button. There are three different initialization methods for these buttons: one that displays a string, one that displays an image, and one to display a predefined system item. The following methods are the different ways to initialize a `UIBarButtonItem`:

```
-initWithTitle:style:target:action:
```

Example 1:

```
UIBarButtonItem *myButtonItem = [[UIBarButtonItem alloc]
                                initWithTitle:@"My button"
                                style:UIBarButtonItemStyleBordered
                                target:self
                                action:@selector(anAction:)];
```

```
-initWithImage:style:target:action:
```

Example 2:

```
UIBarButtonItem *myButtonItem = [[UIBarButtonItem alloc]
                     initWithImage:anImage
                     style:UIBarButtonItemStyleBordered
                     target:self
                     action: @selector(anAction:)];
```

```
-initWithBarButtonSystemItem:target:action:
```

Example 3:

```
UIBarButtonItem *myButtonItem = [[UIBarButtonItem alloc]
          initWithBarButtonSystemItem:UIBarButtonSystemItemAdd
```

```
                          target:self
                          action:@selector(addAction:)];
```

So, you can set the left and right buttons in the same way, like this:

```
self.navigationItem.leftBarButtonItem = myButton;
```

Usually, the left button is used to get back to the previous view controller, and displays a short, appropriate title. To create these kinds of buttons, you should use the method in "Example 1." Whether or not there is a custom left bar button item, the navigation bar displays the item defined in the `backButtonItem` property of the previous view controller. If there is no back button item defined either, the left button is the default button with a title of the previous view controller.

The title and right button can be set to a control defined previously. For instance, the following code snippet shows how to set a segmented control to the right button.

```
- (void) viewDidLoad {

  UISegmentedControl *aSegmentedControl = [[UISegmentedControl
              alloc] initWithItems:[NSArray
      arrayWithObjects:@"Up", @"Down", nil]];

  [aSegmentedControl addTarget:self
              action:@selector(aSegmentAction:)
forControlEvents:UIControlEventValueChanged];

  UIBarButtonItem *aSegmentBarItem = [[UIBarButtonItem alloc]
                      initWithCustomView: aSegmentedControl];

  [aSegmentedControl release];

  self.navigationItem.rightBarButtonItem = aSegmentBarItem;

  [aSegmentBarItem release];
}
```

In the same way the `titleView` can be set directly to a segmented control since they are both views, for instance:

```
self.navigationItem.titleView = aSegmentedControl;
```

20.2.2 Tab Bar Controller

Most iOS apps have a common design as far as how they are controlled. A navigation controller, a tab bar controller, or a combination of both is usually used to control these apps. One common way to implement a complex app is to use a tab bar controller as the root view controller of the main window. A tab bar controller can have several view controllers, some of which can be navigation controllers.

A Tab Bar Application uses a tab bar controller to manage a list of views. It contains a `UITabBarController` object and one or more tabs, each one with an associated view controller. The following example shows how to create a tab bar controller and set the `viewControllers` property to an array of view controllers.

```
- (void) applicationDidFinishLaunching: (UIApplication *)application {
```

```
            UITabBarController *tabBarController =
                        [[UITabBarController alloc] init];
    //create an empty View controller
        UIViewController *firstViewController =
                        [[[UIViewController alloc] init] autorelease];
    //create a custom navigation view controller
        MyNavController *secondViewController =
                        [[[MyNavController alloc] init] autorelease];
    //create a custom View controller
        MyViewController *thirdViewController =
                        [[[MyViewController alloc] init] autorelease];
      NSArray *controllers = [NSArray arrayWithObjects:
                        firstViewController,
                        secondViewController,
                        thirdViewController, nil];
      tabBarController.viewControllers = controllers;
      [window addSubview:tabBarController.view];
    }
```

Each view controller has a `tabBarItem` property that can be created for you. If you do not set this property, the system will set it as a default item with no image and text indicating the title of the view controller. A `tabBarItem` is an instance of `UITabBarItem`, which can be a system item, or contain a title, image, and a tag. Therefore, there are two ways to create it, shown here:

Using `initWithTabBarSystemItem:tag:` method:

```
        - (void) viewDidLoad {
    UITabBarItem *item = [[UITabBarItem alloc]
                        initWithTabBarSystemItem: UITabBarSystemItemRecents
                                            tag: 0];
            self.tabBarItem = item;
            [item release];
        }
```

Or, using the `initWithTitle:image:tag:` method:

```
        - (void) viewDidLoad {
    UITabBarItem *item = [[UITabBarItem alloc]
                        initWithTitle: @"Title"
                                image: anImage
                                  tag: 0];
```

```
            self.tabBarItem = item;
            [item release];
    }
```

Xcode provides some templates that help to create projects using these objects, such as Navigation-based, Tab Bar, and Split View-based (Only for iPad) Applications.

For demonstration purposes, build a tab bar project using the interface builder facilities. First, create a new project in Xcode. Choose "Tab Bar Application" and name it "TabBarExample."

If you run the project you would notice a tab bar controller application by default comes with two tabs connected to two separate `Views`. Now add a third `View` to the project.

Right-click on the project and go to New File. Next select "UIViewController subclass" and make sure the "With XIB for user interface" box is checked, and then name the file "ThirdViewController."

Now that you have a third view, put something in it to distinguish between them. Go to `ThirdViewController.xib` and add a `UILabel` to it and enter the text "Third View."

In order to connect the third view controller to the tab bar controller, open `MainWindow.xib` and add a new tab bar item to the tab bar. Go to the Library and find the "Tab Bar Item" (Figure 20-14) and drag it onto the Tab Bar.

Figure 20-14: Tab Bar item on the Objects Library

Change the Title property of the Tab bar Item from "Item" to "Third." Also, change the class from UIViewController to ThirdViewController. Next, set the NIB Name to "ThirdViewController" (make sure it is spelled exactly the same as the xib file).

Save the changes and then build and Run the project to test it out. As you can see the ThirdViewController created is connected to the Tab Bar Controller. You can select the Third View and see the label on it.

20.3 App Example

Now that you have a basic understanding of the `UITableView` class, `UITableViewCell` class, and controllers, you will create a test project which combines all of them together. This example is divided into four phases in order to get a better understanding and so that you can improve upon the previous example.

20.3.1 Creating a Navigation-based Application

Create a new project in Xcode by selecting "Navigation-based Application" and name it "TableViewExample." Make sure "Use Core Data" and "Include Unit Tests" are both unchecked. In the RootViewController.h file, define an array and then in the RootViewController.m file initialize the array with several objects as follows:

```objc
@interface RootViewController: UITableViewController {
    NSMutableArray *contentsArray;
}
@end


@implementation RootViewController
- (void) viewDidLoad {
   [super viewDidLoad];
   contentsArray = [[NSMutableArray alloc]
                        initWithObjects: @"Banana",
                                         @"Apple",
                                         @"Orange",
                                         @"Pear",
                                         @"Strawberry",
                                         @"Cherry",
                                         @"Grape", nil];
}
```

Now, use the created array to populate the `UITableView` with data. Find the `UITableViewDataSource` methods for `numberOfRowsInSection` and `cellForRowAtIndexPath` and alter the code to look as the example of section 20.1.6.2 Table View Data Delegation. Then, if you build and run you would see a plain `UITableView` setup with the contents of the array.

20.3.2 Creating a table view with various section

Next, make the `UITableView` more advanced, dividing its rows by sections. The sections are classified for three different fruit colors. In the RootViewController.m file update the `viewDidLoad` method as seen below.

```objc
-(void)viewDidLoad {
  [super viewDidLoad];


        self.navigationItem.title = @"Various Fruits";
        NSMutableArray *tempArray1 = [[NSMutableArray alloc]
                                initWithObjects:@"Blackberries",
                                                @"Blueberries",
                                                @"Dried plums",
                                                @"Grapes",
                                                @"Plums",
                                                @"Raisins", nil];
        NSMutableArray *tempArray2 = [[NSMutableArray alloc]
                                initWithObjects:@"Cherries",
```

```objc
                                        @"Cranberries",
                                        @"Grapefruit",
                                        @"Raspberries",
                                        @"Apples",
                                        @"Strawberries", nil];
    NSMutableArray *tempArray3 = [[NSMutableArray alloc]
                        initWithObjects:@"Apricots",
                                        @"Cantaloupe",
                                        @"Lemon",
                                        @"Mangos",
                                        @"Oranges",
                                        @"Peaches",
                                        @"Pineapples", nil];
    NSMutableDictionary *tempSection1 = [[NSMutableDictionary alloc]
                        initWithObjectsAndKeys:
                                        @"Blue/Purple Fruit",
                                        @"Title",
                                        tempArray1,
                                        @"Array", nil];
    NSMutableDictionary *tempSection2 = [[NSMutableDictionary alloc]
                        initWithObjectsAndKeys:@"Red Fruit",
                                        @"Title",
                                        tempArray2,
                                        @"Array", nil];
    NSMutableDictionary *tempSection3 = [[NSMutableDictionary alloc]
                        initWithObjectsAndKeys:
                                        @"Yellow/Orange Fruit",
                                        @"Title", tempArray3,
                                        @"Array", nil];
    contentsArray = [[NSMutableArray alloc] initWithObjects:
                                        tempSection1,
                                        tempSection2,
                                        tempSection3, nil];

    [tempArray1 release];

    [tempSection1 release];

    [tempArray2 release];

    [tempSection2 release];

    [tempArray3 release];

    [tempSection3 release];
}
```

The first line sets the navigation bar title; this can be specific to the view controller, and changes when the view controller changes. In the following lines of code three arrays were created, each one containing fruits of the same color. Then an **NSMutableDictionary** is created for each array, which contains for the color for that group of fruit and the array with the corresponding fruit of that color. Finally, the three dictionaries are added to the contents array.

Next you must implement the `UITableViewDataSource` methods in order to populate the table view with data. Modify the `numberOfSectionsInTableView` and `numberOfRowsInSection` methods, and implement the `titleForHeaderInSection` and `sectionIndexTitlesForTableView` methods as follows:

```
- (NSInteger) numberOfSectionsInTableView: (UITableView *)tableView {
    return [contentsArray count];
}
- (NSInteger)  tableView: (UITableView *)tableView
    numberOfRowsInSection: (NSInteger)section {
    return [[[contentsArray objectAtIndex: section]
                    objectForKey: @"Array"] count];
}
- (NSString *) tableView   : (UITableView *)tableView
    titleForHeaderInSection: (NSInteger)section {
    return [[contentsArray objectAtIndex: section]
                    objectForKey: @"Title"];
}
- (NSArray *)sectionIndexTitlesForTableView:(UITableView *)tableView{
    return [NSArray arrayWithObjects:@"B", @"R", @"Y", nil];
}
```

The `contentsArray` variable now contains multiple dictionaries that have section information, so the size of the `contentsArray` is used to determine the number of sections. In the `numberOfRowsInSection` method, the number of items in each array, within the dictionary, is set as the number of rows, for that particular section. In the `titleForHeaderInSection` method the value for the key "Title" is retrieved from the dictionary corresponding to the section.

Now, the table is mostly set up except you have to update the content of the `UITableViewCell`. So, in the `cellForRowAtIndexPath` method update the code to look as follows.

```
- (UITableViewCell *) tableView: (UITableView *)tableView
         cellForRowAtIndexPath: (NSIndexPath *)indexPath {
  static NSString *myCellIdentifier = @"myCell";
  UITableViewCell *cell = [tableView dequeueReusableCellWithIdentifier:
myCellIdentifier];
  if (cell == nil) {
      cell = [[[UITableViewCell alloc]
              initWithStyle: UITableViewCellStyleSubtitle
              reuseIdentifier: myCellIdentifier] autorelease];
```

```
    }
    cell.textLabel.text = [[[contentsArray objectAtIndex:
                indexPath.section] objectForKey: @"Array"]
                                    objectAtIndex: indexPath.row];
    cell.detailTextLabel.text = [NSString stringWithFormat:
                                    @"Index %i", indexPath.row];
    return cell;
}
```

As you can see, the table view cell style was changed from `UITableViewCellStyleDefault` to `UITableViewCellStyleSubtitle` and the `detailTextLabel` property was used to set the displayed text to the index of the current object. The `textLabel` property will display the correct string depending on the section and row.

Now, the app is ready to run.

20.3.3 Editing in a Table View

This time you will set up the table to allow the deletion of rows. Most of the code that is required should already be present; thus, all that is needed is to uncomment it. Uncomment the `tableView:commitEditingStyle:forRowAtIndexPath:` method. Since you are using a `UITableViewController`, you can just use `self.editButtonItem,` otherwise you would have to create this object manually, handle its events and set the `UITableView` to edit mode. Now, to delete the data from the array, update the `commitEditingStyle` method as follows.

```
-(void)tableView:(UITableView *)tableView
  commitEditingStyle:(UITableViewCellEditingStyle)editingStyle
  forRowAtIndexPath:(NSIndexPath *)indexPath {

    if (editingStyle == UITableViewCellEditingStyleDelete) {

      [[[contentsArray objectAtIndex: indexPath.section]
                  objectForKey: @"Array"]
            removeObjectAtIndex: indexPath.row];

      [tableView deleteRowsAtIndexPaths:[NSArray
                  arrayWithObject: indexPath]
                  withRowAnimation: UITableViewRowAnimationFade];
    }
    else if (editingStyle == UITableViewCellEditingStyleInsert) {

    }
}
```

Finally, add the following line of code to the `viewDidLoad` method:

```
self.navigationItem.rightBarButtonItem = self.editButtonItem;
```

Now when you build and run the app you will see that you can delete rows from the table view. Pressing the "Edit" button in the navigation bar should cause all of the cells to show, on the left, a button (⊖) indicating that a cell can be deleted. You can hide these buttons by clicking on the "Done" button, which replaced the "Edit" button.

Compared to the original `tableView` you now have multiple sections, section headers, a navigation title, and the ability to delete.

20.3.4 Selecting cells in a Table View

Usually when a `UITableViewCell` is clicked in iPhone apps, an action is performed. Commonly, a new ViewController is displayed when the cell is selected. This new ViewController usually displays more information pertaining to the cell that was selected. This is what you are going to attempt to do.

First, add a new view controller by selecting New > File > UIViewController subclass and name it SelectedViewController.

In the SelectedViewController.h file, add two `NSString`'s and a `UILabel` that is an `IBOutlet`. The `NSString`'s will hold the selected text and the section text, so that you can use it to manipulate the view. The `SelectedViewController.h` file should look similar to below.

```objc
@interface SelectedViewController: UIViewController {
    NSString *selectedText;
    NSString *sectionText;
    IBOutlet UILabel *labelSelected;
}
@property (nonatomic, retain) NSString *selectedText;
@property (nonatomic, retain) NSString *sectionText;
@end
```

Now, in the `SelectedViewController.m` file, synthesize the two `NSString`'s, then add the following code to the `viewDidLoad` method:

```objc
- (void)viewDidLoad {
    [super viewDidLoad];
    self.navigationItem.title = self.sectionText;
    labelSelected.text = self.selectedText;
    if ([sectionText rangeOfString:@"Red"].length > 0) {
        self.view.backgroundColor = [UIColor redColor];
        [labelSelected setTextColor:[UIColor greenColor]];
    } else if ([sectionText rangeOfString:@"Blue"].length > 0) {
        self.view.backgroundColor = [UIColor blueColor];
        [labelSelected setTextColor:[UIColor orangeColor]];
    } else if ([sectionText rangeOfString:@"Yellow"].length > 0) {
```

```
            self.view.backgroundColor = [UIColor yellowColor];
            [labelSelected setTextColor:[UIColor purpleColor]];
        }
    }
```

The `navigationItem.title` property is set to the `sectionText` string, so that when this view controller is pushed on to the navigation stack, the navigation bar title will change. The label's text is changed to display the selected text. Afterwards, the `rangeOfString:` method is called to see if "Red," "Blue," or "Yellow" are in the section text. Then the view's `backgroundColor`, and the label text color, is set depending on which section was selected.

Now go to the `SelectedViewController.xib` file and add a `UILabel` and connect it to the `IBOutlet` that was created previously.

In the `RootViewController.m` file, import the `SelectedViewController.h` file. You can now create a `SelectedViewController` to push onto the navigation stack when the user selects a cell. In order to do this, the `didSelectRowAtIndexPath:` method of the `UITableViewDelegate` protocol needs to be implemented, and the following code needs to be added to it:

```
-(void)tableView:(UITableView *)tableView didSelectRowAtIndexPath:
(NSIndexPath *)indexPath {

    SelectedViewController *svController = [[SelectedViewController
            alloc] initWithNibName: @"SelectedViewController"
                                bundle: [NSBundle mainBundle]];

    svController.sectionText = [[contentsArray
                                    objectAtIndex: indexPath.section]
                                    objectForKey: @"Title"];

    svController.selectedText = [[[contentsArray
                                    objectAtIndex: indexPath.section]
                                    objectForKey: @"Array"]
                                    objectAtIndex: indexPath.row];

    [self.navigationController pushViewController:
                                svController animated:YES];

    [svController release];
}
```

The first line of code indicates the creation of an instance of the `SelectedViewController` class, and allocates and initializes it with the XIB file, by using the `initWithNibName:bundle:` method. The values of the section and the cell that were selected are set to the two defined properties of the view controller. Finally, the created `SelectedViewController` is pushed onto the navigation stack.

Build and run your project and you will see that when you select a `UITableViewCell` it will display the `SelectedViewController` with a background color specific to the fruit's color and the name of the fruit selected (Figure 20-15). You will also notice that the navigation controller automatically created the back button.

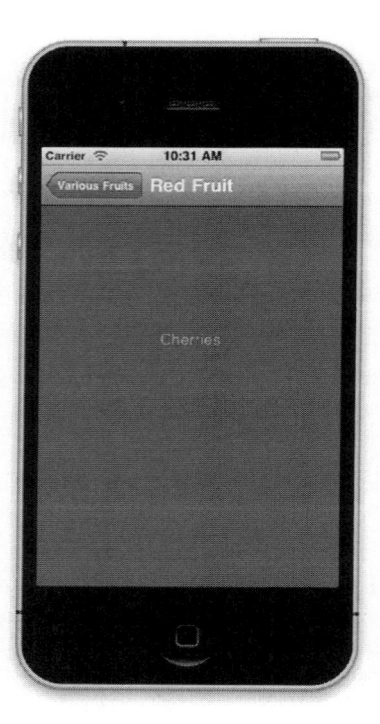

Figure 20-15: Results of the Navigation-based app

Now that you have an understanding of the basics of designing iOS Apps, you should be able to start creating some apps of your own. In the next couple of chapters you will review more advanced topics such as Data Management, Important Frameworks, and User Input.

21

Data Storage and Handling

Data storage and handling is a critical aspect of app development. This chapter discusses the various methods used for storing and handling data in an efficient manner. Basic file input and output, XML, NSUserDefaults, SQLite, and Core Data are covered in this chapter.

21.1 File and Network I/O

File and network input/output is a necessity in today's data driven software. It is especially necessary in iOS apps since users typically run an app for a few seconds at a time; so apps must be able to save state to properly resume later. Also, many apps rely on remote data interaction to function. In this section, the capabilities and limitations of file and network I/O in the iOS environment are discussed.

File input and output on iOS works very similarly to file input and output on Mac OS X. Filenames, directory hierarchies, attributes, etc. are managed in a similar way in both file systems. It is possible to create new files and directories, delete existing files and directories, as well as read from existing files. The iOS file system is case sensitive. An app's data reading and writing abilities are primarily restricted by iOS's sandbox approach to application management.

The Foundation Framework offers four objects that support file input and output directly. These objects include **NSArray**, **NSString**, **NSData**, and **UIImage**. The input and output methods are almost identical for all objects supporting file input and output. You will work primarily with the **NSString** class in this section. For more information on the Foundation Framework see Chapter 18.

21.1.1 Reading Data from a File

Here is a simple and effective way to read a file at a specified path and store the data in an **NSString** object:

```
NSString *path = [[NSHomeDirectory()
            stringByAppendingPathComponent:@"Documents"]
            stringByAppendingPathComponent:@"myFile.txt"];

NSString *myString = [NSString stringWithContentsOfFile:path
            encoding:NSASCIIStringEncoding error:nil];
```

In the above example, you set the directory path to the file being read by creating an **NSString** object defined as **path** and using the function **NSHomeDirectory()** to get the apps home directory. From there, you append the "Documents" directory and finally the filename "myFile.txt". You then create another NSString object defined as myString, and store the data read from the previously defined path.

21.1.2 Writing Data to a File

Here is a simple and effective way to write an NSString to a file at specified path:

```
NSString *path = [[NSHomeDirectory()
                stringByAppendingPathComponent:@"Documents"]
                stringByAppendingPathComponent:@"myFile.txt"];

NSString *myString = @"Hello world!";

[myString writeToFile:path atomically:YES
            encoding:NSASCIIStringEncoding error:nil];
```

The value YES is passed to the atomically parameter to prevent concurrency issues with other threads. In other words, you set atomically to YES to ensure that you do not write to the file at the same time that another thread is writing to the same file. If the file does not exist at the specified path when writing to a file, that file will be created.

21.1.3 File Management with NSFileManager

The NSFileManager class is a powerful tool for manipulating the file system on an iOS device. Even though it may only operate within an apps sandbox, it is still very useful.

First get the NSFileManager object. The defaultManager class method of NSFileManager returns a singleton object used to perform file operations. It is important to note that the defaultManager method returns an NSFileManager that is not thread safe. If you were to use the singleton object of this class in more than one thread at the same time, there will likely be unexpected results or file corruption. It is good practice to use the following code for it to be considered thread-safe:

```
NSFileManager *fileManager = [[NSFileManager alloc] init];
```

Just be sure to release the fileManager object you created when you are done using it.

A quick recap:

Thread-safe Example:

```
NSFileManager *fileManager = [[NSFileManager alloc] init];

[fileManager release];
```

Non-thread-safe Example:

```
NSFileManager *fileManager = [NSFileManager defaultManager];
```

Here is some of what can be done with the NSFileManager class as well as a few commonly used methods:

- Copying an Item
  - -fileManager:shouldCopyItemAtPath:toPath:
  - -copyItemAtPath:toPath:error:

- o - `fileManager:shouldProceedAfterError:copyingItemAtPath:toPath:`
- Removing an Item
    - o - `fileManager:shouldRemoveItemAtPath:`
    - o - `removeItemAtPath:error:`
    - o - `fileManager:shouldProceedAfterError:removingItemAtPath:`
- Creating an Item
    - o - `createDirectoryAtPath:withIntermediateDirectories:attributes:error:`
    - o - `createFileAtPath:contents:attributes:`
- Discovering Directory Contents
    - o - `contentsOfDirectoryAtPath:error:`
    - o - `enumeratorAtPath:`
    - o - `subpathsAtPath:`
    - o - `subpathsOfDirectoryAtPath:error:`
- Determining Access to Files
    - o - `fileExistsAtPath:`
    - o - `fileExistsAtPath:isDirectory:`
    - o - `isReadableFileAtPath:`
    - o - `isWritableFileAtPath:`
    - o - `isExecutableFileAtPath:`
    - o - `isDeletableFileAtPath:`

21.1.4 Network Access to File Input and Output

Typically, classes that support file I/O also support network I/O. For example, to set a string to the contents of a URL, you would do the following:

```
NSURL *myURL = [NSURL URLWithString:
                            @"http://example.com/example.txt"];

NSString *myString = [NSString stringWithContentsOfURL:myURL
                     encoding:NSASCIIStringEncoding error:nil];
```

21.1.5 Managing XML Data

One common way data is stored in today's fast paced world is using XML or existential markup language. The benefits to using XML include its flexibility and wide acceptance in the software development arena. Because of this, XML is used to share data quite often in the IT world.

There are two different approaches to parsing XML, SAX and DOM parsing. SAX, or Simple API for XML, gives access to the data stored in an XML file as a sequence of events while DOM, or the Document Object Model, gives access to the data stored in an XML document by storing the data as a hierarchical object model in memory. The DOM tree-based approach to XML parsing is more useful when handling pure data due to the fact that the order of the data is not critical. The SAX approach is useful when you need to call handler functions when specific nodes are found within the XML document. It is important to understand that no internal representation of the document is stored when using a SAX parser. It is left up to the developer to determine how to best handle the data.

The iOS Foundation Framework gives developers access to the `NSXMLParser` class, which is an Objective-C SAX parser and a member of the `NSXML` cluster of classes. If a SAX parser is required, using `NSXMLParser` is highly recommended for performance. If a DOM parser is required, it is generally easier to use a third-party library to achieve this. The reason for this is that the full `NSXML` cluster of classes is as of this writing currently not available on iOS.

21.2 App Settings

Apps often have settings such as difficulty levels, usernames, or view options. It is convenient to the user to have app settings available inside the settings app of the device. Furthermore, there must be a way for these settings to persist. In this section, you will follow a demonstration on how this is accomplished with the settings bundle.

Apple's guidelines say that an app's settings should be set either in the settings bundle or the app itself, but not both. Here, the former is demonstrated.

When using the settings bundle, the preferences will be on at least one page known as the main page. As the number of preferences increases, it becomes more appropriate to add more pages. These additional pages become child pages of the main page. The user can access them by tapping on a special type of preference that links to another page.

21.2.1 Settings Data Types

The displayed preferences must each have a specific type that defines how the settings application displays that preference. Here's a list of the potential types:

- Text Field – `PSTextFieldSpecifier`
    - Displays an editable `UITextField` and an optional title
    - Can be used for editable strings
- Title – `PSTitleValueSpecifier`
    - Displays a `UILabel` that is not editable
    - Can be used to display information
- Toggle Switch – `PSToggleSwitchSpecifier`
    - Displays a `UISwitch`
    - Typically used for on/off preferences
- Slider – `PSSliderSpecifier`
    - Displays a `UISlider`
    - Can be used to allow the user to select a range of values
- Multi Value – `PSMultiValueSpecifier`
    - Displays a `UITableView` group of options
- Group – `PSGroupSpecifier`
    - Used to organize preferences by groups as seen in the table view
- Child Pane – `PSChildPaneSpecifier`
    - Used to specify child preference pages
    - Can be used to implement hierarchical preferences

21.2.2 Contents

The settings bundle is comprised of a few key components:

- Root.plist
    - The root page of settings that is displayed and can link to subpages of settings
- Additional .plist files
    - Used in a set of hierarchical preferences using child panes
    - Contain the contents for a child pane
- One or more .lproj directories
    - Stores localized string resources for the settings page files
    - Used to provide localized content to display to the user
- Additional images
    - Should be located in the top-level of the bundle directory
    - Can either be a preference icon or images for the slider control

21.2.3 App Settings Icon

Among the app settings is a custom icon. This icon is how your app will be recognized in the device's settings menu. The custom icon must be 29x29 pixels, labeled "Icon-Settings.png," and placed in the top of the app's bundle directory. If this file does not exist then the default icon file for the app will be used.

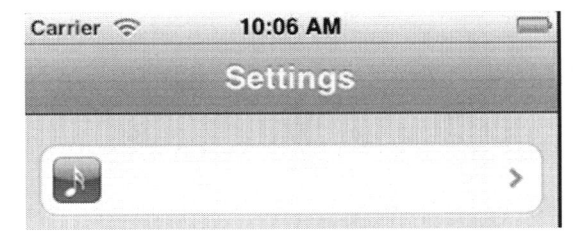

Figure 21-1: Settings

21.2.4 Quick Example

Demonstrated here is how to create a settings bundle which will have your program retrieve the user-entered data. Create a "view-based" iOS project. Now create a `UILabel` in the XIB and wire it up to the view controller. We'll name it `label` in the code.

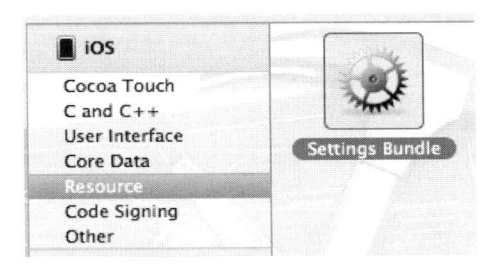

Figure 21-2: New File

Add a new file to the project. Under the iOS section in the resources type option, select *Settings Bundle*. Name it "Settings" - which is the default name. This will create a set of default settings items, including *name_preference* to be used in the rest of the example.

Now you just need to retrieve the value of *name_preference* from the settings bundle. We'll change your label's text to match that of the *name_preference*:

```
label.text = [[NSUserDefaults standardUserDefaults]
              valueForKey:@"name_preference"];
```

21.3 SQLite

As mentioned before, content is king. There must be a way to handle local content efficiently and easily in order for it to be of any use. A great tool to manage local data is SQLite: a single user relational database that uses the very familiar SQL declarative language for queries. This section goes over incorporating SQLite into an iOS app.

SQLite is a relational database management system that may be embedded in an application. In other words, it's not a stand-alone service; the app handles database management itself and only adds about 500 kilobytes to the program size. The binary data is kept on disk as a single, cross-platform file. Since SQLite is designed for single-user access, there aren't any concurrency problems. Besides, the entire database is locked during transactions. Even though it's small and limited by design, it's extremely popular and used by many well-known companies.

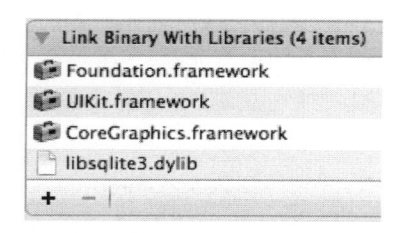

Figure 21-3: Link Binary with Libraries

Before getting started, be sure to include the SQLite library in your project. Select your project in the navigator on the left. In the editor, select your target. Click on the "Build Phases" tab at the top. Under the "Link Binary With Libraries" section, click the plus sign. Add the *libsqlite3.dylib* library. Your project is now ready for SQLite.

21.3.1 Deploying the Database

We've got two options for deploying a SQLite database: using a pre-created DB file or programmatically construct the database.

21.3.1.1 Using a Pre-Created DB File

Shipping with a pre-created database is slightly challenging. Conceptually, you need to include the file in your app's bundle, then once the app has been deployed to the device, the app needs to copy the DB file to the documents directory on the device. It's only practical if the app requires preexisting data. Here's one way to do it:

```
-(NSString*) editableDatabasePathForDB:(NSString*)dbName {
    NSError *errr;
```

```objc
        NSString *resourcePath = [[NSBundle mainBundle] pathForResource:
                        [dbName stringByDeletingPathExtension]
                               ofType:[dbName pathExtension]];
        NSString *documentPath = [[NSHomeDirectory()
                        stringByAppendingPathComponent:@"Documents"]
                        stringByAppendingPathComponent:dbName];
        if([[NSFileManager defaultManager]
                            fileExistsAtPath:resourcePath])
        {
          if (![[NSFileManager defaultManager]
                            fileExistsAtPath:documentPath])
          {
            [[NSFileManager defaultManager] copyItemAtPath:resourcePath
                            toPath:documentPath error:&errr];
          }
        }
        return documentPath;
    }
```

This method creates two paths: one for the document directory that belongs to the device, and the other for the resource folder that is bundled with the app. If the database file exists in the resource folder but does not exist in the documents folder, it copies the database file to the documents folder. This is important because files in the bundle are read-only; to make changes, you must make changes to a copy in the documents directory. Finally, this method returns the path to the editable copy in the document directory.

21.3.1.2 *Programmatically Constructing the Database*

When you need to create, populate, and access a database in an app, then this is the easiest way to deploy it. Please note that this section is reusing the `editableDatabasePathForDB` method defined in 21.3.1.1.

```objc
        NSString *path = [self editableDatabasePathForDB:@"sample.db"];
        sqlite3    *sqlDB;
        char *errorMsg;
        if (sqlite3_open([path UTF8String], &sqlDB) == SQLITE_OK) {
          sqlite3_exec(sqlDB, "DROP TABLE IF EXISTS people", NULL, NULL,    &errorMsg);
          sqlite3_exec(sqlDB, "CREATE TABLE people (firstname TEXT, lastname TEXT)",
NULL, NULL, &errorMsg);
          sqlite3_exec(sqlDB, "INSERT INTO people (firstname, lastname) VALUES
('Johnny', 'Appleseed')", NULL, NULL, &errorMsg);
          sqlite3_exec(sqlDB,"INSERT INTO people (firstname, lastname) VALUES ('Joe',
```

```
  'Average')", NULL, NULL, &errorMsg);
    sqlite3_exec(sqlDB,"INSERT INTO people (firstname, lastname) VALUES ('Susie',
'Sue')", NULL, NULL, &errorMsg);
    sqlite3_exec(sqlDB,"INSERT INTO people (firstname, lastname) VALUES ('John',
'Doe')", NULL, NULL, &errorMsg);
  }
}
sqlite3_close(sqlDB);
```

This code snippet shows the general flow for accessing a SQLite database: open the database, execute queries, then close the database. In this case, you had to create and populate your tables. Basically, the `sqlite3_exec` function lets you execute queries directly.

21.3.2 Getting the Results

You can also execute select queries with the `sqlite3_prepare_v2` function, producing a resulting table. You can step through the table to read results.

```
        NSString *path = [self editableDatabasePathForDB:@"sample.db"];
        sqlite3 *database;
        if (sqlite3_open([path UTF8String], &sqlDB) == SQLITE_OK) {
          NSMutableArray *resultsArray = [NSMutableArray new];
          const char *sqlStatement = "SELECT * FROM people";
          sqlite3_stmt *compiledStatement;
          if(sqlite3_prepare_v2(database, sqlStatement, -1,
                     &compiledStatement, NULL) == SQLITE_OK) {
            int numberOfColumns = sqlite3_column_count(compiledStatement);
            while(sqlite3_step(compiledStatement) == SQLITE_ROW) {
              NSMutableDictionary *rowDictionary = [[NSMutableDictionary
                                      new] autorelease];
              for(int i=0; i<numberOfColumns; i++){
                NSString *ColNam = [NSString stringWithUTF8String:(char *)
                          sqlite3_column_name(compiledStatement, i)];
                NSString *ColVal = [NSString stringWithUTF8String:(char *)
                          sqlite3_column_text(compiledStatement, i)];
                [rowDictionary setValue:ColVal forKey:ColNam];
              }
              [resultsArray addObject:rowDictionary];
            }
          }
```

```
sqlite3_finalize(compiledStatement);
sqlite3_close(database);
```

The same general idea is followed: open the database, execute queries, close the database. In this case, the queries are slightly more complex. You must generate a `sqlite3_stmt struct` (representing a compiled SQL statement) using the `sqlite3_prepare_v2` function. If it's successful, you can start stepping through each row in the table and handle the data accordingly. This example creates an array of column <-> value associations.

21.3.3 Usage and Implementation Considerations

It's generally a good idea to encapsulate `SQLite` operations in another class, or perhaps a set of methods or functions. This way, you gain the benefits of effortlessly accessing your database from the entire project and allow for ease of portability.

Here's an example method for adding a record to your 'people' table made in section 21.3.1.2.

```
- (int) insertIntoDB:(sqlite3*)database
        withFirstName:(NSString*)fName
        andLastName:(NSString*)lName {
  sqlite3_stmt *insertStatement;
  const char *sqlInsertBase = "INSERT INTO people VALUES (?,?)";
  if (sqlite3_prepare_v2(database, sqlInsertBase, -1, &insertStatement, NULL)
        == SQLITE_OK) {
sqlite3_bind_text(insertStatement, 1, [fName UTF8String], -1, SQLITE_TRANSIENT);
sqlite3_bind_text(insertStatement, 2, [lName UTF8String], -1, SQLITE_TRANSIENT);
    if (sqlite3_step(insertStatement) == SQLITE_DONE) {
            int i = sqlite3_last_insert_rowid(database);
            NSLog(@"Insert Succes row index: %i",i);
            sqlite3_reset(insertStatement);
            return i;
    } else {
            sqlite3_reset(insertStatement);
            return -1;
    }
  }
  else {
    return -1;
    }
  }
```

21.4 Core Data

A more Mac-like way of handling data is to use the Core Data framework. Data models are represented by entities, which are instantiated as managed objects. This section shows how to incorporate the Core Data framework in an iOS app.

After learning how to use SQLite effectively, one might not see a reason to use Core Data. Simply, it's just a high level way of doing essentially the same thing. Core Data will manage a SQLite database without the need to actually use any SQL. This saves development time because it saves the developer from writing complex queries and handling the resulting tables as queries. Core Data lets the developer treat data like any other object.

21.4.1 The Anatomy of Core Data

In this section, you will learn about all the different components that make up a functional Core Data implementation.

Data Stores

Core data can store data in four ways on iOS: binary, in-memory, XML, and SQLite. Binary store is not human readable and forces Core Data to load all the data into memory. The in-memory method is reasonably secured and never saved to disk, but is inappropriate for large data sets that would otherwise consume too much memory. XML storage is a standard format and is saved to disk, but requires all data to be loaded into memory at once. Finally, SQLite allows data to be loaded as needed, offering great performance. Generally, SQLite storage is the recommended method.

So regardless of the behind-the-scenes storage method, you still use Core Data as an object representation of a relational data structure, such as records of a table in a database. These managed objects represent the data operated on in an app, which eventually get mapped to a row in a table.

Managed Objects

A managed object context is primarily responsible for handling a collection of managed objects, similar to a table. The managed object context is powerful with a specific purpose in the app, providing the means for object management. When a new managed object is created, it should be inserted into a context. Existing data from the storage medium are fetched into the context as managed objects.

Managed Object Model

The managed object model describes a collection of entities, or managed objects, that are in your app. Its attributes and relationships can be thought of as columns in a table. The model is used to associate the managed objects with the records in the database. If the model is changed then the previous model cannot be read properly.

Persistent Store Coordinator

The persistent store coordinator manages data in a very important way. Even though it's not often directly interacted with when using the framework. It manages a collection of persistent object stores. A persistent store coordinator manages a collection of persistent object stores, thus actually mapping objects in the app to records in the database. Most mobile apps usually contain a single store, while more complex apps can have several. The persistent store coordinator's role is to manage these stores and abstract them from its managed object contexts so that they appear as a singled unified store.

21.4.2 Example Project

Start a new project named "TimeStamper." This app will let the user add the current time to a table view and keep the data in persistent store. This app will save and recall timestamps using Core Data. Go ahead and create it as an Empty Application and make sure "Use Core Data" is checked. Once Xcode generates the project, click on `TimeStamper.xcdatamodeld` in the project navigator.

Figure 21-4: New Project

The editor will now show the data model designer. Create an entity by clicking the Add Entity button at the bottom. A new entity will show up in the list; go ahead and name it `TimeStamp`. Now make sure your new entity is selected under the Entities category on the left. Now in the editor, you will see a new section called "Attributes." Click the plus sign to create a new attribute. Name it `dateTime` and set its data type to `Date`.

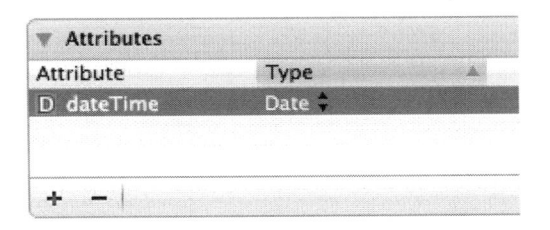

Figure 21-5: Attributes

Now you need to represent your entity in code. To do this, you create a subclass of the `NSManagedObject` class; go to *File > New > New File*. Now in the new window, make sure Core Data is selected under the iOS section. Select the option for `NSManagedObject` subclass. You will be asked to select the data models with entities you would like to manage. Select `TimeStamp` and click the Next button. The next screen will ask you to select the entities for which to generate a class for. Select `TimeStamp` and click Next. Create the file at the project's root.

Chapter 21: Data Storage and Handling   191

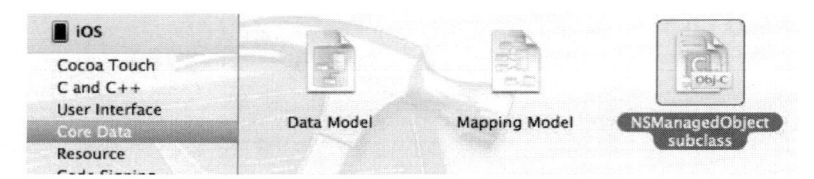

Figure 21-6: Create NSManagedObject subclass

Now create a `UITableViewController` subclass. Call it `TimeStampTableViewController` and skip the XIB, as you will not be using one. Import the `TimeStamp.h` header file. Now in the `TimeStampTableViewController` header file, add the following two member variables and create properties for them:

```
NSManagedObjectContext *managedObjectContext;

NSMutableArray *timeStampList;
```

Also, create these function prototypes:

```
- (void)loadTimeStamps;

- (void)addTimeStamp:(id)sender;
```

Now you implement the controller. In the implementation file, first synthesize the two new properties. Also, add these functions:

```
- (void)addTimeStamp:(id)sender {

    TimeStamp *timeStamp = (TimeStamp*)[NSEntityDescription
            insertNewObjectForEntityForName:@"TimeStamp" =
            inManagedObjectContext:managedObjectContext];
            timeStamp.timeDate = [NSDate date];

    [self.timeStampList insertObject:timeStamp atIndex:0];

    [self.tableView reloadData];

}
- (void)loadTimeStamps
{

    NSEntityDescription *timeStampEntity = [NSEntityDescription entityForName:@"TimeStamp" inManagedObjectContext:managedObjectContext];

    NSFetchRequest *request = [[NSFetchRequest alloc] init];

    [request setEntity:timeStampEntity];

    NSArray *descriptors = [NSArray arrayWithObject:[[NSSortDescriptor alloc] initWithKey:@"timeStamp" ascending:NO]];

    [request setSortDescriptors:descriptors];

    [descriptors release];
```

```objc
        NSMutableArray *fetchedData = [[managedObjectContext
executeFetchRequest:request error:nil] mutableCopy];

    self.timeStampList = fetchedData;

    [fetchedData release];

    [request release];

}
```

At this point, we've implemented a way to create and store `TimeStamp` objects, as well as load them into an `NSMutableArray` so that the table may access the information. The rest is trivial, but it is recommended to perform.

In your `viewDidLoad` method, you need to add some kind of button so the user may trigger your code that adds the current time stamp. Most importantly, you need to load the persistent data back into your table.

```objc
- (void)viewDidLoad
{
    [super viewDidLoad];
  UIBarButtonItem *addButton = [[UIBarButtonItem alloc]
                initWithBarButtonSystemItem:UIBarButtonSystemItemAdd
target:self
                action:@selector(addTimeStamp:)];
  self.navigationItem.rightBarButtonItem = addButton;
  [addButton release];
    self.title = @"Logged TimeStamps";
    [self loadTimeStamps];
}
```

Now you just need to set up your typical table view delegate methods:

```objc
    - (NSInteger)numberOfSectionsInTableView:(UITableView *)tableView
    {
      return 1;
    }
    - (NSInteger)tableView:(UITableView *)tableView
    numberOfRowsInSection:(NSInteger)section
    {
      return [timeStampList count];
    }
```

So now you have finished your table view controller and you just need to tie the loose ends together to get this app running. In the application delegate header file, first make sure we've imported `TimeStampTableViewController.h`, and then add the following member variables.

```
NSManagedObjectContext *managedObjectContext;

NSManagedObjectModel *managedObjectModel;

NSPersistentStoreCoordinator *persistentStoreCoordinator;

UIWindow *window;

UINavigationController *navigationController;
```

Xcode should have already defined read only properties for the first four variables. Make sure to define a property for `navigationController`:

```
@property (nonatomic, retain) UINavigationController *navigationController;
```

In the implementation file, be sure to synthesize the `navigationController` property. In the `application:didFinishLaunching` method:

```
TimeStampTableViewController *timeStampTableController =
  [[TimeStampTableViewController alloc]
   initWithStyle:UITableViewStylePlain];
self.navigationController = [[UINavigationController alloc]
  initWithRootViewController:timeStampTableController];
timeStampTableController.managedObjectContext = self.managedObjectContext;
[window addSubview: [self.navigationController view]];
[window makeKeyAndVisible];
[timeStampTableController release];
return YES;
```

And finally, be sure to release `navigationController` in the `dealloc` method.

22

Important Frameworks

Several public frameworks are available for use in your iOS projects. These frameworks offer methods, data types, or interfaces to increase productivity and functionality. Some are essential while others are supplemental. This chapter investigates which frameworks you should best familiarize yourself with.

22.1   Address Book Framework

The Address Book framework consists of classes that grant the ability to access the information stored in a user's address book. Information such as names, telephone numbers, and addresses is accessible from the address book.

To use this framework first make your view controller conform to the ABPeoplePickerNavigationControllerDelegate protocol as follows:

```
UIViewController <ABPeoplePickerNavigationControllerDelegate>
```

Next create the navigation controller, set its delegate, and then present it modally.

```
ABPeoplePickerNavigationController *picker =
        [[ABPeoplePickerNavigationController alloc]init];
picker.peoplePickerDelegate = self;
[self presentModalViewController:picker animated:true];
```

Run the app and the user's contact list should open as shown in Figure 22-1.

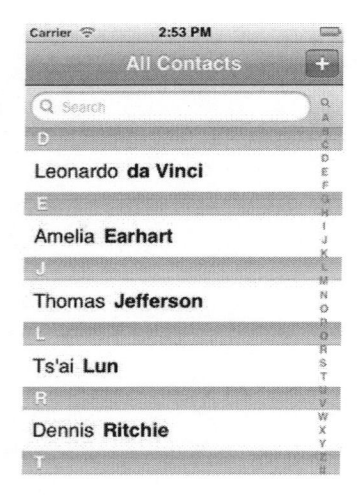

Figure 22-1: Contacts

As the delegate for the navigator, the view controller can implement the following methods:

```
-(BOOL)peoplePickerNavigationController: (ABPeoplePickerNavigationController
*)peoplePicker shouldContinueAfterSelectingPerson:(ABRecordRef)person;

-(BOOL)peoplePickerNavigationController: (ABPeoplePickerNavigationController
*)peoplePicker;
```

Both functions should be implemented to return False. When implementing the first method use the person object to get data for the selected contact. The following is an example of getting values from this record reference.

```
firstName = (__bridge NSString *)ABRecordCopyValue(person,
                        kABPersonFirstNameProperty);

lastName = (__bridge NSString *)ABRecordCopyValue(person,
                        kABPersonLastNameProperty);

ABMultiValueRef multi = ABRecordCopyValue(person,
                        kABPersonPhoneProperty);

number = (__bridge NSString *)ABMultiValueCopyValueAtIndex(multi,
                        0);
```

Note that if the multi object has no numbers then the method ABMultiValueCopyValueAtIndex will cause a crash. To get the number of indices of any ABMultiValueRef object use the following method.

```
ABMultiValueGetCount(var)
```

22.2 Core Data Framework

Core Data is a framework for data management. For more information see Chapter 21.

22.3 Core Location Framework

The Core Location framework grants the ability to locate the device's position and heading by means of GPS. For more information see Chapter 23.

22.4 Event Kit Framework

The Event Kit framework provides access to the user's calendar event information. With Event Kit, events can be created and added to the calendar. Here is an example of how to create an event:

```
EKEventStore *eventDatabase = [[EKEventStore alloc] init];
EKEvent *eventObject = [EKEvent eventWithEventStore:eventDatabase];
eventObject.title = @"Event Title";
eventObject.startDate = [[NSDate alloc] init];
eventObject.endDate = [[NSDate alloc] init];
eventObject.allDay = YES;
[eventObject setCalendar:[eventDatabase defaultCalendarForNewEvents]];
NSError *err;
if(err== noErr){
   UIAlertView *alert = [[UIAlertView alloc] initWithTitle:@"Event
             Created" message:@"" delegate:nil
                   cancelButtonTitle:@"Okay"
             otherButtonTitles:nil];
   [alert show];
   [alert release];
}
```

22.5 AVFoundation Framework

The AVFoundation framework provides interfaces for managing media. These interfaces handle playing or recording audio or video. Here is an example of playing a sound:

```
NSString *filePath = [[NSBundle mainBundle]
             pathForResource:@"mySong" ofType:@"m4a"];
AVAudioPlayer *audioPlayer = [[AVAudioPlayer alloc] initWithContentsURL:[NSURL fileURLWithPath:filePath] error:NULL];
audioPlayer.delegate = self;
[audioPlayer player];
```

This framework can also handle videos. Here is an example of playing a video:

```
NSString *filePath = [[NSBundle mainBundle]
    pathForResource:@"sweatersong" ofType:@"mp4"];
```

```
NSURL *url = [[NSURL alloc]initFileURLWithPath:filePath];
AVPlayer *player = [AVPlayer playerWithURL:url];
AVPlayerLayer *playerLayer = [AVPlayerLayer
              playerLayerWithPlayer:player];
playerLayer.frame = [[self view] bounds];
[self.view.layer addSublayer:playerLayer];
[audio_player stop];
[player play]
```

22.6 Core Audio Framework

Core Audio is a framework for handling audio. Core Audio is very complex because it handles raw audio. Recording, streaming, and storing audio are some of the few uses of this framework. For more information refer to Apple's iOS Developer Library.

22.7 OpenAL Framework

OpenAL is a free, open source, third-party audio API designed for three-dimensional sound projection. For more information visit OpenAL's website at http://connect.creativelabs.com/openal.

22.8 Media Player Framework

The Media Player framework provides an interface for handling video playback. Implementing the movie player is as simple as the following lines of code:

```
MPMoviePlayerController *player = [[MPMoviePlayerController alloc]
                     initWithContentURL: fileURL];
[player.view setFrame:myView.bounds];
[myView addSubview: player.view];
[player play];
```

22.9 Core Animation Framework

The Core Animation framework combines a high-performance compositing engine with animation programming. Animations can be created and assigned to a layer.

Here is an example of how to create an animation:

```
[UIView beginAnimations:nil context:nil];
[UIView setAnimationDelegate:self];
[UIView setAnimationDuration:1.0f];
image.transform = CGAffineTransformMakeScale(0.5f, 0.5f);
[UIView commitAnimations];
```

Note that the animation will not run until the `commitAnimations` message is sent.

22.10 Core Graphics Framework

The Core Graphics framework offers two-dimensional rendering via the Quartz engine. All drawing is done to a `CGContext`, which can be thought of as a canvas. The `CGContext` object can be a window in an app, the contents of a PDF file, or even a printer task.

When drawing with Core Graphics the following method must be implemented in your UIView:

```
-(void)drawRect:(CGRect)rect;
```

This method is where the "drawing" occurs. In order to draw you must get the `CGContextRef`.

```
CGContextRef context = UIGraphicsGetCurrentContext();
```

After you get the context the drawing can begin. The outline of a shape or the line itself is called a stroke and the fill is the area of a shape. To set the stroke color use `CGContextSetStrokeColorWithColor` and for fill colors use `CGContextSetFillColorWithColor`.

```
CGContextSetStrokeColorWithColor(context, [UIColor blackColor].CGColor);

CGContextSetFillColorWithColor(context, [UIColor blackColor].CGColor);
```

To set line width use `CGContextSetLineWidth`.

```
CGContextSetLineWidth(context, 2.0);
```

A path must be made and then stroked to create a line. Move the path's starting point using `CGContextMoveToPoint`.

```
CGContextMoveToPoint(context, 0,0);
```

Next move to the end point of the line.

```
CGContextAddLineToPoint(context, 320, 320);
```

Now you can stroke the path using `CGContextStrokePath`.

```
CGContextStrokePath(context);
```

To draw a rectangle use `CGContextStrokeRect`.

```
CGContextStrokeRect(context, CGRectMake(20, 20, 200, 200));
```

The previous code example will draw a square because the width and height are equal. To draw an ellipse use `CGContextStrokeEllipseInRect`.

```
CGContextStrokeEllipseInRect(context, CGRectMake(20, 20, 200, 200));
```

The previous code example will draw a circle because the width and height are equal.

Drawings can also be clipped in Core Graphics. Clipping will hide anything that is drawn outside the clipped area. To clip first save the graphics state with `CGContextSaveGState`.

```
CGContextSaveGState(context);
```

Next begin drawing a path with `CGContextBeginPath`.

```
CGContextBeginPath (context);
```

Now add the shapes that will form the path. In the example you will add a rectangle.

```
CGContextAddRect(context, CGRectMake(20, 20, 50, 50));
```

After the path is done being formed, close the path using **CGContextClosePath**.

```
CGContextClosePath (context);
```

The next thing to do is to enable clipping using **CGContextClip**.

```
CGContextClip(context);
```

While clipping is enabled draw all the graphics that are to be clipped. In the example a filled rectangle is drawn.

```
CGContextFillRect(context, CGRectMake(45, 45, 50, 25));
```

After all the graphics are drawn restore the graphics state using **CGContextRestoreGState**.

```
CGContextRestoreGState(context);
```

22.11 UIKit Framework

The UIKit framework provides classes for user interface development. For more information see Chapter 20.

User Input

Thanks to their sophisticated hardware, there are many ways in which the user can interact with iOS devices. The capabilities of these devices ensure that they can take in, process, and output a variety of useful information from the user, as well as their environment. In this chapter, you will cover the basics of incorporating user input into your apps through the use of Xcode and your iOS device.

Because the iOS Simulator is limited as far as its user input capabilities, i.e. no accelerometers/gyroscopes, microphone or camera, you will need an iOS device to take full advantage of several of the demonstrations in this chapter.

23.1 Touch

There are four main events that are fired that respond to a user touching the iOS device. They are `touchesBegan`, `touchesMoved`, `touchesEnded`, and `touchesCancelled`.

Create an app that uses these events to get a better understanding of how they work. Open Xcode, create a new `View-Based Application` and name it `TouchApp`. The first event you will work with is the `touchesBegan` event. In the `TouchAppViewController.m` file add the following code.

```
- (void)touchesBegan:(NSSet *)touches withEvent:(UIEvent *)event
{
    UITouch *touch = [touches anyObject];
    CGPoint startLocation = [touch locationInView:self.view];
    NSLog(@"You have touched at location: (%0.0f,%0.0f)", startLocation.x, startLocation.y);
}
```

Build and run your project, and you will see that each time you touch on the screen it tracks where you touched and outputs the location to the console as shown in Figure 23-1. This is because each time the user touches down the `touchesBegan` event is fired.

Chapter 23: User Input   201

```
All Output
GNU gdb 6.3.50-20050815 (Apple version gdb-1518) (Sat Feb 12 02:52:12 UTC 2011)
Copyright 2004 Free Software Foundation, Inc.
GDB is free software, covered by the GNU General Public License, and you are
welcome to change it and/or distribute copies of it under certain conditions.
Type "show copying" to see the conditions.
There is absolutely no warranty for GDB.  Type "show warranty" for details.
This GDB was configured as "x86_64-apple-darwin".Attaching to process 2124.
2011-04-13 15:24:44.813 TouchApp[2124:207] You have touched at location: (138,138)
2011-04-13 15:24:45.156 TouchApp[2124:207] You have touched at location: (202,238)
2011-04-13 15:24:45.556 TouchApp[2124:207] You have touched at location: (66,371)
2011-04-13 15:24:45.892 TouchApp[2124:207] You have touched at location: (122,172)
2011-04-13 15:24:46.309 TouchApp[2124:207] You have touched at location: (209,365)
```

Figure 23-1: Output for touchesBegan

That's how the `touchesBegan` event works; now see how the `touchesMoved` event works. Add the following code below the event youadded previously:

```
- (void)touchesMoved:(NSSet *)touches withEvent:(UIEvent *)event

{

    UITouch *touch = [touches anyObject];

    CGPoint currentLocation = [touch locationInView:self.view];

    NSLog(@"You have moved to location: (%0.0f,%0.0f)", currentLocation.x,
currentLocation.y);

}
```

If you build your project and run it, you will see that when you touch down and drag, it will track your finger's location and output it to the console as shown in Figure 23-2. This is because the `touchesMoved` event is fired if a touch is moved after it has begun and before it is released.

```
2011-04-13 15:43:04.812 TouchApp[2175:207] You have touched at location: (79,119)
2011-04-13 15:43:05.530 TouchApp[2175:207] You have moved to location: (80.119)
2011-04-13 15:43:05.597 TouchApp[2175:207] You have moved to location: (81.119)
2011-04-13 15:43:05.614 TouchApp[2175:207] You have moved to location: (81.120)
2011-04-13 15:43:05.630 TouchApp[2175:207] You have moved to location: (82.121)
2011-04-13 15:43:05.647 TouchApp[2175:207] You have moved to location: (83.121)
2011-04-13 15:43:05.663 TouchApp[2175:207] You have moved to location: (85.123)
2011-04-13 15:43:05.681 TouchApp[2175:207] You have moved to location: (86.125)
2011-04-13 15:43:05.697 TouchApp[2175:207] You have moved to location: (87.126)
2011-04-13 15:43:05.714 TouchApp[2175:207] You have moved to location: (88.127)
2011-04-13 15:43:05.730 TouchApp[2175:207] You have moved to location: (89.128)
2011-04-13 15:43:05.747 TouchApp[2175:207] You have moved to location: (90.129)
2011-04-13 15:43:05.780 TouchApp[2175:207] You have moved to location: (91.130)
2011-04-13 15:43:05.814 TouchApp[2175:207] You have moved to location: (91.131)
2011-04-13 15:43:05.830 TouchApp[2175:207] You have moved to location: (92.131)
```

Figure 23-2: Output for touchesMoved

The third event is `touchesEnded`, which is fired when the user ends a touch. In order to see how this works, add the following code below the previous event:

```
- (void)touchesEnded:(NSSet *)touches withEvent:(UIEvent *)event

{

    UITouch *touch = [touches anyObject];

    CGPoint endLocation = [touch locationInView:self.view];

    NSLog(@"You have stopped at location: (%0.0f,%0.0f", endLocation.x,
endLocation.y);

}
```

Now when you run your project you will see your location is tracked when you touch, drag, and release the touch as shown in Figure 23-3.

```
2011-04-13 16:57:23.322 TouchApp[2332:207] You have touched at location: (76,95)
2011-04-13 16:57:23.360 TouchApp[2332:207] You have moved to location: (76,96)
2011-04-13 16:57:23.377 TouchApp[2332:207] You have moved to location: (76,97)
2011-04-13 16:57:23.393 TouchApp[2332:207] You have moved to location: (77,97)
2011-04-13 16:57:23.410 TouchApp[2332:207] You have moved to location: (77,99)
2011-04-13 16:57:23.427 TouchApp[2332:207] You have moved to location: (82,107)
2011-04-13 16:57:23.444 TouchApp[2332:207] You have moved to location: (89,118)
2011-04-13 16:57:23.460 TouchApp[2332:207] You have moved to location: (96,129)
2011-04-13 16:57:23.477 TouchApp[2332:207] You have moved to location: (97,131)
2011-04-13 16:57:23.497 TouchApp[2332:207] You have stopped at location: (97,131
```

Figure 23-3: Output for touchesEnded

Each time you touch the screen, you create a UITouch object. The UITouch class has some important properties and methods. The properties are phase, timestamp, and tapCount. The methods locationInView: and previousLocationInView: are important as well. First you will examine what each of the properties represents and then you will review what the two methods do.

- *phase:* Tells you the current phase of the touch. There are several values that UITouchPhase can be. They are: UITouchPhaseBegan, UITouchPhaseMoved, UITouchPhaseStationary, UITouchPhaseEnded, and UITouchPhaseCancelled.

- *timestamp:* This is the time when the touch changed its phase.

- tapCount: This represents the number of taps the user made on the screen. For example, touching the same spot 5 times would result in a tapCount of 5.

- *LocationInView:* This returns the location of the touch in a given view.

- *previousLocationInView:* Returns the previous location of the touch in a given view.

23.2 Multi-Touch

Multi-Touch is very useful and powerful and is almost as easy to implement as Single-Touch. You will find that it is a major part of developing iPhone apps. Create an example to learn how to work with Multi-Touch. You will create a new View-Based Application and name it MultiTouchApp.

Once you have your project you need to add a new UIView subclass. Do this by right-clicking on the MultiTouchApp folder, selecting "New File," then select Objective-C class and make sure the subclass is set to UIView. Name it MultiTouchView, and then save.

Figure 23-4: New Objective-C Class Icon

Now you will setup your `UIView` subclass to allow you to draw lines on it. Go to the `MultiTouchView.h` header file and update it to look like the following:

```
@interface MultiTouchView : UIView {
    CGPoint firstTouch;
    CGPoint secondTouch;
}
@property (nonatomic) CGPoint firstTouch;
@property (nonatomic) CGPoint secondTouch;
@end
```

The two location properties will hold the values of the first and second `touch` to allow you to draw lines between them. You did not retain your properties because they are not `objects`; they are `structs`.

Now there are several things you must do in the `MultiTouchView.m` file. First you must synthesize your properties if you haven't already. Next, you must implement the `isMultipleTouchedEnabled` method so that your `UIView` subclass can detect multiple `touches`, like so:

```
@implementation MultiTouchView
@synthesize firstTouch, secondTouch;
- (BOOL)isMultipleTouchEnabled
{
    return YES;
}
```

Next you must implement the `touchesBegan:withEvent:` method like you did before, except this time you will do it differently. You will create an array with your `touches` object and check to see if you have more than one touch so that you will know if it is multi-touch or not. Afterwards, you will update the location variables you created with the location of the `touch/touches`. Lastly, you will call the `setNeedsDisplay` method, which tells the view to redraw itself.

```
- (void)touchesBegan:(NSSet *)touches withEvent:(UIEvent *)event
{
```

204 Chapter 23: User Input

```objc
    NSArray *allTouches = [touches allObjects];
    if ([allTouches count] > 1)
    {
        NSLog(@"Multi-Touch");
    self.firstTouch = [[allTouches objectAtIndex:0] locationInView:self];
        self.secondTouch = [[allTouches objectAtIndex:1] locationInView:self];
        NSLog(@"First Touch at: (%0.0f,%0.0f) Second Touch at: (%0.0f,%0.0f)",
self.firstTouch.x, self.firstTouch.y, self.secondTouch.x, self.secondTouch.y);
        } else {
    NSLog(@"Single-Touch");
        self.firstTouch = [[allTouches objectAtIndex:0] locationInView:self];
        NSLog(@"Touch at (%0.0f,%0.0f)", self.firstTouch.x, self.firstTouch.y);
    }
    [self setNeedsDisplay];
}
```

Now you need to implement the `touchesMoved:withEvent:` method which will just call the `touchesBegan` method so you do not need to duplicate code. The last method you have to implement is the `drawRect:` method, which is called after doing a `setNeedsDisplay`. In this method you will get a pointer to the current graphics context which you will use to set the line color, starting point, add a line to the second point, and finally, draw the line.

```objc
- (void)touchesMoved:(NSSet *)touches withEvent:(UIEvent *)event
{
    [self touchesBegan:touches withEvent:event];
}
- (void)drawRect:(CGRect)rect
{
    CGContextRef context = UIGraphicsGetCurrentContext();
    CGContextSetStrokeColorWithColor(context, [UIColor blueColor].CGColor);
    CGContextMoveToPoint(context, self.firstTouch.x, self.firstTouch.y);
    CGContextAddLineToPoint(context, self.secondTouch.x, self.secondTouch.y);
    CGContextStrokePath(context);
}
```

Go to the `MultiTouchAppViewController.xib` file, and in the `Identity Inspector` change the view class from `UIView` to `MultiTouchView`, which is the custom class as shown in Figure 23-5.

Figure 23-5: Custom Class set in Identity Inspector

Now you can build your project and run it. Pressing and holding the alt key while clicking simulates Multi-Touch. You should see a blue line between the two touches as shown in Figure 23-6. You should also see the location of the two touches being output in the console as in the following Figure 23-7.

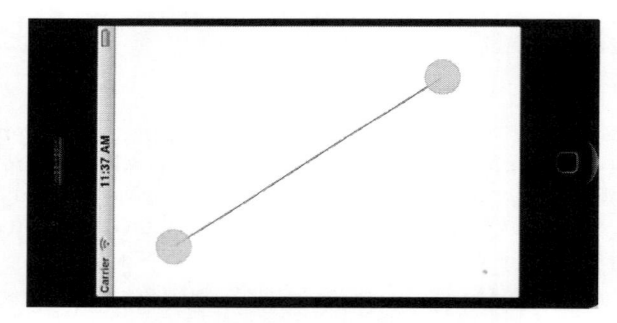

Figure 23-6: Multi-Touch on iPhone Simulator

Figure 23-7: Output for Multi-Touch

23.3 Gestures

Now that you have basic understanding of touch and multi-touch, you can use this knowledge to implement gestures like pinch and swipe. Alternatively, you can use the `UIGestureRecognizer` class to create objects that will recognize gestures like swipe, tap, pinch, rotation, pan, and long press. It is important to know that the `UIGestureRecognizer` needs to be connected to the `UIView` that is receiving the touch events.

There are two types of gestures, those that are discrete, and those that are continuous. The difference is that a discrete gesture occurs only once, whereas a continuous gesture will take place over a period of time and ends when the user ceases touching the device. This is important to know because when a discrete gesture is recognized, a single action message is sent to the gesture recognizer's target. A continuous gesture, however, sends multiple action messages until the gesture is ended. The tap and swipe gestures are both discrete gestures, while the pinch, pan, rotate, and long press gestures are all continuous.

23.3.1 Tap Gesture

The tap gesture is useful when you want to allow the user to select something. If you wanted to select something but want a different behavior than the regular single tap you could use a double tap gesture. For example a single tap could select a `UITableViewCell` and a double tap could delete it. The double tap gesture is handled with the class `UITapGestureRecognizer`, which is a subclass of `UIGestureRecognizer`. This class is also the same class used to handle a single tap gesture. The difference is when using the `UITapGestureRecognizer` to handle a double tap the `numberOfTapsRequired` property should have an integer value of 2.

To implement the double tap gesture, first you have to create an instance of the `UITapGestureRecognizer` as follows:

```
UITapGestureRecognizer *doubleTap = [[UITapGestureRecognizer alloc]
        initWithTarget:self action:@selector(handleDoubleTap:)];
doubleTap.numberOfTapsRequired = 2;
[self.view addGestureRecognizer:doubleTap];
[doubleTap release];
```

First the `UITapGestureRecognizer` is allocated and initialized with the action selector `handleDoubleTap:` which will be implemented next. After the gesture recognizer is created, the integer value 2 is assigned to the `numberOfTapsRequired` property. The gesture recognizer is then added to the view that will receive the touches with the `addGestureRecognizer` method of the `UIView` class. Finally, the `doubleTap` object is released since now it is retained by the view.

Now that the gesture recognizer is added to the view that will receive the touches, the action method for handling the gesture must be implemented similar to as follows:

```
- (IBAction)handleDoubleTap:(UITapGestureRecognizer *)sender
{
    CGPoint point = [sender locationInView:sender.view];
    NSLog(@"DoubleTap performed at x:%f y:%f", point.x, point.y);
}
```

As you can see, the method is relatively simple and straightforward. The `UIGestureRecognizer` passed to the method is used to determine where in the view the double tap occurred. The x and y coordinates are then displayed using the `NSLog` function.

23.3.2 Swipe Gesture

The swipe gesture is used very often in iOS apps and is useful for scrolling and paging between views. The gesture recognizer used to handle the swipe gesture is a subclass of `UIGestureRecognizer` called `UISwipeGestureRecognizer`.

To implement the swipe gesture, an instance of the `UISwipeGestureRecognizer` class must first be created and initialized as follows:

```
UISwipeGestureRecognizer *swipeGesture = [[UISwipeGestureRecognizer alloc]
                          initWithTarget:self action:@selector(handleLeftSwipe:)];
swipeGesture.direction = UISwipeGestureRecognizerDirectionLeft;
[self.view addGestureRecognizer:swipeGesture];
[swipeGesture release];
```

The gesture recognizer is initialized with the action selector `handleLeftSwipe:`, which must be implemented next. Once the gesture recognizer is created, if no direction is assigned to it, the default direction will be right. However, you are going to assign it the value `UISwipeGestureRecognizerDirectionLeft`, which will detect left swipes. The gesture recognizer is then added to the view and then released since the view now retains it.

Lastly, you will implement the `handleLeftSwipe:` method to handle the actions that should occur when a left swipe is performed. The `handleLeftSwipe:` method will look as follows:

```
- (IBAction)handleLeftSwipe:(UISwipeGestureRecognizer *)sender
{
    CGPoint point = [sender locationInView:sender.view];
    NSLog(@"LeftSwipe occurring at x:%f y:%f", point.x, point.y);
}
```

In this method, the point at which the left swipe occurred is obtained by calling the `locationInView:` method. Then, using the `NSLog` function, the coordinates are displayed.

23.3.3 Pinch Gesture

The pinch gesture is used very often in iOS apps in order to zoom in or out. The `UIPinchGestureRecognizer` class is used to handle the pinch gesture. The pinch gesture is a continuous gesture, which means that its gesture recognizer sends multiple action messages to its target until the gesture ends.

Implementing the pinch gesture consists of first creating an instance of the `UIPinchGestureRecognizer` class, and initializing it similar to the following:

```
UIPinchGestureRecognizer *pinchGesture = [[UIPinchGestureRecognizer alloc]
                          initWithTarget:self action:@selector(handlePinch:)];
[self.view addGestureRecognizer:pinchGesture];
```

```
[pinchGesture release];
```

As you can see, an instance of the `UIPinchGestureRecognizer` class is allocated and initialized with the action selector `handlePinch:`, which must be implemented as well. The `pinchGesture` object is then added to the view that will receive the touches. Finally, the `pinchGesture` object is released because it is now retained by the view to which it was added.

What's left is to implement the `handlePinch:` method, which should look like this:

```
- (IBAction)handlePinch:(UIPinchGestureRecognizer *)sender
{
    if (sender.state == UIGestureRecognizerStateEnded)
    {
        NSLog(@"Pinch occurred with scale:%f", [sender scale]);
    }
}
```

Because the pinch gesture is a continuous gesture, when implementing the `handlePinch:` method, you have to take into consideration that it will be called multiple times. Of all the times that the method is called, you are probably only concerned with the state of the gesture when it began or ended. It is for that reason that you are checking to see if the state of the `UIPinchGestureRecognizer` is the same as `UIGestureRecognizerStateEnded`, so that you know the pinch gesture has ended. Once it has ended, the `scale` (or scale factor relative to the two touches) of the pinch gesture is printed out with the `NSLog` function.

23.3.4 Pan Gesture

The pan gesture is used most often to move a view around, for example when viewing an image that is zoomed in to a resolution that is larger than the resolution of the device. The gesture recognizer used to handle the pan gesture is the `UIPanGestureRecognizer` class.

To use the pan gesture, an instance of the `UIPanGestureRecognizer` class must first be created and initialized similar to the following:

```
UIPanGestureRecognizer *panGesture = [[UIPanGestureRecognizer alloc]
                    initWithTarget:self action:@selector(handlePan:)];

[self.view addGestureRecognizer:panGesture];

[panGesture release];
```

As you can see, an instance of the `UIPanGestureRecognizer` class is allocated and initialized with the action selector `handlePan:` that will be implemented next. Then the gesture recognizer is added to the view that will receive the touches using the `addGestureRecognizer:` method of the `UIView` class. Finally, the `panGesture` object is released because now the view it was added to retains it.

Now the `handlePan:` method must be implemented to handle what should be done when a pan gesture occurs. The `handlePan:` method will look as follows:

```
- (IBAction)handlePan:(UIPanGestureRecognizer *)sender
{
    if (sender.state == UIGestureRecognizerStateEnded)
    {
        CGPoint point = [sender translationInView:sender.view];
        NSLog(@"Panned a distance of x:%f y:%f", point.x, point.y);
    }
}
```

Since the pan gesture is a continuous gesture, the `handlePan:` method is called multiple times until the gesture ends. It is for this reason that the previous code has a conditional to only perform what is inside of the statement when the gesture recognizer's state is `UIGestureRecognizerStateEnded`, which tells you that the gesture has finished. When the gesture has finished, the `translationInView:` method of the `UIPanGestureRecognizer` class is called which returns the change in the x-y coordinates of the view that is panning. This change in coordinates is then output to the console with the `NSLog` function.

23.3.5 Rotate Gesture

The rotate gesture can be used to rotate or flip a view between different orientations. The gesture recognizer used to handle the rotate gesture is the `UIRotationGestureRecognizer` class.

To implement the `UIRotationGestureRecognizer` class, first an instance of it must be created as follows:

```
UIRotationGestureRecognizer *rotateGesture = [[UIRotationGestureRecognizer
            alloc] initWithTarget:self action:@selector(handleRotate:)];
[self.view addGestureRecognizer:rotateGesture];
[rotateGesture release];
```

As you can see, an instance of the `UIRotationGestureRecognizer` class named `rotateGesture` is initialized with the `handleRotate:` method as its action selector. Once initialized, the `rotateGesture` object is added to the view that will receive the touches. After it is added to the view, it is ok to release the `rotateGesture` object since now the view retains it.

Now all that is left is to implement the `handleRotate:` method, which is implemented as follows:

```
- (IBAction)handleRotate:(UIRotationGestureRecognizer *)sender
{
    if (sender.state == UIGestureRecognizerStateEnded)
    {
        CGFloat radianAngle = [sender rotation];
```

```
        CGFloat degreeAngle = radianAngle * 180 / M_PI;
        NSLog(@"Rotating with angle:%f", degreeAngle);
    }
}
```

This method checks to see when the state of the gesture recognizer matches `UIGestureRecognizerStateEnded`, meaning that the gesture has ended. When it is determined that the gesture has ended, the `rotation` property of the gesture recognizer is accessed to determine the angle that the rotation has changed. The value returned by this property is in radians, so in order to get the angle in degrees, the value is converted in the next line. Once the angle in degrees is obtained, it is output to the console with the `NSLog` function.

23.3.6 Long-Press Gesture

The long-press gesture is useful when you want to select something, yet perform a different behavior than that of a single tap. The gesture recognizer that is used to handle a long-press gesture is the `UILongPressGestureRecognizer` class.

In order to implement the long-press gesture, first an instance of the `UILongPressGestureRecognizer` class must be created as follows:

```
UILongPressGestureRecognizer *longPressGesture = [[UILongPressGestureRecognizer
            alloc] initWithTarget:self action:@selector(handleLongPress:)];
[self.view addGestureRecognizer:longPressGesture];
[longPressGesture release];
```

An instance of the `UILongPressGestureRecognizer` class named `longPressGesture` was allocated and initialized with the `handleLongPress:` method as its action selector. After creation, the `longPressGesture` object is added to the view that will be receiving the touches by using the `addGestureRecognizer:` method of the `UIView` class. Once added to the view, the `longPressGesture` object is released since now the view retains it.

After adding the gesture recognizer to the view, you need to implement the action selector that it was created with. The `handleLongPress:` method should look as follows:

```
- (IBAction)handleLongPress:(UILongPressGestureRecognizer *)sender
{
    if (sender.state == UIGestureRecognizerStateBegan)
    {
        startTime = [[NSDate date] retain];
    }
    if (sender.state == UIGestureRecognizerStateEnded)
    {
        endTime = [NSDate date];
```

```
            NSLog(@"Press had duration of %f seconds", [endTime
    timeIntervalSinceDate:startTime]);
            [startTime release];
        }
    }
```

Since the long-press gesture is a continuous gesture, you are checking to see when the gesture began and ended by checking to see if its state matches either `UIGestureRecognizerStateBegan` or `UIGestureRecognizerStateEnded`. When the gesture's state indicates that it has begun, the date is stored at that point with the `startTime` object. Once the gesture has ended, the date is again stored with the `endTime` object. The time interval between the two dates is then output to the console with the `NSLog` function using the `timeIntervalSinceDate:` method of the `NSDate` class.

23.4 Camera

To use the built-in camera, you must first create a `UIImagePickerController` and then present that modally. However it is important to first make sure that the `SourceType` is available on your device. Using the `isSourceTypeAvailable` class method of `UIImagePickerController` with the parameter `UIImagePickerControllerSourceTypeCamera` will return true if a camera is available on the iOS device. So in your `viewDidLoad` method, you should have something similar to the following:

```
- (void)viewDidLoad
{
    [super viewDidLoad];
    if ([UIImagePickerController
isSourceTypeAvailable:UIImagePickerControllerSourceTypeCamera] == NO) {
        lbl.textColor = [UIColor redColor];
        lbl.text = @"No Camera Available";
        btn.hidden = YES;
    }
    else {
        lbl.textColor = [UIColor blueColor];
        lbl.text = @"Camera Available";
        btn.hidden = NO;
        imgPicker = [[UIImagePickerController alloc] init];
        imgPicker.sourceType = UIImagePickerControllerSourceTypeCamera;
        imgPicker.allowsEditing = NO;
        imgPicker.delegate = self;
```

                }
        }

As you see in the preceding code, when the camera is not available, the text and text color of a label is changed to indicate this, and a button is hidden. When the camera is available, the label is changed and the button is also changed to be visible. In addition to this, the `UIImagePickerController` is initialized and configured to work with the camera.

Before you go any further, it is important to show you what the XIB file looks like, so that when you are going through this example you will know how the `IBOutlets` and `IBActions` are wired up. As you can see in Figure 23-8, there is a very simple `UILabel` that will change to tell you whether or not the camera is available, a `UIButton` that will launch the camera interface, and a `UIImageView` that will display the image that was taken with the camera.

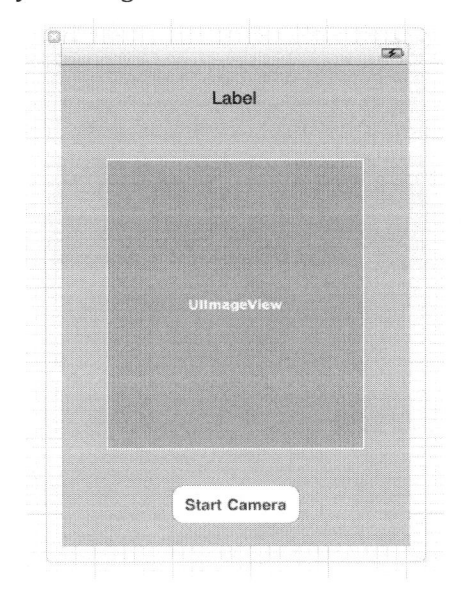

Figure 23-8: CameraAppViewController.xib

When using the `UIImagePickerController` class, it is important to make sure that your view controller implements the `UIImagePickerControllerDelegate` protocol as well as the `UINavigationControllerDelegate` protocol. Your view controller's header file should look similar to the following:

```objc
#import <UIKit/UIKit.h>

#import <MobileCoreServices/UTCoreTypes.h>

@interface CameraAppViewController : UIViewController
<UIImagePickerControllerDelegate, UINavigationControllerDelegate> {

    UIImagePickerController *imgPicker;

    IBOutlet UILabel *lbl;

    IBOutlet UIButton *btn;
```

```objc
    IBOutlet UIImageView *imgView;
    UIImage *pickedImage;
}

@property (nonatomic, retain) UIImage *pickedImage;
- (IBAction)startCamera;
@end
```

The `startCamera` method is just a simple method that is wired up to a UIButton that, when pressed, presents the `UIImagePickerController` modally, and looks as follows:

```objc
- (void)startCamera
{
    [self presentModalViewController:imgPicker animated:YES];
}
```

Now that you have a method to display the `UIImagePickerController`, you need to implement some delegate methods to handle what happens when the user hits cancel, or when they are done taking the picture.

```objc
#pragma mark - UIImagePickerController delegate
- (void)imagePickerControllerDidCancel:(UIImagePickerController *)picker
{
    [self dismissModalViewControllerAnimated:YES];
}
- (void)imagePickerController:(UIImagePickerController *)picker
didFinishPickingMediaWithInfo:(NSDictionary *)info
{
    NSString *mediaType = [info objectForKey:UIImagePickerControllerMediaType];
    // Handle a still image capture
    if (CFStringCompare ((CFStringRef) mediaType, kUTTypeImage, 0) ==
kCFCompareEqualTo)
    {
        pickedImage = (UIImage *) [info
objectForKey:UIImagePickerControllerOriginalImage];
        imgView.image = pickedImage;
    }
    [self dismissModalViewControllerAnimated:YES];
}
```

When you run your app and take a picture, you should end up with something similar to Figure 23-9.

Figure 23-9: Picture taken with app

23.5 Microphone

When it comes to accessing the microphone to record sound, the easiest way is to use the `AVAudioRecorder` class. However, before you do that, you need to setup your XIB so that it has a button for record and a label that will tell you when it is recording. It should look similar to Figure 23-10.

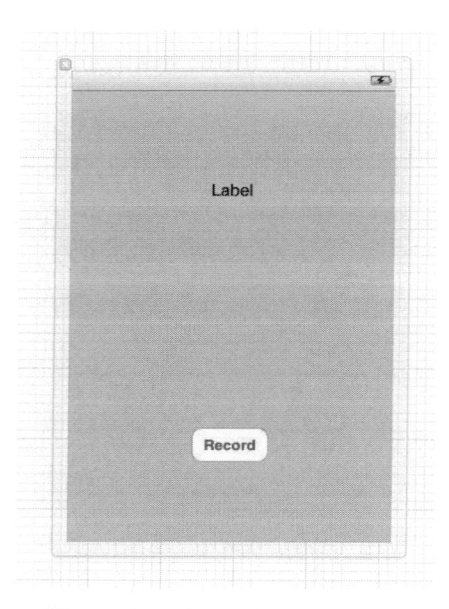

Figure 23-10: MicrophoneAppViewController.xib

Now that you have the XIB file set up, you can set up the audio recorder. First you have to make sure your interface implements the **AVAudioRecorderDelegate** protocol and the **AVAudioSessionDelegate** protocol. Your header file should look similar to the following:

```
#import <UIKit/UIKit.h>

#import <AVFoundation/AVAudioRecorder.h>

#import <AVFoundation/AVAudioSession.h>

@interface MicrophoneAppViewController : UIViewController
<AVAudioRecorderDelegate, AVAudioSessionDelegate> {

    NSURL *soundFile;

    BOOL isRecording;

    IBOutlet UILabel *lbl;

    AVAudioRecorder *audioRecorder;

}

- (IBAction)recordOrStop:(UIButton *)sender;

@property (nonatomic, retain) NSURL *soundFile;

@property (nonatomic, retain) AVAudioRecorder *audioRecorder;

@end
```

As you can see, **AVAudioRecorder.h** and **AVAudioSession.h** were imported so that you can use them. Other things to note are that you have an **IBOutlet** for the **UILabel**, and an **IBAction** called **recordOrStop** for the **UIButton** you set up (These can be wired up now).

216　Chapter 23: User Input

There is an NSURL property for the sound file that you will be recording, and an AVAudioRecorder property for the audio recorder that you will use. The last thing of note here is the BOOL named recording that will be used to determine whether or not the device is currently recording.

So now that you have the interface set up, it is time to work on the implementation part. The first thing to do is set up your soundFile object and an instance of the AVAudioSession class in the viewDidLoad method. Your viewDidLoad method should be very similar to the following:

```objc
- (void)viewDidLoad
{
    [super viewDidLoad];
    NSString *tempDir = NSTemporaryDirectory();
    NSString *filePath = [tempDir stringByAppendingString:@"recording.caf"];
    NSURL *tempURL = [[NSURL alloc] initFileURLWithPath:filePath];
    self.soundFile = tempURL;
    [tempURL release];
    AVAudioSession *audioSession = [AVAudioSession sharedInstance];
    audioSession.delegate = self;
    [audioSession setActive:YES error:nil];
    isRecording = NO;
}
```

As you can see, in the viewDidLoad method you first create a temporary directory and then append the file name and extension to it. This file path is then used to initialize an NSURL, which is then assigned to the soundFile property that you created earlier. You then create a sharedInstance (singleton object) of the AVAudioSession class, set its delegate property to self, and then activate the audio session. Finally, you set the isRecording flag to NO because the default state is not recording.

The next step is to set up the record button so that when it's pressed it will start or stop recording depending on whether the device is currently recording or not. To do this, you set up the recordOrStop method as follows:

```objc
- (void)recordOrStop:(UIButton *)sender
{
    if (recording) {
        [audioRecorder stop];
        isRecording = NO;
        self.audioRecorder = nil;
        lbl.text = @"Stopped Recording";
```

```objc
        [sender setTitle:@"Record" forState:UIControlStateNormal];
        [sender setTitle:@"Record" forState:UIControlStateHighlighted];
        [[AVAudioSession sharedInstance] setActive: NO error: nil];
    }
    else {
        [[AVAudioSession sharedInstance]
setCategory:AVAudioSessionCategoryRecord error: nil];
        AVAudioRecorder *newRecorder = [[AVAudioRecorder alloc]
initWithURL:soundFile settings:nil error:nil];
        self.audioRecorder = newRecorder;
        [newRecorder release];
        audioRecorder.delegate = self;
        [audioRecorder prepareToRecord];
        [audioRecorder record];
        isRecording = YES;
        lbl.text = @"Recording";
        [sender setTitle:@"Stop" forState:UIControlStateNormal];
        [sender setTitle:@"Stop" forState:UIControlStateHighlighted];
    }
}
```

As you can see, there is a lot going on in there, so it's important to break it down to get a better understanding of what is going on. First, you will review what happens when you press the button for the first time. Since there is no audio currently being recorded, the isRecording flag is set to NO, which brings you into the else part of the statement. In here you use the AVAudioSession singleton object to set the audio session's category to AVAudioSessionCategoryRecord, which sets up the device for recording. Then you create an instance of AVAudioRecorder and initialize it with the sound file that you set up in viewDidLoad. Next, you assign it to the audioRecorder property, which then has its delegate property set to self so that the view controller is the delegate of the audioRecorder. Then the prepareToRecord method is called followed by the record method, which starts the audio recording. Since the audio has begun recording, you have to set the isRecording flag to YES, as well as change the label and the button to indicate that the device is recording.

Now that you know what happens when the device is not recording, you need to understand what happens when the device *is* recording. When the button is pressed, since the device is currently recording, the isRecording flag currently has the value YES, which brings you into the first part of the if statement. Since the desired behavior is to stop the recording, the first thing to be done is to send the stop message to the audioRecorder object. Since the recording is stopped, the isRecording flag is set to NO to indicate as much. The audioRecorder object is set to nil since it is no longer required, and the label and button are both changed accordingly to indicate that the recording has ceased. Lastly, the `AVAudioSession` is set to inactive since it is no longer being used.

23.6 Core Location

The Core Location framework lets you access the device's location information. In the case of an iPhone 3GS and later, you may access GPS coordinates and a compass heading. When these features are not available in hardware, the API uses other means to provide an estimate. For example, an iPod touch may use ISP information through its Wi-Fi connection to approximate a location. Unfortunately, there is no way to approximate heading on a device that does not have a magnetometer (compass).

First, you set up the project to use Core Location. You will start with a blank `View-Based project` for iPhone. The framework needs to be added, so go to the `Build Phases` tab under your project's target settings. In the `Link Binary with Libraries` section, go ahead and add the `CoreLocation.framework` item. To use the framework, be sure to `#import <CoreLocation/CoreLocation.h>` in the code.

For the sake of these examples, go ahead and set up a label in the XIB and take the necessary steps to wire it up with the code. Make it fairly large and have 3 lines of text. You will use the name `label` to refer to this label.

23.6.1 Positional Data

Core Location is accessed through a manager object of the class `CLLocationManager`. So the first thing you need to do is create a member object for your view controller class. Call it `locationManager`:

```
CLLocationManager *locationManager;
```

Now set up your `locationManager` object inside of your view controller's `viewDidLoad` method.

```
locationManager = [[CLLocationManager alloc] init];

locationManager.delegate = self;

locationManager.desiredAccuracy = kCLLocationAccuracyBestForNavigation;

[locationManager startUpdatingLocation];
```

Now all that's left is implementing the delegate method that gets called each time Core Location updates its location:

```
- (void)locationManager:(CLLocationManager *)manager
    didUpdateToLocation:(CLLocation *)newLocation
           fromLocation:(CLLocation *)oldLocation
{
    if(newLocation.horizontalAccuracy >= 0) {
        CLLocationCoordinate2D coord = newLocation.coordinate;
```

```
NSString *textLabel = [NSString stringWithFormat:@"LAT: %f \nLON: %f",
  coord.latitude, coord.longitude];
label.text = textLabel;
  }
}
```

Go ahead and build and run the app. Your current latitude and longitude will be displayed in the label as soon as Core Location reports an update.

23.6.2 Compass Heading

Following the previous example, this should be incredibly simple. After setting up the `locationManager` object in `viewDidLoad`, you just need to call its `startUpdatingHeading` method. Then you just need to implement the appropriate delegate method:

```
- (void)locationManager:(CLLocationManager *)manager
didUpdateHeading:(CLHeading *)newHeading
{
        NSLog(@"%f\n", newHeading.trueHeading);
}
```

Here, you were shown how to access the compass heading. You just did a basic `NSLog`, but feel free to get creative. Keep in mind that this will not work in the simulator.

23.7 Core Motion - Gyroscope

The Core Motion framework lets you take advantage of the device's gyroscope and accelerometer sensors. Following similar steps described in 12.6, add the `CoreMotion` framework to the project. The proper import directive is `#import <CoreMotion/CoreMotion.h>`. Core Motion does not use delegates; instead, it uses blocks (also known as closures). We'll create an object of the class `CMMotionManager` and call it `motionManager`.

To initialize the `motionManager` object, first `alloc` then `init` the object. Then before using it, you must check to see if the gyroscope hardware exists and make sure it's not already in use:

```
if(motionManager.gyroAvailable && !motionManager.gyroActive) {
```

So once we've determined that the gyroscope is in fact available for you to use, you first need to set its update interval. Keep in mind that this is a floating-point value depicting the interval, in seconds, between data updates. To specify the frequency, set the interval to 1 divided by the desired frequency. In this example, the frequency is 2 Hz, or 2 times per second:

```
motionManager.gyroUpdateInterval = 1.0/2.0;
```

And finally, you just need to tell `motionManager` what to do when it updates the data. This is where things get a little weird. Instead of using delegate methods, you will pass a block (also known as a closure) as a parameter. In this example, you will also print out the gyroscope acceleration values to the console:

```
[motionManager startGyroUpdatesToQueue:[NSOperationQueue mainQueue]
```

```
withHandler:^(CMGyroData *gyroData, NSError *error) {
        NSLog(@"%2.2f\t%2.2f\t%2.2f\n", gyroData.rotationRate.x,
            gyroData.rotationRate.y, gyroData.rotationRate.z);
}];
```

23.8   The Accelerometer

The accelerometer API is given to you by the UIKit framework, which is already included by default in your project. Unlike Core Motion, the accelerometer does use delegate methods, so your view controller must implement the UIAccelerometerDelegate protocol.

To get started, set up the accelerometer in viewDidLoad:

```
UIAccelerometer *accelerometer = [UIAccelerometer sharedAccelerometer];

accelerometer.updateInterval = 1.0/30.0;

accelerometer.delegate = self;
```

Lastly, implement the delegate method:

```
- (void)accelerometer:(UIAccelerometer *)accelerometer
didAccelerate:(UIAcceleration *)acceleration
{
  NSLog(@"%2.2f, %2.2f, %2.2f\n", acceleration.x, acceleration.y,
        acceleration.z);
}
```

In the delegate method you are printing out the acceleration of each coordinate when the accelerometer accelerates.

This concludes the chapter on the different ways of accepting user input. Now you should be able to recognize Touch and Multi-Touch events, Gestures, Microphone Input, GPS coordinates as well as compass heading, and finally Gyroscope and Accelerometer input.

Instruments

Apple's Xcode 4 comes with a set of tools, called Instruments, which can be used to profile your app. You will learn how to implement them in this chapter, as well as using them to optimize an app by finding common problems like memory leaks.

The Instruments application that comes with Apple's Xcode Tools download is a very useful application that can be used to dynamically trace and profile your iOS code. This application has a large set of tools that can be used for a wide variety of things for iOS and Mac OS X code. However, the tools used to profile Mac OS X code fall outside the scope of this book, therefore only the tools that apply to iOS will be presented.

There are two separate sets of instrument templates for iOS code. One set is for use with the iOS simulator. The other is for use with the iOS device. This makes sense because it would not be possible, for example, to use the Energy Diagnostics Instruments to track energy usage on the iOS simulator because there is no battery on the iOS simulator. This tool would have to be used with an iOS device.

In this chapter you will be going over some of the instrument templates and showing briefly how to use them. For the sake of simplicity, you will only touch on some of the tools and focus on the instruments that are the most useful. For testing purposes, create a simple View-based application, and name it "InstrumentTest." This will be a simple app with a button and a label within the main view. When you run the project it should look something like this.

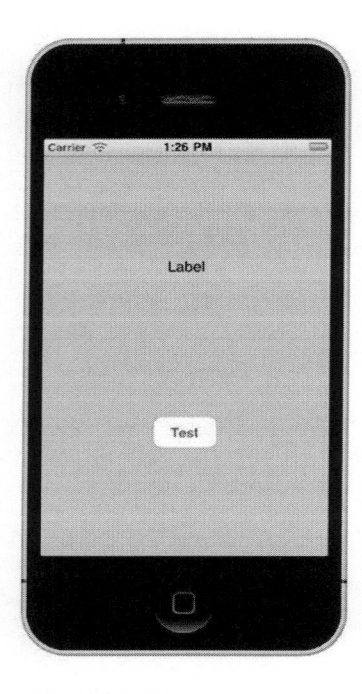

Figure 24-1: InstrumentTest App

24.1 Instruments for the iOS Simulator and iOS Device

In this section, you will have a brief overview of the instrument templates that can be used with either the iOS simulator or the iOS device. These instruments include: Allocations, Leaks, Activity Monitor, Time Profiler, and Automation.

24.1.1 Allocations

This template adds two instruments, one that tracks memory allocations and another that tracks virtual memory usage. It's useful when you are interested in object-allocation patterns and monitoring memory.

Before you can use this instrument template, you first need to add some code to your "InstrumentTest" project. You should already have a `UILabel` and a `UIButton` in the `.xib` file, for now all you are concerned with is the `UIButton`. Add an `IBAction` in the "InstrumentTestViewController.h" file named `testBtnPressed` and wire it up to the button. Add the following code to the method.

```
- (void)testBtnPressed
{
    NSString *tempString2 = [[NSString alloc] initWithUTF8String:"Test String"];
}
```

This will make it so that each time the button is pressed it will allocate memory for a new `NSString` object. By doing this, you have an easily accessible and traceable object allocation.

Now that you have your code set up, it is time to launch the Allocations instrument template. To do this, go to the top left corner of Xcode, where the Run button is, then click the drop down and select Profile.

Figure 24-2: Profile button

The Instrument template window will appear, asking you to choose a template or open a recent document. Select the Allocations template.

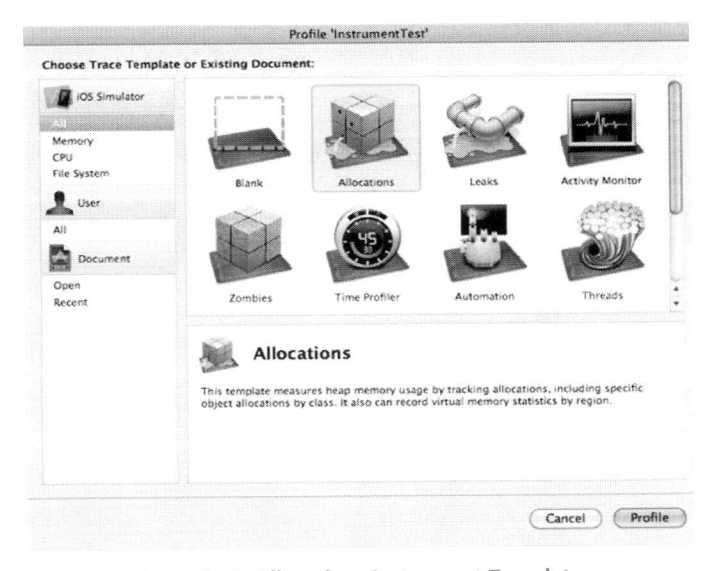

Figure 24-3: Allocations Instrument Template

Once you choose the instrument template, it will load up the Allocations and VM Tracker instruments along with your program in the iOS Simulator. Now you can see all of the memory allocations occurring in your app.

Chapter 24: Instruments   225

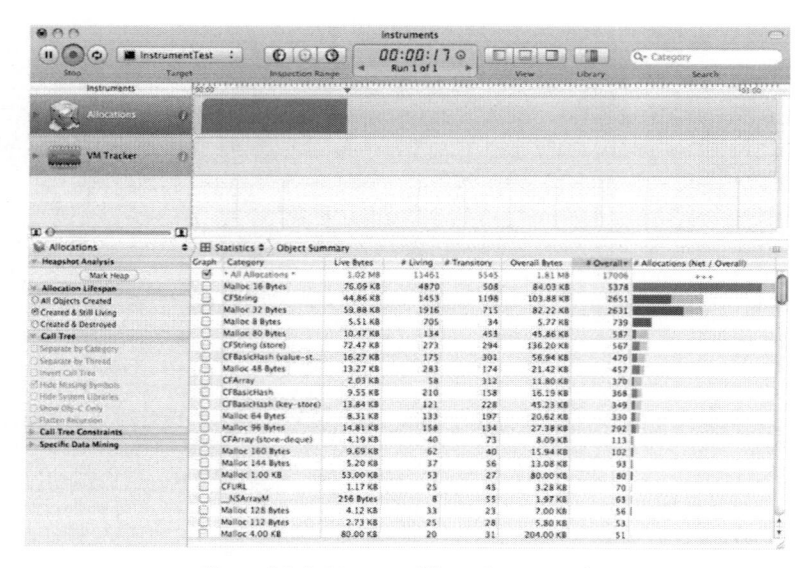

Figure 24-4: Memory Allocations sample

As you can see, there is a lot of information to sort through. What's shown is memory that is allocated upon startup and loading of your app. What you want to see is the memory allocation that occurs when you click the button. In order to do this, you need to use the "Mark Heap" button to take a "Heapshot."

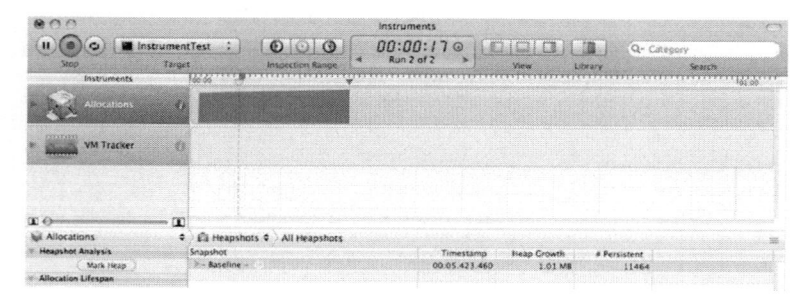

Figure 24-5: Allocations sample with Heapshot

When the "Mark Heap" button is pressed, it creates a small red marker at the time at which it was pressed. This also creates a "Baseline" of all the current allocations so that you can separate out the initial memory allocations that occur when the app loads. For the purpose of this section, they are not as important as the memory allocations that will be created during the testing process. Finally, click the button in the app once and then click the "Mark Heap" button in the instrument window once more. Doing so will create a "Heapshot" which can be expanded to see the memory allocation of the `CFString` that was created in the `testBtnPressed` method.

[Table showing Heapshots data]

Figure 24-6: String created from button press

If you toggle the Extended Detail View pane at the top, it will display a pane along the right side of the window with the stack information showing where the `CFString` was allocated.

[Extended Detail panel showing Stack Trace]

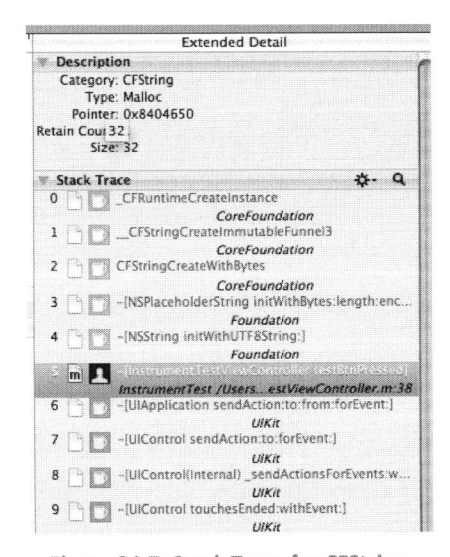

Figure 24-7: Stack Trace for CFString

As you can see, it shows the method `testBtnPressed`, which is where your allocation originated. If you double click it, you will be able to see where it is in code.

```
- (void)testBtnPressed
{
    NSString *tempString2 = [[NSString alloc] initWithUTF8String:"Test String"];
}

- (void)dealloc
{
    [super dealloc];
}
```

Figure 24-8: Line of code where allocation originated

Chapter 24: Instruments   227

One more thing to note is that you can change the Track Pane to display "Allocation Density" instead of the default "Current Bytes," which is useful because with it you can see a graph of the memory allocations when you are clicking on the button. To do this, click the small i button in the Instrument Pane next to the Allocations instrument. Then, under style change it to "Allocation Density." Now when you click on the test button, it should display the memory allocations graphically.

Figure 24-9: Allocations Instrument graph options

In order to see the memory allocations more easily, uncheck the "All Allocations" box and check the CFString box. Now the CFString memory allocations will be the only thing being tracked and displayed graphically. You may have to pause the tracking and then resume it for the display to change.

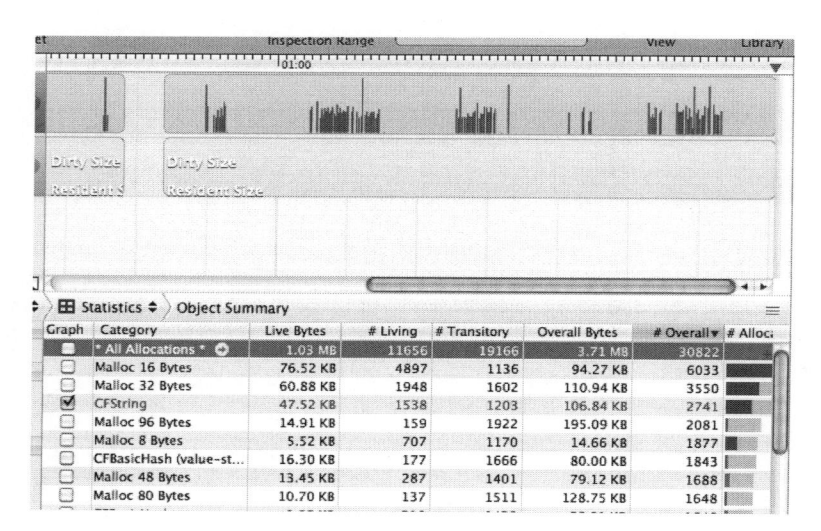

Figure 24-10: Allocation Density Graph

The blue spikes you see in Figure 24-10 are the CFString memory allocations that occurred when the test button was pressed.

This concludes the overview of the Allocations template. Next you will review memory leaks and the Leaks template, which can be very useful.

24.1.2 Leaks

The Leaks instrument template is very useful for detecting memory leaks in your projects. This template adds the Allocations instrument and the Leaks instrument to the document. You are already familiar with the Allocations instrument from the previous section, now you are going to see how the Leaks instrument works.

Before you go into detail about how this instrument works, it's worth explaining what a memory leak is, and how it occurs. Basically, a memory leak occurs when a program consumes memory but it is unable to release it back to the operating system. For you, this means that somewhere in your code – where you called alloc on an object – you forgot to send the `release` message. If you recall from the "InstrumentTest" project, in the `testBtnPressed` method you called `alloc` on an `NSString` object, however you did not call `release` anywhere. This is a simple example of a memory leak. Memory leaks are bad, especially in iOS apps, because memory cannot be freed until the user quits the app. This can cause the app to run slower, be unable to complete an operation, and in the worst case, crash. That's why it's a good idea to optimize your app by checking to see if any memory leaks occur.

In order to use the Leaks instrument, you will use the same code from the previous example. Open up the "InstrumentTest" project from before, and since you already have a memory leak in the `testBtnPressed` method, all you have to do is select the Profile option again. Then, in the Instrument template window, select the Leaks template.

Figure 24-11: Leaks Instrument template

Once selected, it should load up the Allocations instrument and the Leaks instrument. There are a few things worth mentioning at this point. For one, you will notice that if you click on the Leaks instrument, below in the detail pane there is an option that says "Snapshot Interval (sec)," and a timer underneath it. This interval is the amount of time in seconds that the instrument will wait to take an automatic snapshot to detect memory leaks; the default is 10 seconds. The timer, as you might have guessed, shows the countdown until the next snapshot occurs. There is also a button that allows you to take snapshots manually.

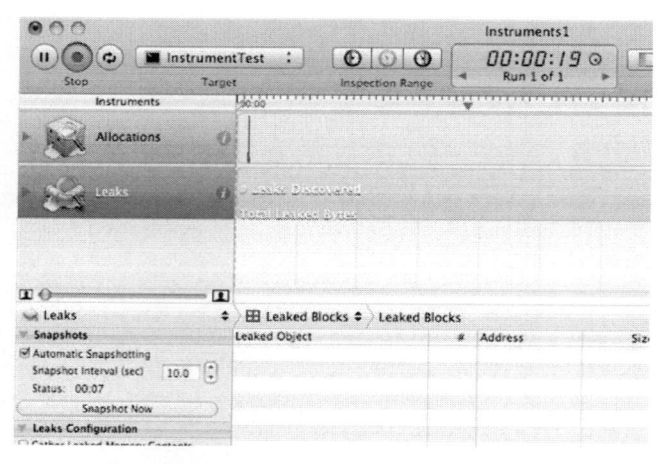

Figure 24-12: Sample with no Leaks discovered

As you have probably noticed, the Leaks instrument isn't detecting any memory leaks even though it is taking Snapshots every 10 seconds. This is because you have not created the memory leak yet. Click on the test button in your app to create a memory leak. Now, in the Leaks instrument, you will see a memory leak detected upon the next Snapshot being taken similar to Figure 24-13.

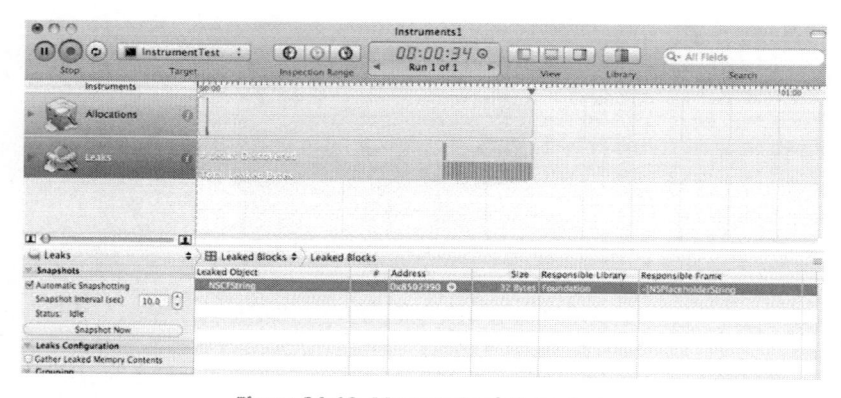

Figure 24-13: Memory Leak Detected

You can see that in the bottom pane it is displaying leaked blocks of memory and there is only one, which is an `NSCFString`. If you add the Extended Detail Pane, you will be able to see the Stack Trace, which will show you where the leak occurred.

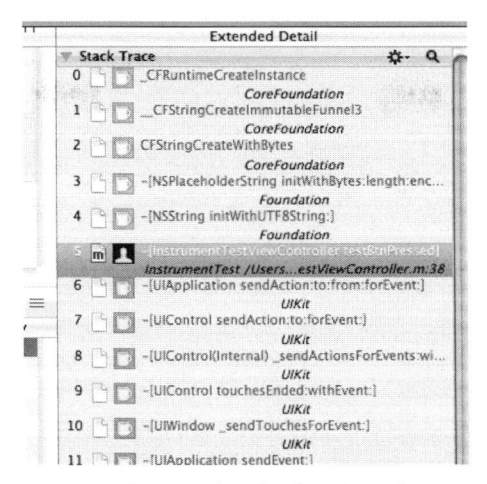

Figure 24-14: Stack Trace showing location of memory leak

You can see that it is occurring in the `testBtnPressed` method on line 38. If you double click on it, the "InstrumentTestViewController.m" file will appear, and the line will be highlighted.

Figure 24-15: Line of code where memory leak occurs

This concludes the coverage of the Leaks instrument; next the Activity Monitor instrument template will be discussed.

24.1.3 Activity Monitor

The Activity Monitor instrument is used when you want to measure the load on the system against the size of the virtual memory. Even though this instrument can be used with the iOS Simulator, it is not as useful as when it is used in conjunction with an iOS Device. This is because it will compare how it runs on the iOS simulator against the computer you are using, which most likely will have a faster processor and more memory than an iOS Device.

To demonstrate this instrument, you can just use the "InstrumentTest" project as is. Go to Profile and then select the Activity Monitor Instrument template.

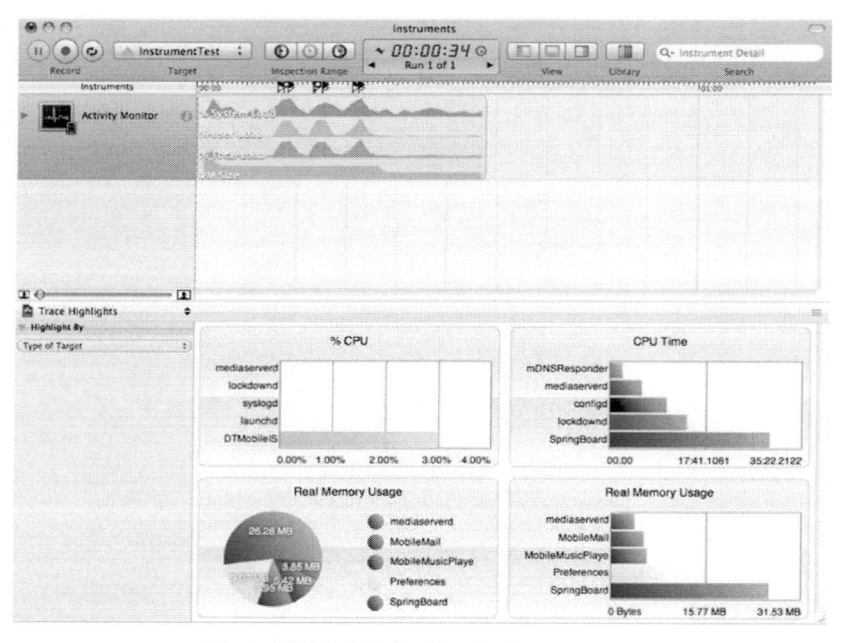

Figure 24-16: Activity Monitor Instrument

As you can see in Figure24-16, the Activity Monitor Instrument is preloaded and is configured by default to track the System Load, User Load, Total Load, and Virtual Memory Size. Above are 4 graphs that show by process: the percentage of CPU utilization, the amount of CPU Time used, and Real Memory Usage - which is displayed both as a bar graph and a pie chart. In the trace window above you can see there are 3 "peaks" where the load spiked. Opening and closing the app on the iOS Device while the trace was recording produced this. This concludes the brief explanation of how to use the Activity Monitor instrument.

24.1.4 Time Profiler

The Time Profiler instrument is useful when you want to know where most of the execution time is being spent in your program. It does this by recording stack trace information for each thread in the program at predefined intervals. By knowing where most of the execution time is being spent in your program, you can use that information to optimize your code. For example, say you have a program with 20 functions, if one of those functions has 80% of the execution time it would be better to optimize that one function instead of optimizing the other nineteen.

In order to see how this instrument works you will have to add a bit of code to your "InstrumentTest" project. Before adding any code, you must make a couple of changes to the XIB so that it looks similar to below.

Figure 24-17: InstrumentTest project with 3 buttons

Now you need to wire all of the buttons up to the `testBtnPressed` method. To distinguish between the three different buttons, you need to set the tags for Button 1, Button 2 and Button 3 to 1, 2 and 3 respectively. Now to change the `testBtnPressed` method to work with the tags, add the following code:

```objc
- (void)testBtnPressed:(UIButton *)sender
{
    switch (sender.tag) {
        case 1:
            testLbl.text = @"Button 1";
            [self buttonOneFunction];
            break;
        case 2:
            testLbl.text = @"Button 2";
            [self buttonTwoFunction];
            break;
        case 3:
            testLbl.text = @"Button 3";
```

```objc
        [self buttonThreeFunction];
        break;
    }
}
```

This sets up the `testBtnPressed` method to call one of three methods depending on which button is pressed. Now you have to implement those methods. To do this, add the following code to the "InstrumentTestViewController.m" file:

```objc
- (void)buttonOneFunction
{
    NSMutableArray *tempArray = [[NSMutableArray alloc] initWithCapacity:10];
    for (int i = 0; i < 10; i++)
    {
        NSNumber *tempNum2 = [[NSNumber alloc] initWithInt:i];
        [tempArray addObject:tempNum2];
        NSLog(@"i=%i and tempNum=%i", i, [tempNum2 intValue]);
        [tempNum2 release];
    }
    [tempArray removeAllObjects];
    [tempArray release];
}


- (void)buttonTwoFunction
{
    NSMutableArray *tempArray = [[NSMutableArray alloc] initWithCapacity:100];
    for (int i = 0; i < 100; i++)
    {
        NSNumber *tempNum2 = [[NSNumber alloc] initWithInt:i];
        [tempArray addObject:tempNum2];
        NSLog(@"i=%i and tempNum=%i", i, [tempNum2 intValue]);
        [tempNum2 release];
    }
    [tempArray removeAllObjects];
    [tempArray release];
}
```

```objectivec
- (void)buttonThreeFunction
{
    NSMutableArray *tempArray = [[NSMutableArray alloc] initWithCapacity:1000];
    for (int i = 0; i < 1000; i++)
    {
        NSNumber *tempNum2 = [[NSNumber alloc] initWithInt:i];
        [tempArray addObject:tempNum2];
        NSLog(@"i=%i and tempNum=%i", i, [tempNum2 intValue]);
        [tempNum2 release];
    }
    [tempArray removeAllObjects];
    [tempArray release];
}
```

As you can see, the three functions are essentially the same. However, the size of the array is increased tenfold, and thus the number of iterations in the for loop. The reason behind this is so that you can see the difference in execution time being taken between the three very similar methods. If you haven't already modified the "InstrumentTestViewController.h" file, then do so now. It should look similar to the following:

```objectivec
#import <UIKit/UIKit.h>
@interface InstrumentTestViewController : UIViewController {
    IBOutlet UILabel *testLbl;
}
- (IBAction)testBtnPressed:(UIButton *)sender;
- (void)buttonOneFunction;
- (void)buttonTwoFunction;
- (void)buttonThreeFunction;
@end
```

Now that you have your code set up, you can run the Time Profiler instrument. Select Profile and then the Time Profiler instrument template. The trace will start running, but the first thing you are going to want to do is change a few options to make it easier to see the data. In the bottom left corner of the Instruments window, check the options to "Hide System Libraries" and "Show Obj-C Only."

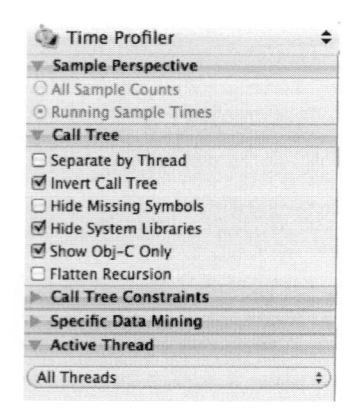

Figure 24-18: Call Tree options for Time Profiler Instrument

Once the trace is setup and running, press each button one time and watch how the graph changes. It should look similar to Figure 24-19.

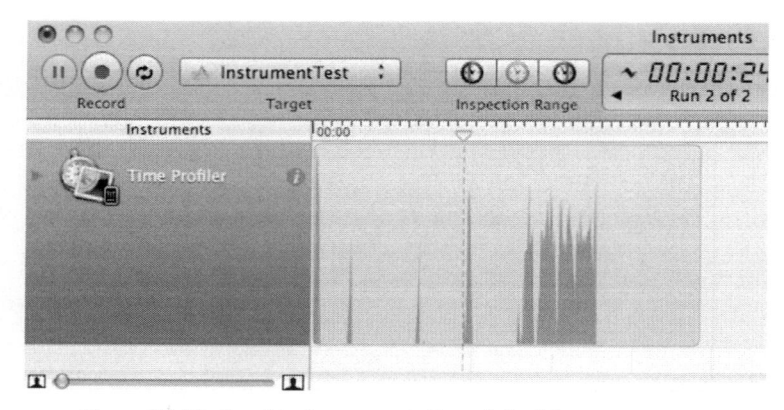

Figure 24-19: Graphical representation of the 3 button presses

The first two spikes were from loading the app. The third spike was from pressing "Button 1." As you might have guessed, the fourth and much higher spike was from pressing the second button. Finally, the third button being pressed resulted in the cluster. This is as expected, because the third button is working with ten times as much data as the second and 100 times more than the first. In the bottom pane you can see the Call Tree, which lists the methods that were called and the amount of time spent on each. According to the data, the `buttonThreeFunction` method took 86.8% of the execution time or 1145ms to execute. The `buttonTwoFunction` method took 105ms or 7.9% of the execution time, while the `buttonOneFunction` method took only 40ms or 3.0% of the execution time. This information can be very useful when trying to figure out what is causing your app to run slow or if you are just trying to speed it up a bit. This concludes the tutorial on the Time Profiler instrument.

24.1.5 Automation

The Automation instrument is used to automate user interface tests of your iOS app. This can be very useful for several reasons. It allows you perform more comprehensive testing, develop repeatable regression tests, and minimize procedural errors, all while freeing up your time and allowing you to perform other tasks.

In order to use the Automation instrument, you need to create a test script. These automated tests are written in JavaScript and use the user interface automation API to perform actions in your app. Details on the API can be found on Apple's website under *UI Automation Reference Collection*.

To create a new automated test script, add a new file in the "InstrumentTest" project and select Empty. Name the file "AutomationTest.js", and make sure you do not forget the `.js` extension so that it will be recognized as a JavaScript file.

Figure 24-20: Create empty file

Before you write the test script, you have to set the Accessibility label of the UI controls in order to access them during the tests. This is easily done by going to the `.xib` file in Interface Builder, selecting the control, and then going to the Identity Inspector. Make sure you set the label to a unique value for the view. For "Button 1" set the label to "buttonone," and set the label for the other two buttons similarly.

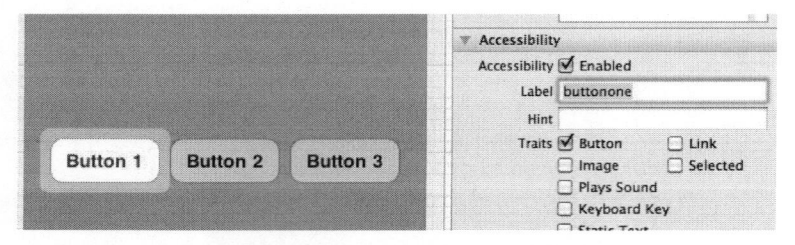

Figure 24-21: Configure Accessibility labels for buttons

Now that you have the UI controls set up for testing, add the following code to the AutomationTest.js file:

```javascript
var window = UIATarget.localTarget().frontMostApp().mainWindow();
var buttons = window.buttons();
var target = UIATarget.localTarget();
//Check number of buttons
UIALogger.logStart("Button Check");
if (buttons.length !=3)
{
  UIALogger.logFail("Fail: Invalid number of buttons");
}
else
{
  UIALogger.logPass("Pass: Correct number of buttons");
    target.delay(3);
}
//TESTCASE_001: Check for UIButton "Button 1" on screen and press it
UIALogger.logStart("TESTCASE_001");
if(buttons["button1"] == null || buttons["button1"].toString() == "[object UIAElementNil]")
{
    UIALogger.logFail("Fail: UIButton 'Button 1' not found.");
}
else
{
    buttons["buttonone"].tap();
    UIALogger.logDebug("Button 1 pressed.");
```

```
        target.delay(3);

        UIALogger.logPass("Pass: UIButton 'Button 1' found.");
}
//TESTCASE_002: Check for UIButton "Button 2" on screen and press it
UIALogger.logStart("TESTCASE_002");
if(buttons["buttontwo"] == null || buttons["buttontwo"].toString() == "[object
UIAElementNil]")
{
    UIALogger.logFail("Fail: UIButton 'Button 2' not found.");
}
else
{
    buttons["buttontwo"].tap();
    UIALogger.logDebug("Button 2 pressed.");
    target.delay(3);
    UIALogger.logPass("Pass: UIButton 'Button 2' found.");
}
//TESTCASE_003: Check for UIButton "Button 3" on screen and press it
UIALogger.logStart("TESTCASE_003");
if(buttons["buttonthree"] == null || buttons["buttonthree"].toString() ==
"[object UIAElementNil]")
{
    UIALogger.logFail("Fail: UIButton 'Button 3' not found.");
}
else
{
    buttons["buttonthree"].tap();
    UIALogger.logDebug("Button 3 pressed.");
    target.delay(10);
    UIALogger.logPass("Pass: UIButton 'Button 3' found.");
}
```

Save the test script, then go to Profile and select the Automation instrument template. In the bottom left corner, select the script and click "Start Script" once. You should see results similar to Figure 24-22.

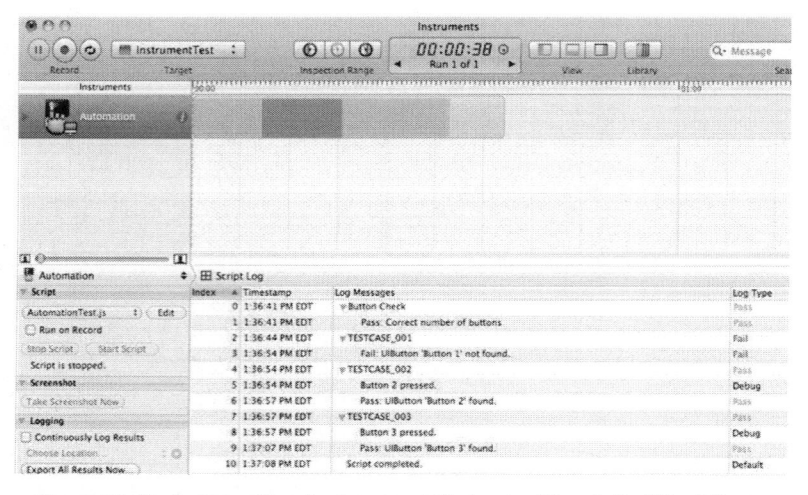

Figure 24-22: Automation Instrument Output with a failed Test Case

As you can see, part of the graph is red, and in the Script Log there is a Fail logged there. The red part of the graph corresponds to the test case that failed. In the code you added, you used `button1` instead of `buttonone`, causing the button to not be found. It is also worth noting that when accessing items using bracket references by element name, JavaScript interprets numbers as an index, which yields incorrect results. To rectify this, rename the button appropriately and the tests should be successful and look similar to Figure 24-23.

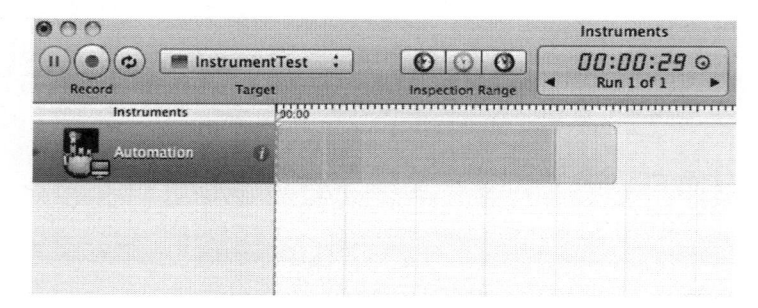

Figure 24-23: Script Log with a Pass for all Test Cases

In addition to the *UI Automation Reference Collection* API used to test the system, there is another use for the Automation instrument. It can be used in conjunction with another instrument, like the previous Time Profiler instrument, for example. All you need to do is drag the desired instrument from the library onto the trace window, and start recording. If you use the Time Profiler instrument, this can help you determine what is taking up CPU cycles and execution time. As you can see in Figure 24-24 at `TESTCASE_003`, when the third button is being pressed, the CPU usage increases greatly.

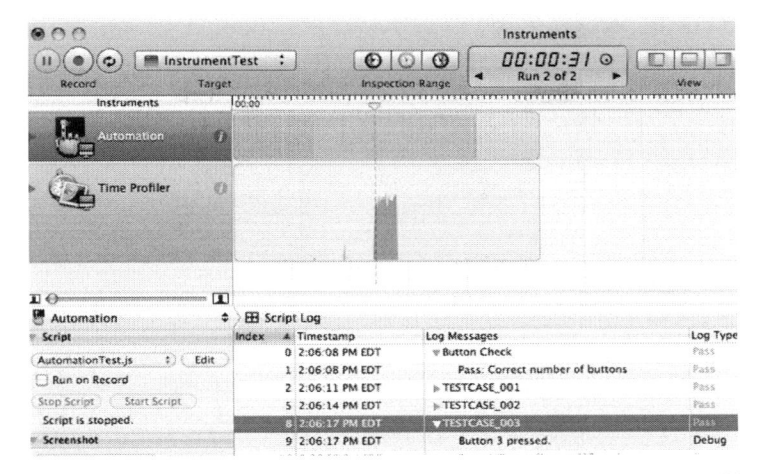

Figure 24-24: Automation Instrument used in conjunction with Time Profiler

This concludes the tutorial on the Automation instrument. Next, you will go into detail on iOS Simulator specific instruments.

24.2 iOS Simulator-Specific Instruments

In this section you will briefly go over the instrument templates that are specific to the iOS simulator. They are: Zombies, Threads, and File Activity.

24.2.1 Zombies

The Zombies instrument is useful if you are getting the dreaded EXEC_BAD_ACCESS error. This error can occur when you are sending a message to an object that has already been released. Using the Zombies instrument allows you to see exactly where the source of the error is because it tracks the alloc, retain, release, and autorelease history of objects.

In the "InstrumentTest" project, you need to add a button and a method named printZombieString and wire it up to the new button. Update the printZombieString method and the viewDidLoad method to look as follows:

```
- (void)printZombieString
{
    NSLog(@"Zombie String %@", zombieString);
}


- (void)viewDidLoad
{
    [super viewDidLoad];
    zombieString = [[NSString alloc] initWithUTF8String:"Zombie String"];
```

```
        [zombieString release];
}
```

Now when you select Profile and choose the Zombies instrument template, it adds the Allocations instrument to the trace pane configured to work with `NSZombie` objects. If you were to click on the button that triggers the `printZombieString` method, you would have results similar to Figure 24-25.

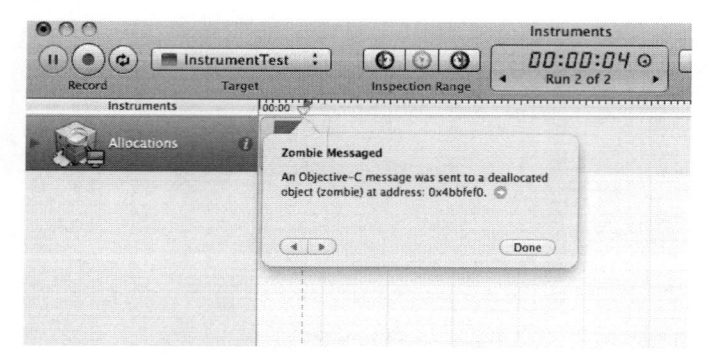

Figure 24-25: Zombie object messaged

Clicking on the arrow next to the memory address of the zombie object will take you to the `alloc`, `retain`, and `release` history of the object. In this history, you can see where the object is being messaged after it was deallocated. You can then examine the stack trace in the extended detail pane to see where in code this was occurring.

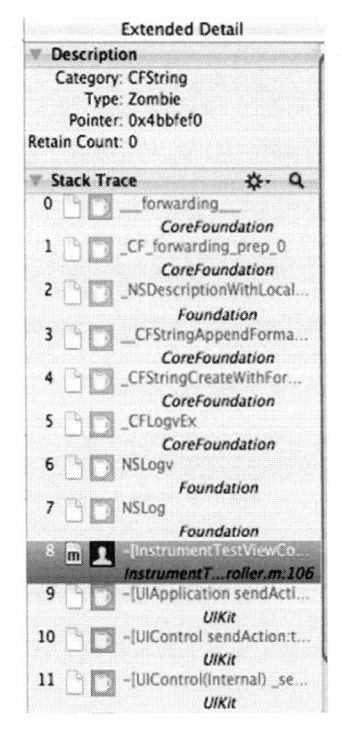

Figure 24-26: Stack Trace of Zombie object

After clicking on the part in the stack that represents the InstrumentTestViewController.m file, it will show you where in the code the Zombie object was messaged. In this case it was in the `printZombieString` method as shown in Figure 24-27.

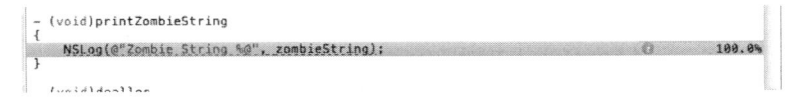

Figure 24-27: Line of code that Zombie object was messaged

This concludes the tutorial for the Zombies instrument. Next, the Threads instrument will be discussed.

24.2.2 Threads

The Threads instrument template adds the Thread States instrument to the track pane. This instrument is useful for analyzing thread state transitions within a process.

Using the "InstrumentTest" project, go to Profile and then select the Threads template. You will see that Thread States instrument is recording but no data is being graphed. If you click the home button on the simulator and then click back on the "InstrumentTest" application, you should see results similar to Figure 24-28.

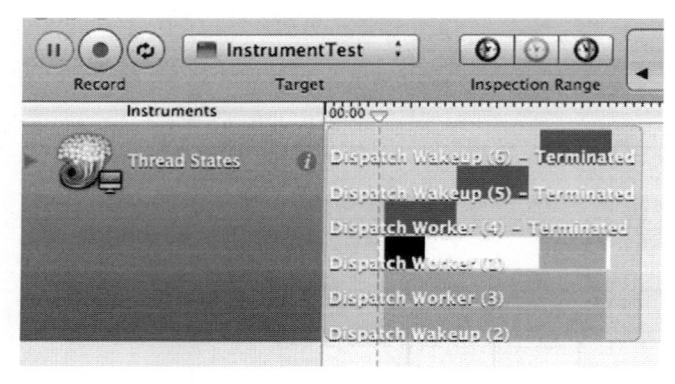

Figure 24-28: Thread States Instrument

24.2.3 File Activity

The File Activity instrument template adds the File Activity, Reads/Writes, File Attributes, and Directory I/O instruments to the track pane. This is useful when you want to analyze file usage patterns in the system. These instruments will allow you to monitor changes in the file system, as well as monitor open, close, read, and write operations on files.

24.3 iOS Device-Specific Instruments

In this section, you will briefly review the instrument templates that are specific to the iOS device. Those instruments are as follows: Energy Diagnostics, System Usage, Core Animation, OpenGL ES Driver, and OpenGL ES Analysis.

24.3.1 Energy Diagnostics

The Energy Diagnostics instrument template adds the Energy Usage, CPU Activity, Display Brightness, Sleep/Wake, Bluetooth, WiFi, and GPS instruments to the track pane. This template is useful if you need to get diagnostic information regarding energy usage in iOS devices. In order to use this template effectively, the device needs to be disconnected from external power. This is because devices behave differently when operating on external power than they would while operating on battery power.

To use this template properly while having the device connected, launch either Xcode or Instruments. In your case you will be testing it against your "InstrumentTest" project. Enable power logging on the device by going to Settings > Developer, this will enable your device to log all the information for these instruments. After that is enabled, the device should be disconnected and the tests should be performed.

In your case you performed a simple test consisting of launching the "InstrumentTest" application, pressing buttons 1 through 3 in order, then turning off the screen and waiting a few seconds, turning it back on, then repeating the button pressing sequence once more.

Once the tests have been performed, the device needs to be reconnected to the computer. Instruments should then be run and the Energy Diagnostics template chosen. You will notice it will start recording automatically. This should be stopped. Going to File > Import Energy Diagnostics from Device should load the data that was recorded on the device. You should have similar results to Figure 24-29.

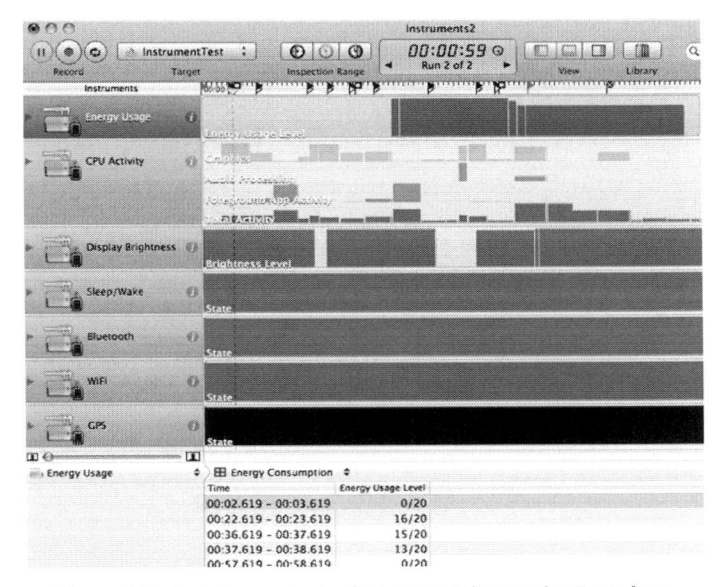

Figure 24-29: Instruments in the Energy Diagnostics Template

As you can see, there are several instruments with various kinds of information. The first instrument is showing the energy usage on a scale from 0-20. The second instrument is recording the CPU Activity, and the data is split amongst four categories; Graphics, Audio Processing, Foreground App Activity, and Total Activity, in that order. The Graphics activity was highest when the app was loading or when the screen was being turned back on. Audio Processing was pretty much non-existent since the app did not use any audio. Foreground App Activity corresponds to when the buttons were pressed the largest red block was from "Button 3," which is the most CPU intensive. Total Activity represents the total percentage of CPU Activity. The third instrument is showing the changes in brightness level, which is generally only when the screen was turned off and back on. Instruments 4 through 7 display either a red or black band indicating the state of the device. The Sleep/Wake instrument is showing a red band meaning that the device is running. The Bluetooth and Wi-Fi instruments both have a red band, showing that Bluetooth and Wi-Fi are both active. Lastly, the GPS instrument is displaying a black band, which tells you that GPS is off.

24.3.2 System Usage

The System Usage instrument template adds the I/O Activity instrument to the track pane. This is used to record calls to functions that operate on files in a process running on an iOS device, such as read, write, open, and close functions.

24.3.3 Core Animation

The Core Animation instrument template adds the Core Animation and Sampler instruments to the track pane. This is useful if you want to measure the number of Core Animation frames per second for your app while it's running on an iOS device. The Core Animation instrument also has options to provide visual hints to help you understand how content is rendered on the screen. Using the "InstrumentTest" project, select Profile, then select the Core Animation template. Check the three "Debug Options," then press the buttons and see how they react. Your trace pane should look somewhat like Figure 24-30.

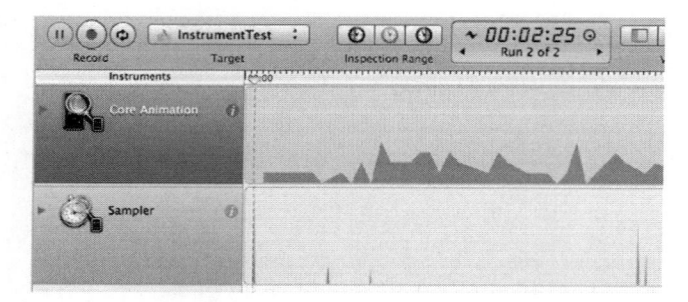

Figure 24-30: Core Animation instrument template

In Figure 24-30 you can see some of the visual hints were enabled, namely the ones that are labeled: "Color Blended Layers," "Color Misaligned Images," and "Flash Updated Regions." After enabling those options, the app should look similar to Figure 24-31.

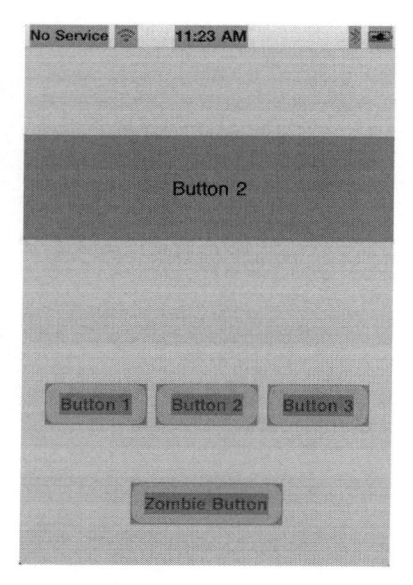

Figure 24-31: InstrumentTest app with visual hints enabled

You will also notice that when you select a button it flashes yellow and so does the label, this occurs because these are the updated screen regions. There are seven rendering hints that can be selected. They are as follows:

- *Color Blended Layers*: Layers drawn without blending have a green overlay and layers that were drawn with blending enabled are red.

- *Color Copied Images:* Makes images that were copied by Core Animation have a cyan overlay.

- *Color Immediately:* Removes the 10 ms wait after performing a color-flush operation.

- *Color Misaligned Images:* Images whose source pixels are not aligned to destination pixels have a magenta overlay.

- *Color Offscreen-Rendered Yellow:* Offscreen-rendered content has a yellow overlay.

- *Color OpenGL Fast Path Blue:* Content that is detached from the compositor gets a blue overlay.

- *Flash Updated Regions:* Updated screen regions will flash yellow.

Lastly, the Sampler instrument is similar to the Time Profiler instrument from previously in this chapter. It will show CPU usage in the track pane, and underneath will display methods and their execution time.

24.3.4 OpenGL ES Driver

The OpenGL ES Driver instrument template adds the OpenGL ES Driver and Sampler instruments to the track pane. This is used to determine how efficiently you are using OpenGL and the GPU on the iOS Device. There are several pieces of information gathered by the OpenGL ES Driver instrument. They are as follows:

- Context Count
- Command Buffer Allocated Bytes
- Command Buffer Submitted Bytes
- Command Buffer Submit Count
- Command Buffer Render Count
- Command Buffer Transfer Count
- Command Buffer Swap Count
- Renderer Utilization %
- Tiler Utilization %
- Device Utilization %
- Tiled Scene Bytes
- Split Scene Count
- Resource Bytes
- Resource Count
- Core Animation Frames Per Second

The GPU hardware in Apple's iOS devices has two components: a tiler and a renderer. The OpenGL ES Driver instrument captures information on the utilization of these two components.

This is important because it can be useful for determining bottlenecks which could greatly slow down performance of an app.

In order to use the OpenGL ES Driver instrument you will need an app that uses OpenGL. Create a new project in Xcode and select the OpenGL ES Application template. Name the project "OpenGLTest."

Figure 24-32: OpenGL ES Application template

If you run the project now, you will see a simple colored square moving up and down on the screen similar to Figure 24-33.

Figure 24-33: Basic OpenGL ES Application

Now that you have a functioning OpenGL ES Application, you can run instruments to analyze it. In the instrument pane select the OpenGL ES Driver instrument template. As you can see, by default Core Animation Frames Per Second (FPS) and Resource Bytes are displayed in the track pane. It should look similar to Figure 24-34.

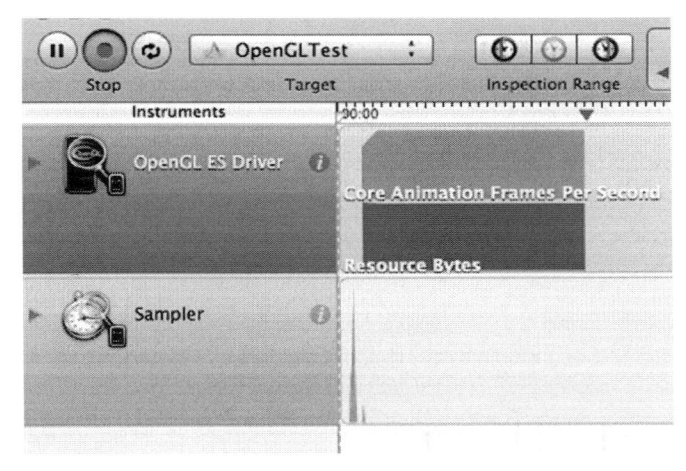

Figure 24-34: OpenGL ES Driver Instrument template

The instrument can be configured to show more than just those two sets of data. To do this, click on the small "i," then click on "Configure." Select three more options from the list, "Device Utilization %," "Renderer Utilization %," and "Tiler Utilization %." After clicking "Done," make sure the three options that were just selected are checked under the "Statistics to Observe" section. Now, running the instrument should yield similar results as seen in Figure 24-35.

Figure 24-35: Device, Tiler, and Renderer Utilization displayed by percentage

As you can see, the graphs display the percentage of utilization of the renderer and tiler components. Since this is a very basic OpenGL ES app, the tiler and renderer are only utilized at about 5-6 percent. In a more complex app those numbers may be much larger.

24.3.5 OpenGL ES Analysis

The OpenGL ES Analysis instrument template adds the OpenGL ES Analyzer and OpenGL ES Driver instruments to the track pane. This is very useful for measuring and analyzing OpenGL ES activity in your apps. It's also helpful for finding and addressing problems because it offers relevant solutions to problems based on best practices and knowledge of Apple's hardware and software.

Using the "OpenGLTest" project, you can see how this instrument works. Go to Profile and select the OpenGL ES Analysis template. Letting the app run for about 30 seconds will give results similar to Figure 24-36.

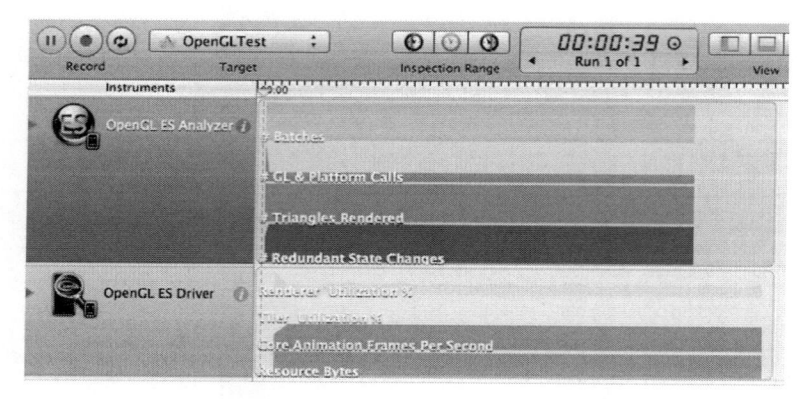

Figure 24-36: OpenGL ES Analysis template

As you can see, there is a lot of information being collected by both instruments. The second instrument was already discussed in the previous section. The first instrument is displaying several data in the track pane, which can be seen in more detail by selecting "Frame Statistics" in the pane below.

If "Expert" is selected, the bottom pane will show problems and recommended solutions. As you can see in Figure 24-37, it found several "problems" in the app as well as recommended changes. These recommended changes usually appear when best practices aren't followed, resulting in a less efficient app. Among the list are several redundant calls, two recommendations to use vertex buffer objects (VBO's), a recommendation to use indexed primitives for lines and triangles, and a recommendation to compact the FBO color attachment, renderbuffer.

Figure 24-37: Expert system recommends fixes based on best practices

Clicking on the arrow next to the "Category" will bring you to a list of the unique occurrences. These can then be selected to show where they are on the stack, as well as the file and lines of code responsible for them.

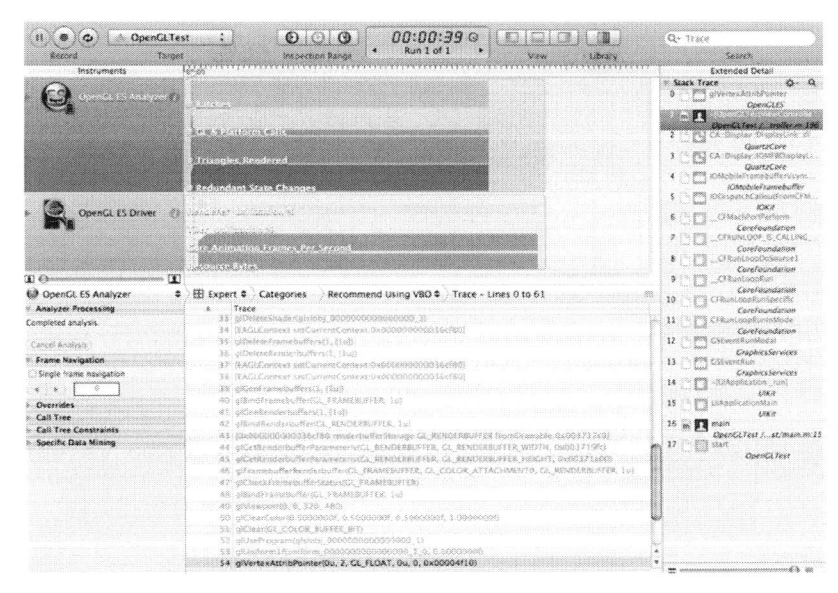

Figure 24-38: Stack trace for a recommended problem fix

This concludes the tutorial for the OpenGL ES Analysis instrument template. Next you will review how to create custom instrument templates.

24.4 Custom Instruments and Building Your Own Instrument Template

It is possible in the Instruments application to create your own set of instruments to track your app. Selecting the Blank template when Instruments is launched will create an empty Trace Document on which you can add as many recording instruments as you would like.

Figure 24-39: Blank Instrument template

After you have launched your blank document, it should look like Figure 24-40.

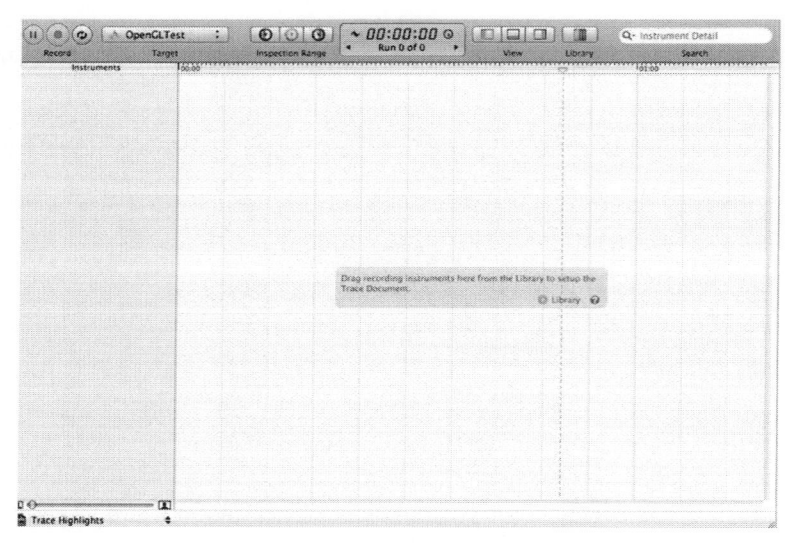

Figure 24-40: Blank document with no Instruments

Now that you have a blank template, you can drag whichever instruments you want from the library onto the empty document. For example, if you were to drag the Time Profiler, CPU Monitor, and CPU Activity instruments into the document, the result would look similar to Figure 24-41.

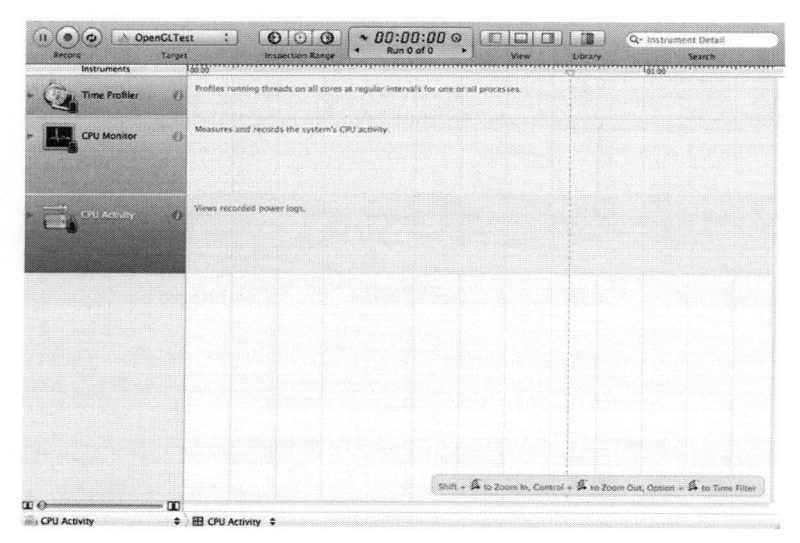

Figure 24-41: Document after adding several Instruments

Another option that provides some flexibility in analyzing apps using Instruments is the option to create a custom instrument with DTrace. DTrace is a dynamic tracing facility, created by Sun and ported to Mac OSX v10.5, which taps into the operating system kernel. Many of the built-in instruments are based on DTrace.

For more information or instructions on how to create a custom instrument, Apple has a complete guide on creating custom instruments on their developer website at the URL:

http://developer.apple.com/library/ios/#documentation/DeveloperTools/Conceptual/InstrumentsUserGuide/Introduction/Introduction.html

SECTION IV

ALTERNATIVE DEVELOPMENT OPTIONS

Going beyond the core development tools and utilities, provide here are step-by-step instructions, or "walkthroughs," for each individual iOS development option as described in the prior sections and continue to learn additional tools and utilities for developing apps for publication.

Each walkthrough guides you through the entire process – from installing the development environment all the way to creating a simple app, such as the traditional "Hello World."

You also examined the use and practice of 3D game editors to which enable you to create 3D console quality games (and even publish to them). This book will help guide you through this process using easy to follow instructions, code examples and images of desired output.

25

Unity 3D

Unity3D is a game development tool that can be used to create 3D games for multiple platforms. It consists of a game engine and an editor. The platforms that are supported with Unity are the following:

- Microsoft Windows
- Mac OS X
- Web via Unity Web Player plugin
- Adobe Flash Player
- iOS
- Android
- Nintendo Wii
- Xbox 360
- PlayStation 3

One of the pros of Unity is that you can develop apps for multiple platforms concurrently. When creating an app, you have the ability to switch its target platform instantly. This allows you to develop for iOS, Android, Web, PC/Mac and consoles all at the same time. This gives you the ability to easily build an app for multiple platforms without starting from scratch for each one.

Unity 3D has a free evaluation edition available for PC, Mac, and Web. There are several fully paid licenses available, supporting Apple iPhone, Google Android, and others. Committing to the fully licensed Unity cross-platform environment for development can be a substantial investment for some developers with a low budget.

25.1 Installation

Before you can start making games with Unity 3D, you have to download and install it. To download the program, go to the Unity 3D website and click the download link or go directly to the address http://unity3d.com/unity/download.

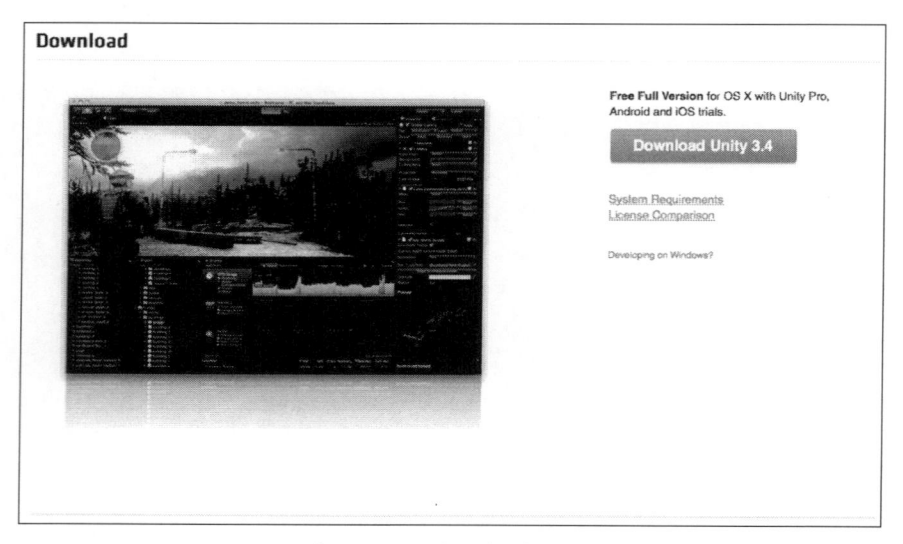

Figure 25-1: Download Unity

When the download completes, open the file and run the installer. You should see the following screen.

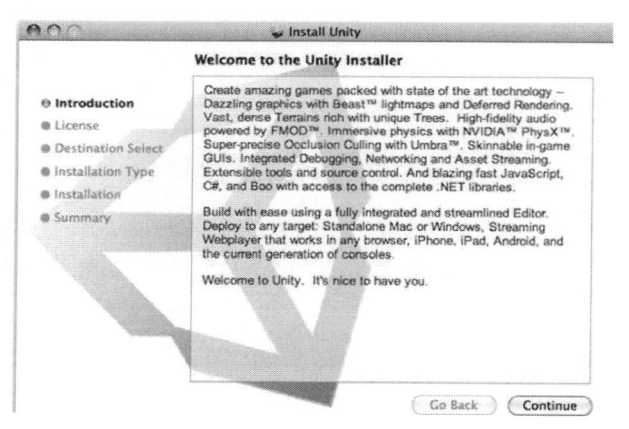

Figure 25-2: Unity Installer

Continue the installation by reading through and agreeing to the license agreement and choosing the installation destination. The installation should finish after a couple minutes.

Figure 25-3: Unity Installation Complete

After you install Unity and run it for the first time, a pop-up window similar to Figure 25-4 will appear asking you to register in order to continue.

Figure 25-4: Unity Registration

You have two registration options: Internet activation or Manual activation. For the purposes of this tutorial you will choose Internet activation.

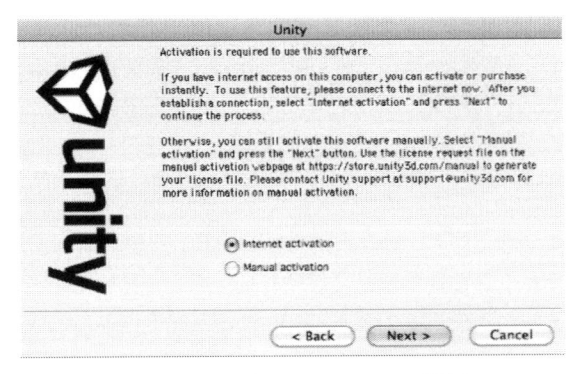

Figure 25-5: Activate your copy of Unity

Chapter 25: Unity 3D   259

Choosing Internet activation will bring you to their website where you will be required to register with an email address and have the option of choosing either the free license or starting the 30-day Pro/iOS trial. For the purposes of this demo, you will select the "Free" license.

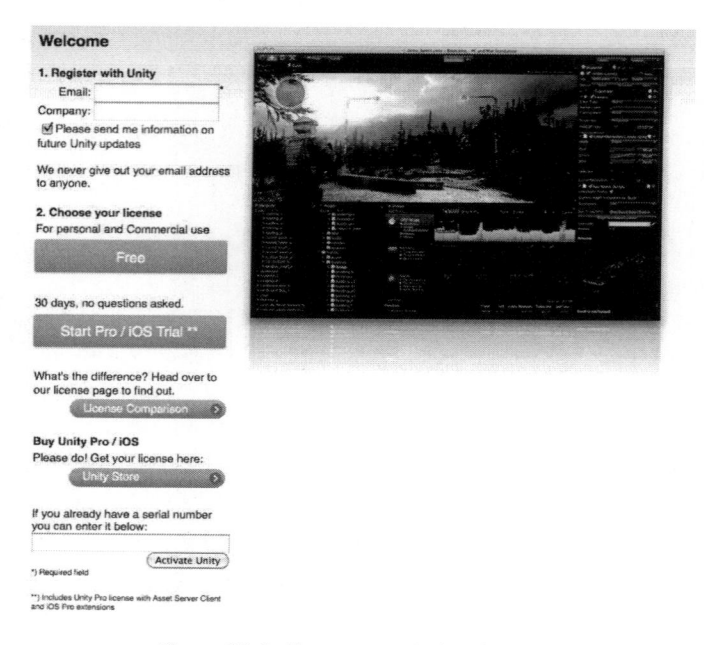

Figure 25-6: Choose your Unity License

After entering in the required information and selecting the appropriate license, the copy of Unity should be authorized. A window indicating this will appear, as in Figure 25-7.

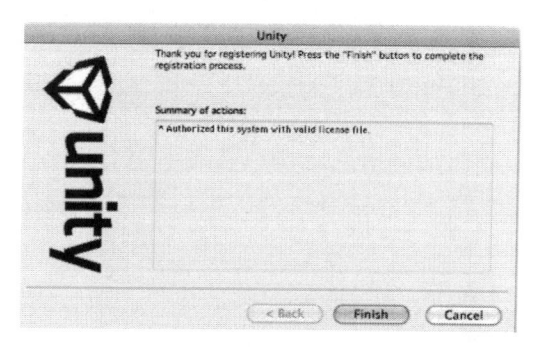

Figure 25-7: Unity Registration Complete

Afterwards the program should launch and you will see a welcome window containing useful links to several Unity-related resources.

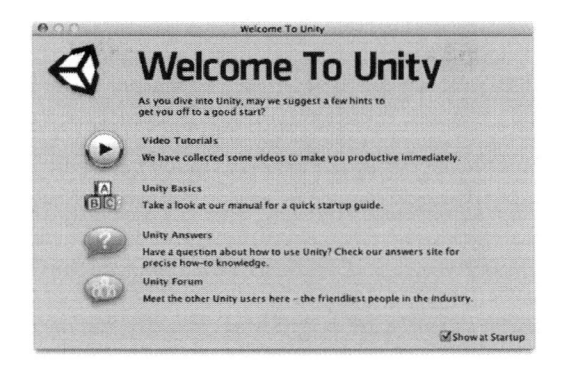

Figure 25-8: Unity Welcome Screen

25.2 Unity 3D Walkthrough

In this section you will walk through the Unity 3D development environment and create a simple app to showcase the basic functions of Unity 3D. In this app you will create a simple plane and a sphere that will bounce on the plane.

25.2.1 Creating a new project

To start creating your app, go to File > New Project, select where you want to save it, and name it "Hello World."

Now the project should be open and look similar to Figure 25-9.

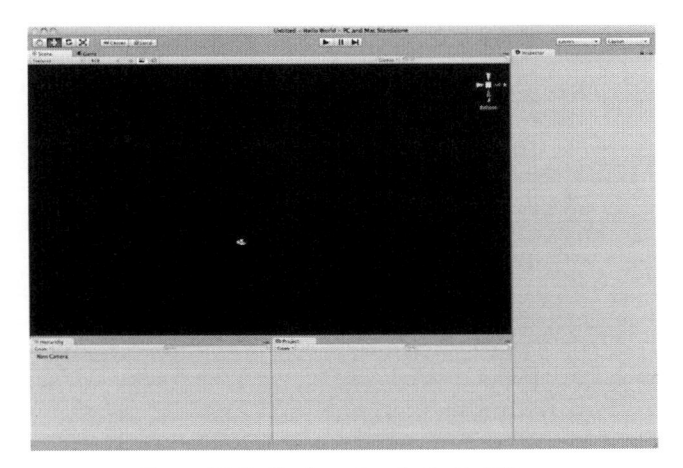

Figure 25-9: Blank project in Unity Editor

Configure the different views so that you can see all of them including the Game View. The Unity Editor should look similar to Figure 25-10.

Chapter 25: Unity 3D   261

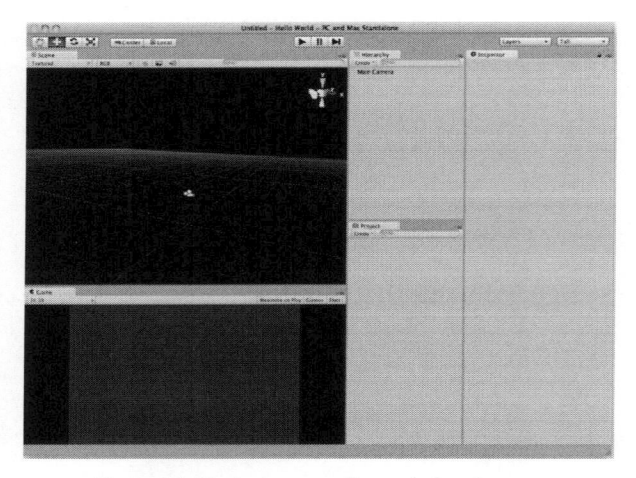

Figure 25-10: Custom configured view layout

25.2.2 Creating your first game

Now that you have an empty project the sphere and plane GameObjects can be added. A GameObject is basically a container that holds different Components. Components can be a variety of thing from the object's position to its collider, which handles collisions with other GameObjects.

Before adding the GameObjects, first align the Main Camera's view with the Scene's view. To do this select the Main Camera in the Hierarchy view then go to GameObject > "Align with View" or press Shift+Command+F. This will ensure that the objects you will create will be in front of the camera.

To create a plane, go to GameObject > Create Other > Plane and a plane will be created. With the selected you can click and drag on the different colored arrows and the plane will be moved in different x, y, and z directions. Try and position it in the center of the view like in Figure 25-11.

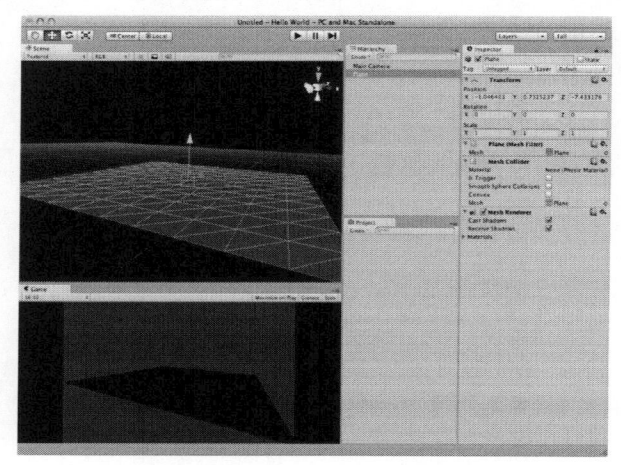

Figure 25-11: Position Plane GameObject

262   Chapter 25: Unity 3D

Now that you have a plane, you need to create a sphere. To do this go to GameObject > Create Other > Sphere and a sphere will be created. Again use the colored arrows to drag the sphere around and try and position it above the plane. Once positioned correctly it should look similar to Figure 25-12.

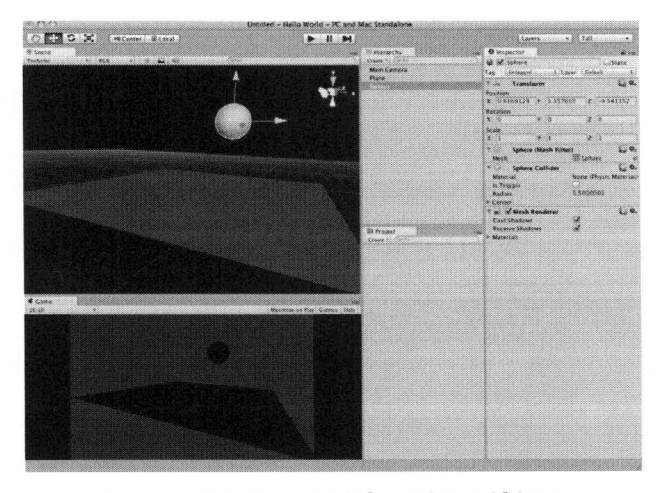

Figure 25-12: Position Sphere GameObject

Once the sphere and plane are positioned correctly, you will notice that they are very difficult to see in the Game View. That is because a light source is needed. To do this you will use a Directional Light, which casts light on everything based on which direction it is facing, similar to the sun. To add a Directional Light select GameObject > Create Other > Directional Light and after it appears rotate the light source with the rotation tool in the upper left hand corner. With the light source set up the Game view should look similar to Figure 25-13.

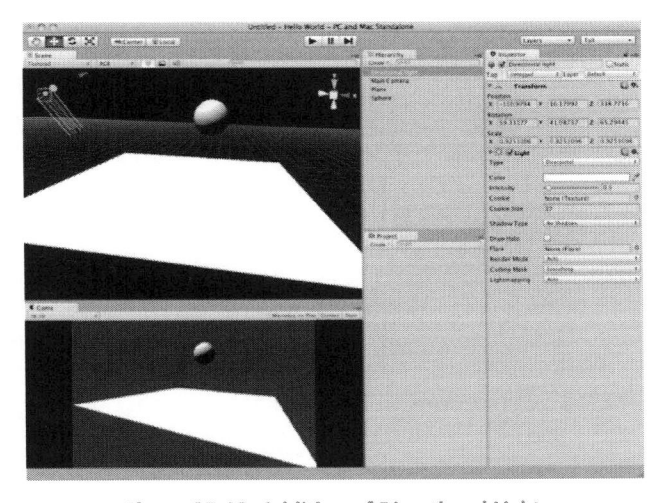

Figure 25-13: Addition of Directional Light

Now that you have your scene set up with all of your GameObjects you need to configure them so that the ball will bounce on the plane. If you were to run the game right now nothing would happen. This is because you have not set up the physics on the GameObjects. First select the sphere, then go to Component > Physics > Rigidbody in order to add the Rigidbody component to the sphere. This will give the GameObject a mass thus allowing it to interact with gravity. Press play and you will see the sphere will now fall and land on the plane.

Now that the sphere has a mass and reacts to gravity it needs to be set up so that it will bounce when it collides with the plane. This is done by adding a Physic Material to the Sphere, which is done by first selecting Assets > Create > Physic Material. After creating the Physic Material, rename it to Bounce Material so that way it is easily distinguishable. Rename the Physic Material by selecting it in the Project view, clicking on the name so that it becomes editable, and then type in the new name. Configure the Bounce Material to allow the sphere to bounce when it collides with another GameObject: select it and change the Bounciness property to 1 and the Bounce Combine property to Maximum. The Bounciness property tells you how bouncy a surface is, a value of 1 means that the object will bounce without losing energy. The Bounce Combine property tells how the bounciness of two objects that are colliding is combined. Setting these two values as such ensures maximum bounciness. After the changes are made the values for the Physic Material should look similar to Figure 25-14.

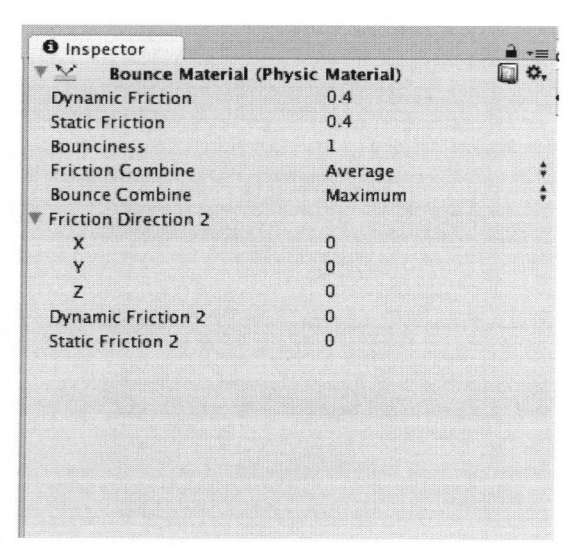

Figure 25-14: Bounce Material Properties

Once the Bounce Material is set up, add it to the Sphere GameObject. To do this, select the Sphere GameObject and in the Sphere Collider Component drag the Bounce Material to the material property. When finished the Sphere Components should look similar to Figure 25-15.

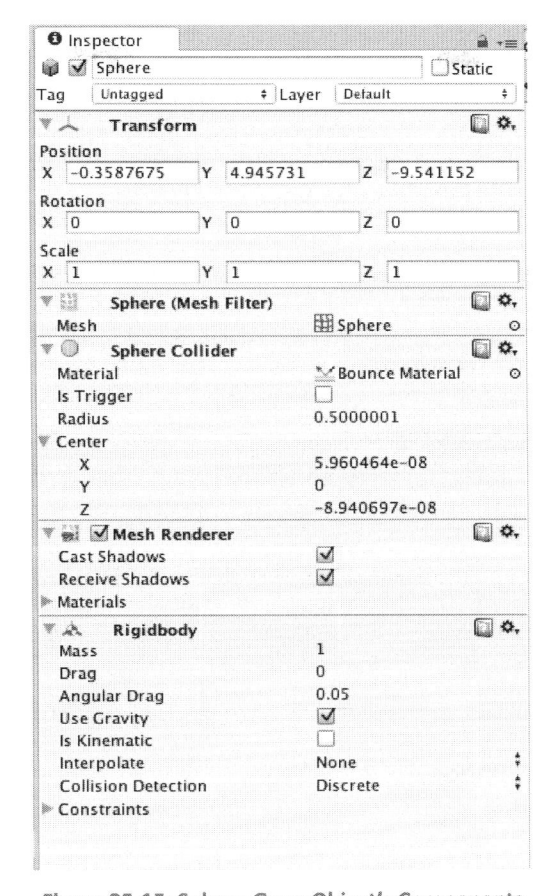

Figure 25-15: Sphere GameObject's Components

If you were to play the game right now you would see the sphere fall and continuously bounce off the plane.

25.2.3 Writing your first script

To add a little more complexity to the game you can make it so that when the sphere bounces it changes colors and after several bounces it will display "Hello World." In order to accomplish this you will need to do some scripting. Unity supports scripting with both JavaScript and C#, but for the purposes of this tutorial you will be using JavaScript.

First create a new script by going to Assets > Create > JavaScript and name it "HelloWorldScript." Select the script and click the "Open" button in the Inspector view. A new Monodevelop window should appear with an empty script file containing only a function named `Update()`.

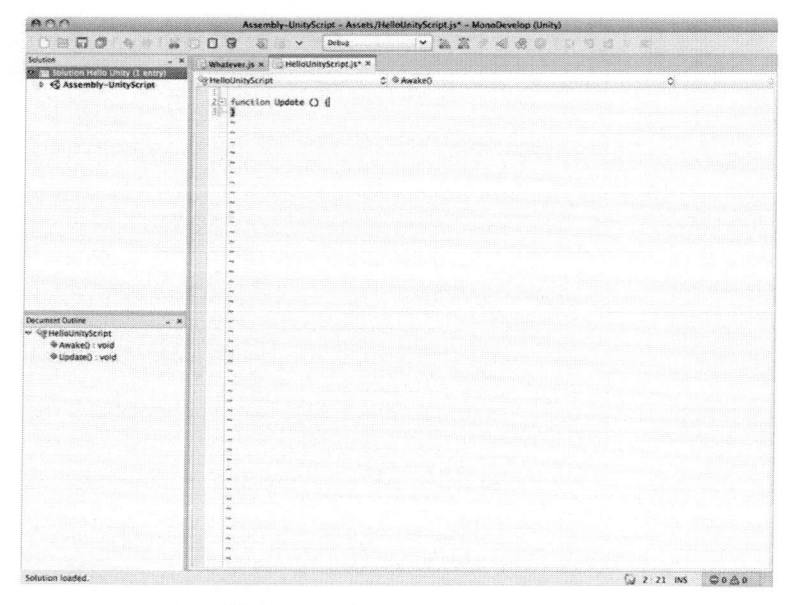

Figure 25-16: The script opens in Mono Develop

Since you want the sphere to change colors when it bounces, you will have to randomly choose a different color for the sphere each time it collides with the plane. However, before you can assign the sphere a random color, you have to have an array of colors that you can use to get a random color. Outside of the function create a variable named `colorArray` that is an array of `Color` objects.

```
var colorArray : Color[];
```

The `OnCollisionEnter` function is called when two colliders/rigidbodies begin touching. Inside this function is where the code to change the color of the sphere should go. Inside of the function place the following code.

```
function OnCollisionEnter()
{
    renderer.material.color =
    colorArray[Random.Range(0, colorArray.length)];
}
```

This will change the color of the sphere to a random color in the array each time a collision occurs. The script is set up, but there are still other things that need to be taken care of, so for now save the script.

First, the script needs to be attached to the Sphere GameObject, which is easily accomplished by dragging the script from the Project view to the Sphere in the Hierarchy view. Now, when you select the sphere, there should be a Script component that looks similar to Figure 25-17.

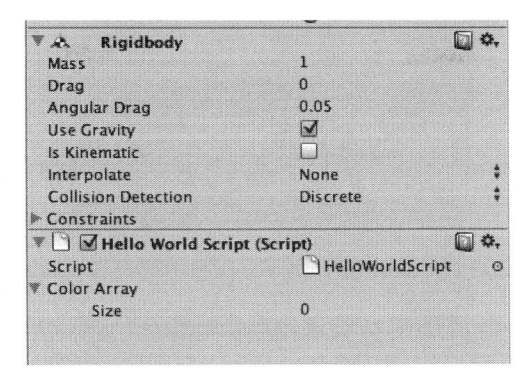

Figure 25-17: Script Component of Sphere

As you can see, the script has a Color Array with a Size of 0. Change the size to 6 and you will see 6 Elements appear where you can manually select the color for each one. Go through each element and change the color to a random color of your liking but different from the previous colors so that color changes will be easily distinguishable. When you have finished it should look something like Figure 25-18.

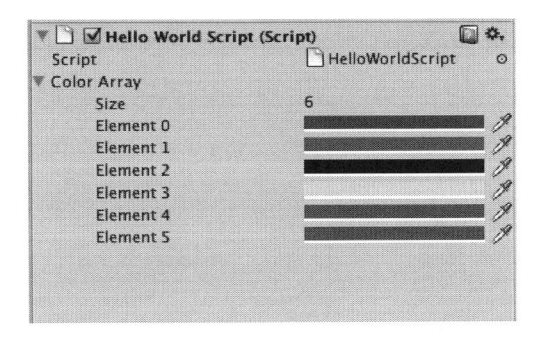

Figure 25-18: Color Array from Script

If you run the project, you will see that each time the sphere bounces off the plane it changes colors. Next you will add a GUI Text GameObject that can be used to display the "Hello World!" string. To do this, go to GameObject > Create Other > "GUI Text" and you will see the text in the Game view, but it must be moved by its x and y coordinates in the Scene view. Try and move the GUI Text towards the top and center of the view. Once positioned change the font size to 24 so it is easier to see.

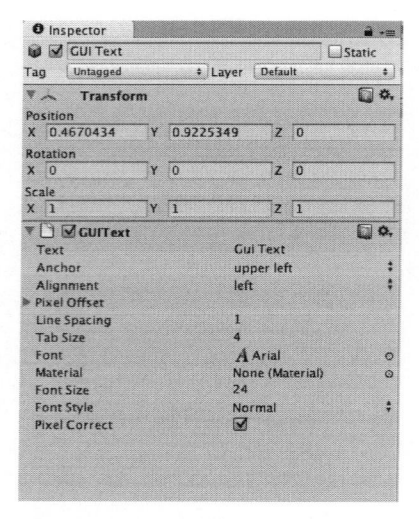

Figure 25-19: GUIText Properties

The script now needs to be modified so that "Hello World!" is displayed slowly while the sphere bounces. To do this, new variables need to be added to the script. Add the following variables:

```
var counter : int = 0;
var helloText : GUIText;
var helloWorld = "Hello World!";
```

The first variable is a counter that will keep track of the number of bounces that have occurred. The second variable will be a reference to the GUIText GameObject. The final variable is the desired text to be displayed.

Lastly the `OnCollisionEnter` method needs to be updated so that each time a collision occurs a new character from the "Hello World!" string will be displayed. To do this update the method to look as follows:

```
function OnCollisionEnter()
{
    renderer.material.color =
    colorArray[Random.Range(0, colorArray.length)];
    counter++;
    if(counter <= helloWorld.length)
    {
    helloText.text = helloWorld.Substring(0, counter);
    }
}
```

As you can see, each time that a collision occurs, the counter is incremented. After being incremented the counter is then checked to see if it is less than or equal to the size of the "Hello World!" string. This is done so that once the number of bounces has increased past the size of the string; the inner code will not be executed anymore. Inside of the conditional statement the text of the GUIText GameObject is set to a substring of the "Hello World!" string, increasing in size with each collision until the entire string is displayed.

Now that the script is complete the sphere needs to be configured with the GUIText GameObject that was placed earlier. To do this, select the sphere and in the Script Component of the Inspector view change the "Hello Text" element from "None (GUIText)" to "GUI Text (GUIText)" by dragging the GUI Text GameObject onto it or selecting it from the list by clicking the circle. This signifies that the GUIText GameObject is now connected to the variable `helloText` in the script. The Script Component should look like Figure 25-20.

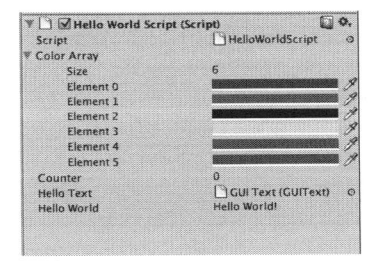

Figure 25-20: Variables for Script

If everything was done correctly, when the project is run, the sphere will change colors each time it collides with the plane it should change colors and a new character from the "Hello World!" string will appear. You should see results similar to what you see in Figure 25-21.

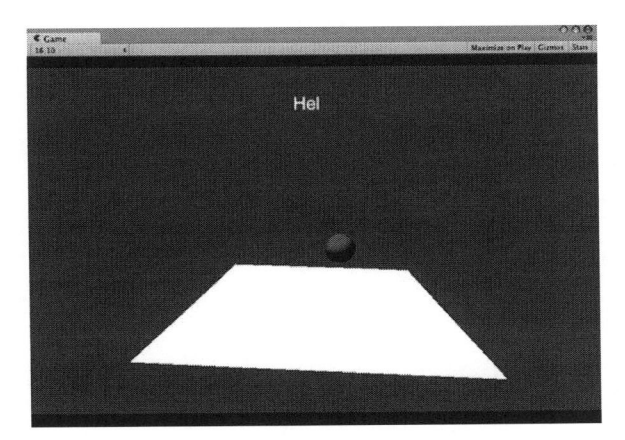

Figure 25-21: Working Hello World Game

After several bounces you should see the entire string and the ball should still be changing colors.

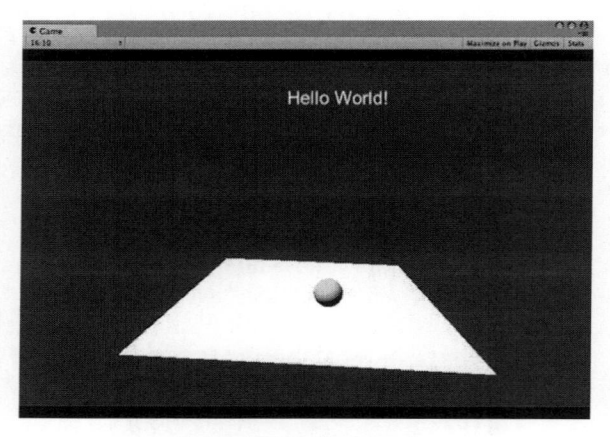

Figure 25-22: Final Result

Now that you have completed your first app in Unity, you should be familiar with the process of creating 3D games in this program.

25.3 Publishing Builds

After having created your app, you will want to build it for distribution. However, before you can build the app for certain platforms, you will need to purchase a Unity license. Once purchased the following steps will walk through the proper way to do build the project for iOS distribution.

1. First you will have to save the scene that you created in the previous section so that it can be included in the build. Go to File > "Save Scene as …" and save the Scene as "HelloUnity" in the Assets folder as seen in Figure 25-23.

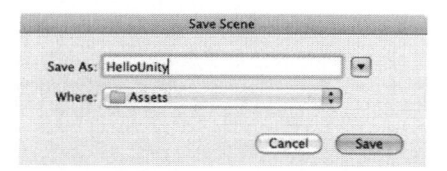

Figure 25-23: Saving a Scene

2. Once the scene is saved, you can include it in your build. Go to File > "Build Settings" and a window will pop up shown in Figure 25-24.

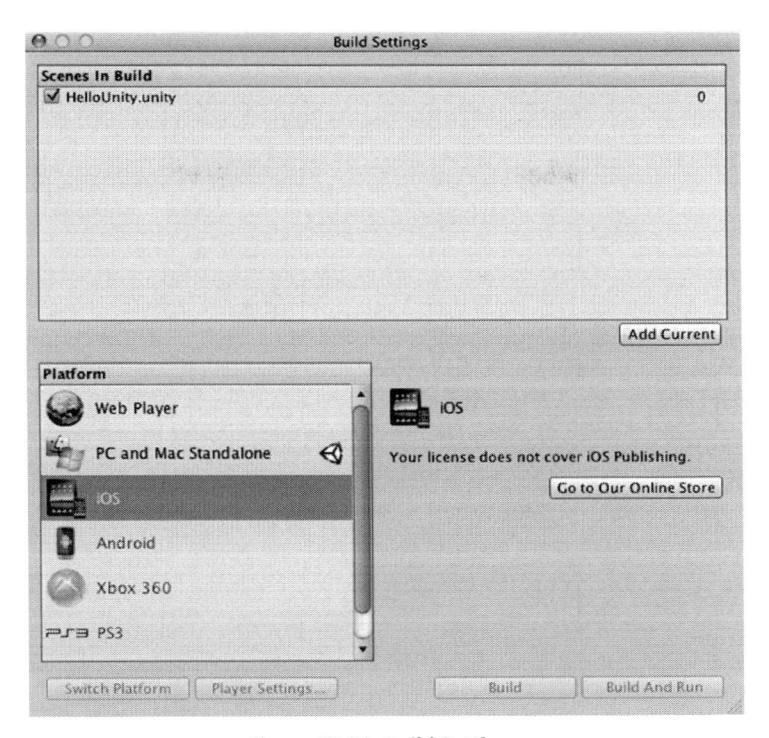

Figure 25-24: Build Settings

3. Select the platform and scenes that you wish to include in the build. In this case you would select iOS for the platform and the "HelloUnity" Scene that was just saved.

If you do not see the "HelloUnity" Scene then you can either drag and drop it into the list or use the "Add Current" button to add the current scene to the list - which in this case is the "HelloUnity" Scene.

As you can see in Figure 25-24 the "Build" button along with others are grayed out, this is because you are currently on the free version of Unity. If the iOS license was purchased then those options should be available to you.

4. Click the player settings button and the inspector will change as shown in Figure 25-25.

Figure 25-25: Player Settings

5. Change the product name (this is what shows up on the device) to whatever you desire.

6. Put your icon image in the "Default Icon" spot (if you do not put one here it will be a default Unity icon).

7. Under Per-Platform Settings click on the tiny iPhone icon as shown in Figure 25-25 in order to display Settings for iOS.

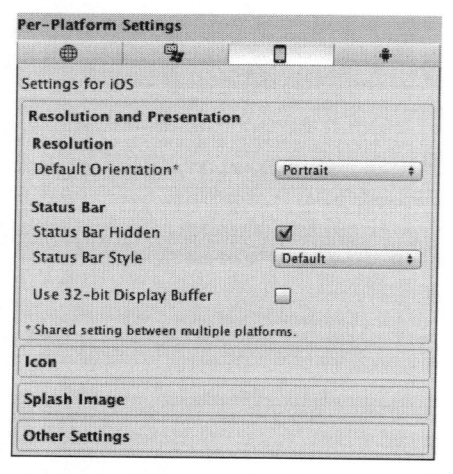

Figure 25-26: Resolution and Presentation iOS Settings

8. In the "Resolution and Presentation" section shown in Figure 25-26, select which orientation you want the game to be in.

9. In the "Other Settings" section shown in Figure 25-27, fill out the bundle identifier and select the target device (iPhone, iPad, or both).

Figure 25-27: Other Settings for iOS

10. Click the "Build" button to build the desired file (in this case an Xcode project for iOS).

11. Now that you have an Xcode project you can follow the instructions in Section 5 to get your app on the App Store.

Congratulations on completing your first game in Unity3D. Using the knowledge gained from this walkthrough and the documentation at www.unity3d.com/support/documentation/ you will be on your way to create and sell the game you always wanted to make.

Chapter 25: Unity 3D   273

ShiVa3D

ShiVa3D consists of a multi-platform 3D game engine, a 3D editor, an Authoring tool, and a MMO server. The ShiVa3D Editor is a WYSIWYG (What You See Is What You Get) editor that is used to develop games. The ShiVa3D engine is a multi-platform game engine, which will handle things like physics and shading. The ShiVa3D Authoring Tool compiles the projects created in the editor into executables for the supported platforms such as: Windows, Mac, Linux, iOS, Android, WebOS, Marmalade, BlackBerry QNX, and Nintendo Wii. The ShiVa3D Editor supports built-in Lua Scripting. C, C++, and Objective-C are also supported, though some steps must be taken for them to work with it. Further information on how to do this can be found on the developer's website at www.stonetrip.com.

One of the advantages to using ShiVa3D is that you can create 3D games for multiple platforms much more easily than if you built them using native code for each platform. It is compatible with many of the major 3D software such as: 3ds Max, Maya, XSI, Blender, and many more.

One of the disadvantages of ShiVa3D is that the documentation for it is not user friendly and there aren't as many tutorials available compared to other options. As a beginner it can be quite frustrating trying to get a simple game to work. But depending on your level of experience with 3D editors and the Lua scripting language, you may find ShiVa3D fairly simple to use.

26.1 Installation

Before you can start creating 3D games you need to download and install ShiVa3D Suite on a Windows based PC.

IMPORTANT NOTE FOR DEVELOPING FOR iOS: You will need an Intel based Mac with OS X 10.6 or above installed. You will also need Xcode and the iOS SDK installed. Instructions on how to do this can be found in Chapter 16.

This section will be a walkthrough of the process of downloading and installing ShiVa3D Suite. You can click on the download link that is on the ShiVa3D website or go directly to the URL http://www.stonetrip.com/download.html.

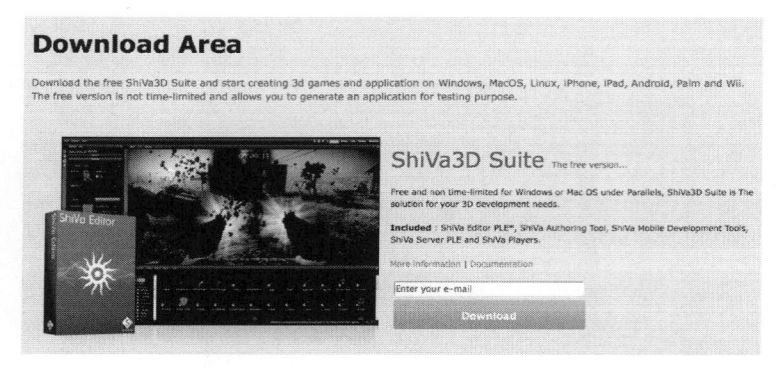

Figure 26-1: Download ShiVa3D

1. After downloading open the executable file to start the ShiVa3D Setup Wizard shown in Figure 26-2.

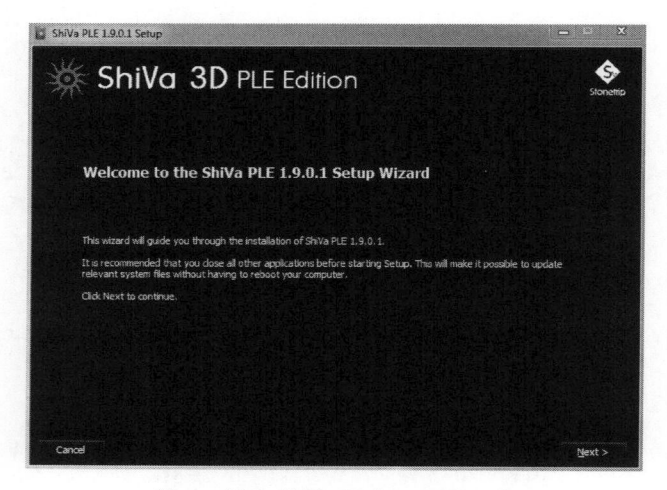

Figure 26-2: ShiVa 3D Setup Wizard

2. Step through the wizard, and install the ShiVa Editor, ShiVa Authoring Tool, and Device Development Tools, as well as the rest of the components.

3. When the installation is complete launch the ShiVa Editor PLE, shown in Figure 26-3.

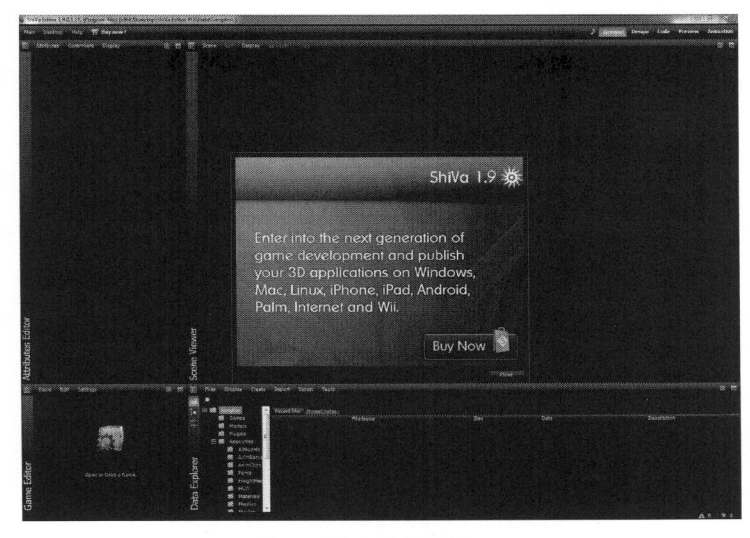

Figure 26-3: ShiVa Editor

You are now ready to create the "Hello ShiVa" example.

26.2 The "Hello ShiVa" Example

In this section you will create a simple ShiVa3D app, which will make a ball bounce.

1. Open the ShiVa Editor if it is not currently open from the previous section.
2. Go to Main > Projects to set up a new workspace. This will open the window shown in Figure 26-4.

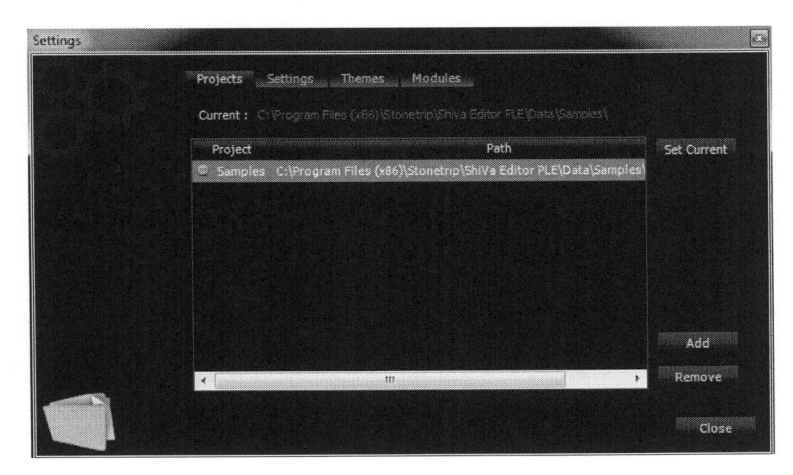

Figure 26-4: Project Workspace

Chapter 26: ShiVa3D   277

3. Click on the button labeled "Add" and then choose the directory where you wish your project to be located. Once chosen, this will automatically set this directory as the current project workspace.

4. Once the project workspace is determined, the next step is to create a Game. This is done by going to Game > Create in the Game Editor window depicted in Figure 26-5. When prompted set the name as "HelloShiVa."

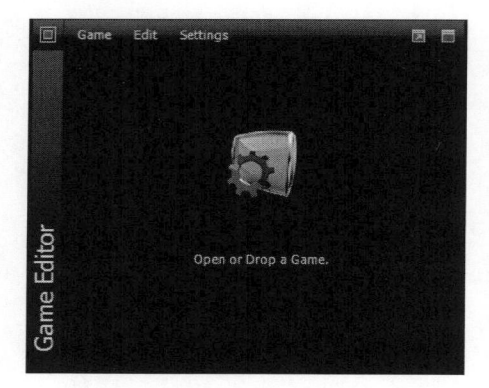

Figure 26-5: Game Editor

5. Once the empty Game is created, the next step is to create a scene for the Game. This is done in the Game Editor by going to Edit > Scene > Create... and then choosing a name for the scene. Choose the name "MainScene" and then click "OK." The new Scene should appear in the Game Editor as shown in Figure 26-6.

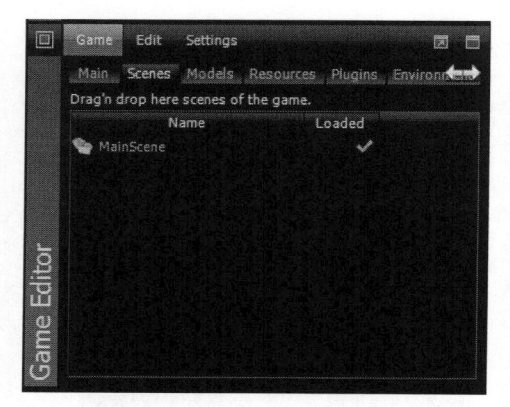

Figure 26-6: Scene added to Game

6. Now that you have a Scene, you need to create an AIModel. This is done in the Game Editor similar to creating a Scene. Go to Edit > UserMainAI > Create... and name it "MainAI."

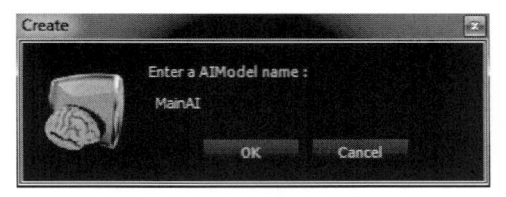

Figure 26-7: Create an AIModel

7. To access the "Code Desktop," click on the "Code" tab in the top right corner. Code Desktop contains the Script Editor, AIModel Editor, and Scene Viewer.

8. In the AIModel Editor click on "Add Handler..." twice and select onInit to create the for onInit handler.

9. Add the following code to the onInit handler.

```
application.setCurrentUserScene ( "MainScene" )
local Camera = application.getCurrentUserActiveCamera ( )
object.setTranslation (Camera, 20, 20, 20, object.kGlobalSpace)
```

This code sets the current scene to "MainScene" which was created earlier and also sets the Camera to the point (20, 20, 20).

Next, you will add some objects to the scene.

1. First add a sphere to the scene. To do this select Create > Model > Shape > Sphere in the Data Explorer. This will prompt you to name the model. Name it "Ball."

2. You will be prompted to create a Sphere mesh as shown in Figure 26-8. Set the number of segments to 32.

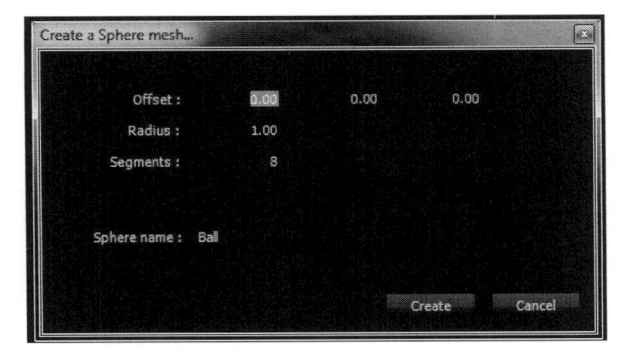

Figure 26-8: Create the Sphere mesh for the Ball object

3. Create a floor so that when the sphere falls it will land somewhere and be able to bounce. Go to Create > Model > Shape > Box in the Data Explorer. Enter "Floor" for the name of the model. Normally a Plane model would be used, but currently ShiVa is having

Chapter 26: ShiVa3D   279

some problems with the use of physics and planes.

4. You will be prompted to create a Box mesh. Set the Size values to 20, 1, and 20 as shown in Figure 26-9.

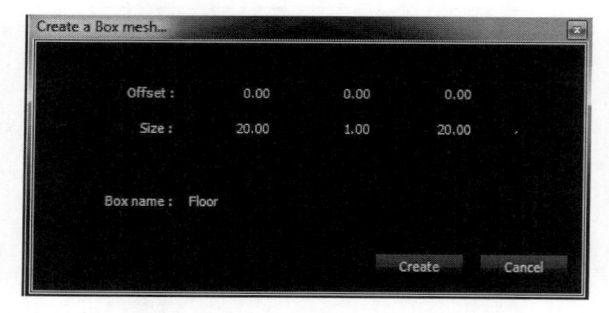

Figure 26-9: Create the Box Mesh for the Floor object

5. In the Data Explorer, click on the Models folder and drag the Ball and Floor models out into the Scene Viewer one at a time so that the ball is positioned above the floor as seen in Figure 26-10.

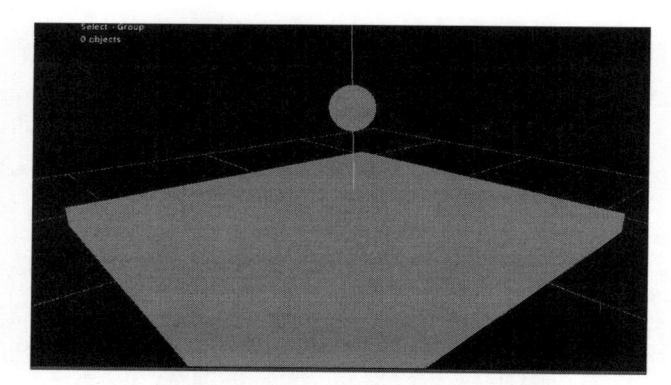

Figure 26-10: Scene Viewer with Ball and Floor objects

Now that the models are in the Scene, you can add some materials to them to change the way they look.

1. In the Data Explorer go to Create > Resource > Material and create a material named "Ball." Repeat this to create another material and name it "Floor."

2. Drag the Ball material onto the Ball model and the Floor material onto the Floor model. The previously pink models should now be a shade of grey as shown in Figure 26-11.

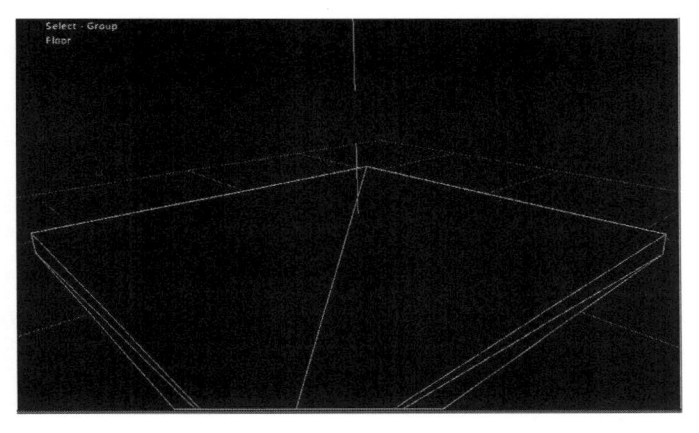

Figure 26-11: Ball and Floor objects with materials

3. Now that both models have materials they need to be edited to suit preference. To do this click on the Design dashboard in the upper right corner to access the Material Editor.

4. In the Material Editor go to Material > Open and choose the Ball Material when prompted. With the Ball Material selected in the editor, click on the section labeled "Lighting" and change the color slider under the Emissive section so that the color emitted from the material is a shade of red as shown in Figure 26-12.

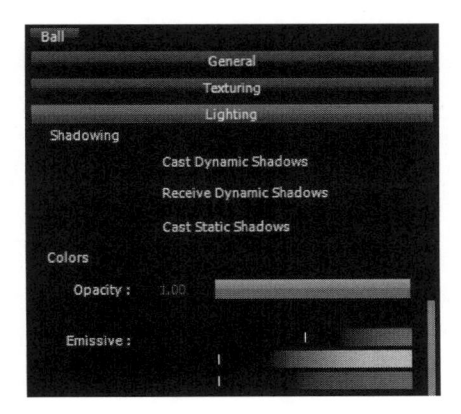

Figure 26-12: Modify the Emissive color in the Material Editor

5. Click on the section labeled "Texturing" and change the section titled "Effect Map 0" to the "DefaultChecker" pattern as shown in Figure 26-13.

Chapter 26: ShiVa3D

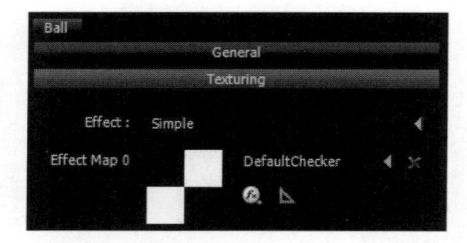

Figure 26-13: DefaultChecker texture for Ball material

6. The material for the Ball model is finished; the Floor material needs to be edited as well. In the Material Editor go to Material > Open and choose the Floor material.

7. In the Floor material change the Emissive lighting to a green color. Then change the texture to the "DefaultChecker" pattern as well. Afterwards you should have results similar to what is shown in Figure 26-14.

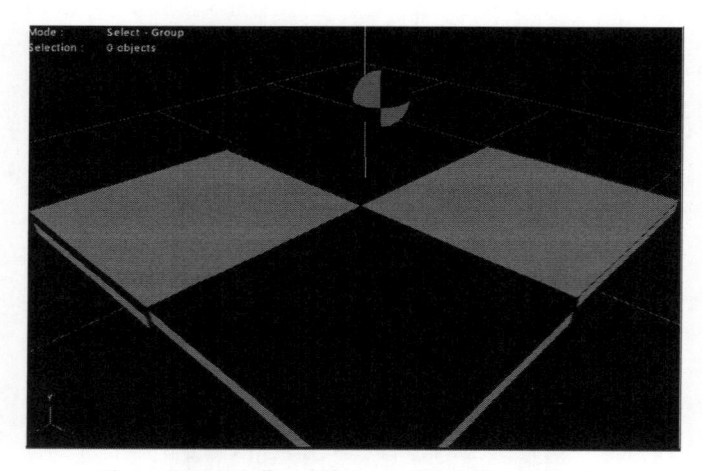

Figure 26-14: Ball and Floor models with materials

Now that the two models have materials on them, the Ball model needs a Dynamics Controller so that it will have mass and interact with the physics engine.

In the Scene Viewer select the Ball model and in the Attributes Editor go to Controllers > Dynamics > Create Controller and this will create a Dynamics Controller for the Ball model.

If you were to play the game, you would now see the ball drop and fall through the Floor model created earlier. This is because the Floor model does not currently have a Collider set and the Dynamics Controller does not have Collisions enabled by default.

1. To set the Floor as a Collider select it first, then go to the Attributes Editor and select Attributes > Collider > Set as Collider.

2. In the Scene Editor select the ball again and edit its Dynamics Controller so that the value for Bounce is 1 as shown in Figure 26-15. Also make sure to check the "Enable Collisions" flag.

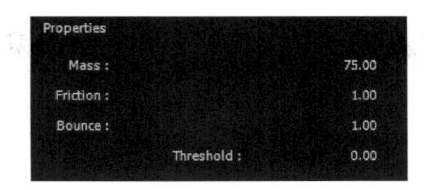

Figure 26-15: Set Bounce value to 1

When you play the game, by either pressing F9 or clicking on the play button, you should see the ball drop and bounce off the floor as depicted in Figure 26-16.

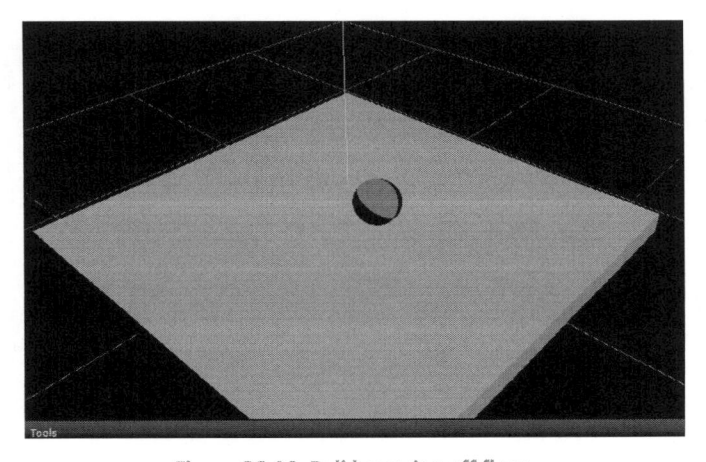

Figure 26-16: Ball bouncing off floor

Congratulations you have built your first game with ShiVa3D!

26.3   Building for Distribution

Now that you have a working game you need to build it for distribution. This is accomplished through the use of the ShiVa Authoring Tool. However, when building for iOS there are some extra steps required before you can make a build to put on the App Store. You will need a Mac with OS X Snow Leopard (10.6) or above. In addition, you will need to install Xcode and the iOS SDK, instructions on how to do this can be found in Chapter 16. Also, the ShiVa Authoring Tool for Mac OS X needs to be installed. Once these prerequisites are met, you can follow the steps below to build your game.

1. First, you need to export your game, which will convert all your scripts into C++ files. Before this can be done the game needs to be compiled to make sure there are no errors or warnings with the scripts. Going to Game > Compile in the Game Editor will compile all the scripts in the game.

2. Once all the errors and warnings are resolved, the exporting process can begin. Go to Game > Export and an export window will appear. In the window choose the folder you want to export to and make sure "Runtime Package (.stk and .cpp source)" is selected as seen in Figure 26-17.

3. Click "Export" to begin the export process. This can take a while depending on the size of the project and number of scripts.

4. Take the created files and transfer them to your Mac OSX computer where the ShiVa Authoring Tool is installed.

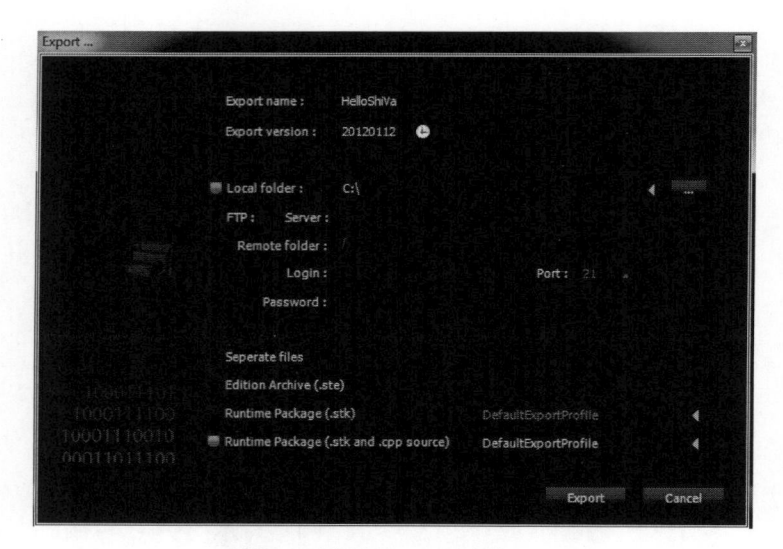

Figure 26-17: Export window

5. Open the ShiVa3D Authoring Tool in Mac OSX and choose the iPhone icon as seen in Figure 26-18.

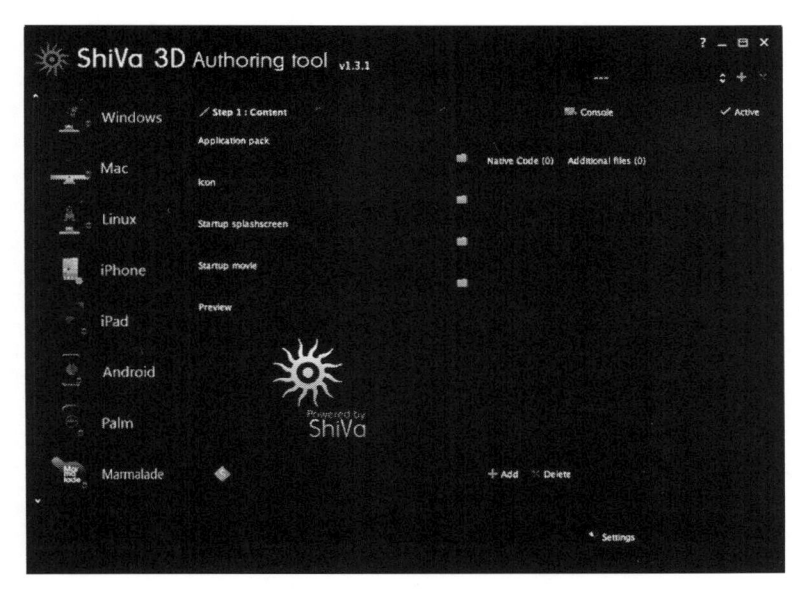

Figure 26-18: ShiVa3D Authoring Tool iPhone option

6. In the section titled "Application pack," load the .stk file that was created earlier during the export process.
7. Add the native code files one by one by clicking on the "Add" button on the right-hand side underneath the Native Code section. Once all the code is in click "Next" to choose whether to build an .app file for upload on the App Store or put on an iPhone, or build an Xcode project.
8. To build for the App Store select Application under "Authoring type" if not already selected, as shown in Figure 26-19.

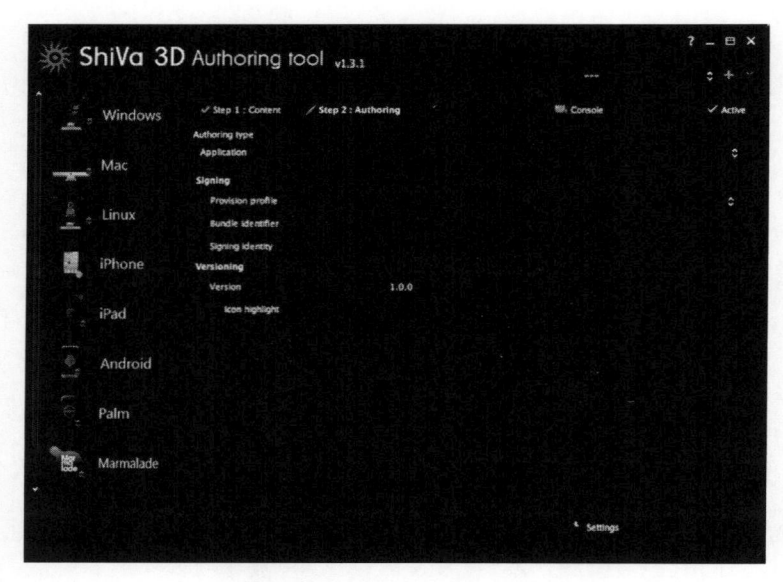

Figure 26-19: Authoring Tool

9. Next choose the provisioning profile, bundle identifier, and signing identity and click the "Build" tab to choose the final build settings.

For help with setting up provisioning profiles and certificates please refer to Chapter 9. Now that you have a built .app file you can upload it to the App Store by following the directions in Chapter 40: Submitting Apps Using Application Loader.

PhoneGap

PhoneGap offers an easy way for those familiar with programming in HTML5, CSS, and JavaScript to program for iOS devices, as well as several other mobile platforms such as Android, WebOS, Windows Phone 7, Symbian, bada, and Blackberry. With PhoneGap, developers can take the source code from web apps that they have already written and convert them into stand-alone iOS apps.

Because PhoneGap integrates closely with Xcode, it supports the use of the iPhone's advanced functionalities such as the compass, gyroscope, and GPS. However, it's still not wise to use it for large, graphics-heavy games.

27.1 Installation

Before installing PhoneGap, be certain you have Xcode installed. For help on how to install Xcode, please refer to Chapter 16.2.

The latest release of PhoneGap is free and available on their website at http://www.phonegap.com/. The example shown here is version 2.0.0.

1. Click on the download image shown in Figure 27-1.

Figure 27-1: Download PhoneGap

2. After downloading unzip the file and open the resulting folder.
3. In that folder go to lib > ios and run the installer it will have a .dmg file extension.
4. Next, click on the installer, shown in Figure 27-2, to install PhoneGap's template into Xcode.

Figure 27-2: The PhoneGap Installer package

27.2 Hello World!

To better understand how PhoneGap works, you will slightly modify the default PhoneGap app, which is created whenever you start a new PhoneGap-based app. The app simply displays static text, over which pops an alert box. You will modify the static text to read "Hello, World!" upon opening the app.

1. Open Xcode, click "Create new Xcode project" in the next window, select "Cordova-Based Application," from the available templates, then click "Next."

Figure 27-3: The PhoneGap template icon as it appears in Xcode

2. In the box next to "Product Name" type "HelloWorld," then click "Next."

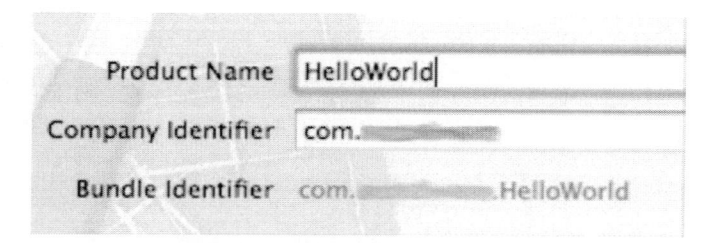

Figure 27-4: The app name and company identifier make up the bundle identifier

3. You will be asked where you want to create your project. Select a location; click "Create" and Xcode will create a folder at that location with the title "HelloWorld."

4. When Xcode displays your new project, make sure the Scheme is set to "iPhone X.X Simulator," where "X.X" is the version of iOS on which you want to test your program.

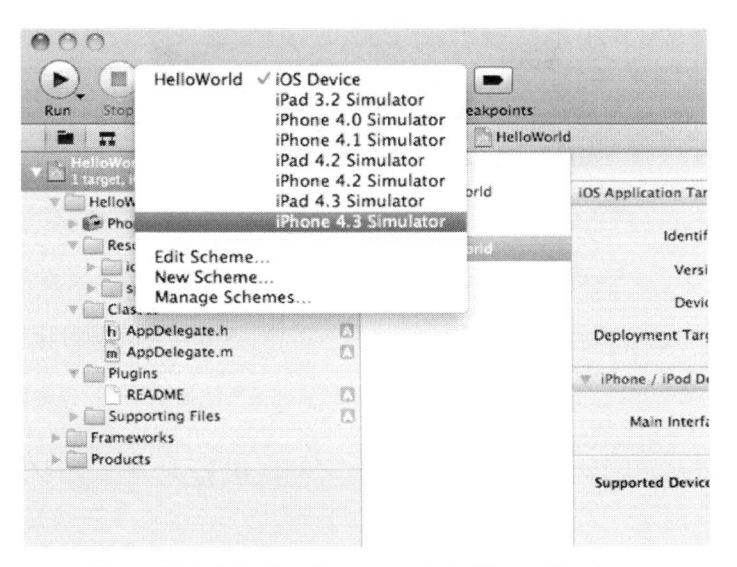

Figure 27-5: Selecting the appropriate iPhone Simulator

5. Click "Run" in the top left corner. This will put the basic JavaScript files and directory structure for your app into the "HelloWorld" folder you created previously. PhoneGap will search for an .html file to initialize the building of the app. Since you have not created one yet, you should get a message saying that "www/index.html" was not found. PhoneGap will now create an "index.html" file for you to use. Close the iOS Simulator for now.

6. Right-click on the project in the left part of the window and select "Show in Finder." This will open up the directory that Xcode created in step 5, where there should be a new folder called "www."

7. With both the Finder window you just opened, and Xcode visible, drag the "www" folder from the Finder window ***directly into*** Xcode, dropping it on top of the project file in the left side of the Xcode window.

Chapter 27: PhoneGap   289

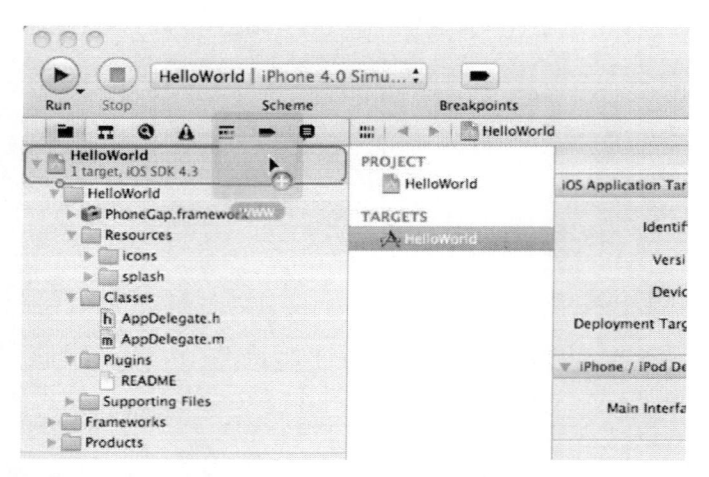

Figure 27-6: Drag and drop the "www" folder from the Finder window directly onto the HelloWorld project

8. In the window that opens, make sure "Create folder references for any added folders" is selected, and that there is a check in the box next to "HelloWorld." Next, click "Finish" to add the folder to the project.

9. Inside Xcode, click the arrow next to the "www" folder that you just added. Here you will find the HTML and JavaScript files that were just created to help begin building the app. Click the "index.html" file and view its contents in Xcode's main window. Inside the `<body></body>` tag, type `<h1>Hello, World!</h1>`, as in the example below, then save the file. This is a very basic HTML file that, upon launching, runs the `onBodyLoad()` function and displays some header text.

```
    </script>
  </head>
  <body onload="onBodyLoad()">
        <h1>Hello, World!</h1>
  </body>
</html>
```

290 Chapter 27: PhoneGap

Figure 27-7: The basic app running on iOS Simulator

10. Go back to Xcode and click "Run" again, you should see the simple app that you just created, along with an alert box to let you know PhoneGap is running properly.

11. If you look a little further up from the </script> tag in the index.html file, you can see where the onBodyLoad() function is told to display the alert box. To prevent this box from popping up, simply comment out that line of code, as seen below:

```
function onBodyLoad()
...
{
  // do your thing!
  //navigator.notification.alert("PhoneGap is working")
}
```

If you save and run the app again, you will notice the alert box does not pop up, but the "Hello, World!" text still shows up because you commented out the alert box function, but not the <h1></h1> tags.

To test the app on a device, simply connect a compatible device to your computer, change the Scheme in Xcode to the device name, and click "Run" again.

Chapter 27: PhoneGap   291

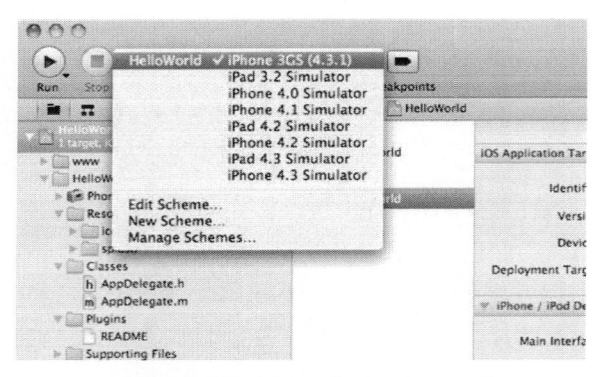

Figure 27-8: Selecting a device for testing

27.3 Building for Distribution

This was just a quick exercise to familiarize you with the general process of developing an app with PhoneGap. When building your actual apps, simply build everything as you would a normal web app, keeping all your files in the "www" folder. After you have built and tested your app, build the app for distribution and upload to the App Store using Apple's Application Loader or directly through Xcode 4, as described in Chapters 39 and 40.

27.3.1 PhoneGap Build Service

PhoneGap recently introduced a new service called PhoneGap Build that offers you the ability to not use Xcode at all. To use it, you will still need to have your certificate and provisioning profile from Apple. But instead of starting the project in Xcode, you will build your app using JavaScript, HTML, and CSS with a text editor or web design software. Upon completion, you will compress it, and then upload it to PhoneGap through the PhoneGap Build portal at https://build.phonegap.com/. PhoneGap will compile your app and make it instantly available for a variety of mobile operating systems.

To test on a device or distribute your app on The App Store, you will have to obtain a distribution certificate and provisioning profile from Apple. PhoneGap will guide you through uploading these, as well as the compressed file, that contains your app's files, to their website. After completing these steps, PhoneGap Build will compile your project and send you back a project file that will be ready for testing or distribution without ever having to open Xcode.

28

MonoTouch

The MonoTouch SDK is based on the Mono framework, which is a software platform that allows .NET applications to be run on multiple platforms. Using MonoTouch, you can create iOS apps using C# and .NET. With the MonoTouch SDK you can use the iPhone APIs as well as reuse code and libraries that have been built for .NET. When developing with MonoTouch, MonoDevelop is the IDE that is used. MonoDevelop supports Linux, Windows, and Mac OSX platforms and multiple languages such as, C#, Visual Basic.Net, C/C++, and Vala; these other languages do not apply when creating iOS apps with the MonoTouch SDK

One of the main advantages to using MonoTouch is that developers who are already experienced with C# and .NET do not need to spend time learning how to use Objective-C. However, the developer will still need to spend time learning the C# APIs that relate to Cocoa Touch. Another advantage is the fact that the Mono Runtime comes with a garbage collector meaning that memory is managed automatically. Another one of the major advantages is that .NET libraries are available in MonoTouch. These libraries are more extensive and have more support than their Objective-C counterparts for things like XML, Web Services, and String and Date manipulation.

One of the disadvantages of MonoTouch is that currently you need to use Interface Builder along with MonoDevelop to create an app. This is unfortunate because the latest version of Xcode has Interface Builder built in and this version is incompatible with the MonoDevelop IDE. You are still able to download Xcode 3, which does not have Interface Builder integrated into it. However, it is not known how long this version will be supported. Another disadvantage is that an app created with MonoTouch is going to be much larger than one created with Objective-C and native APIs, because it will have the Mono Framework included in it which adds around 5MB to the file size.

Another thing of note is that there is an SDK called Mono for Android. This is similar to MonoTouch however it is used to develop apps for the Android operating system.

IMPORTANT NOTE: While the walkthrough in this chapter uses Xcode 3 for the latest version of MonoTouch now offers support for Xcode 4 so the workflow for integration between MonoDevelop and Xcode are somewhat different. If you are interested in using Xcode 4 instead of or after learning how to use Xcode 3 as shown in this chapter, there is a great guide on Xamarin's website at the following URL:
http://docs.xamarin.com/ios/tutorials/transitioning_from_xcode_3_to_xcode_4

28.1 Installation

Before you can create apps using MonoTouch you first have to download and install four separate components in a specific order.

1. Download and install Apple's iOS SDK and Xcode 3. This can be obtained from Apple's iOS Dev Center.
2. Download and install the latest version of Mono for OS X. This can be obtained from http://mono-project.com/downloads. (Version 2.6.7 is used here as an example.)

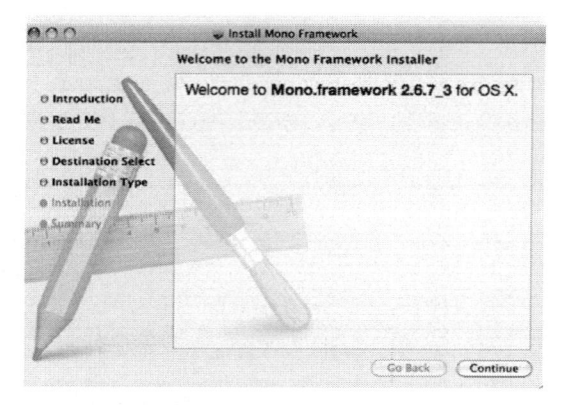

Figure 28-1: Mono Framework Installer

3. The next item on the list is the MonoDevelop IDE for OS X. This can be obtained from http://monodevelop.com/download.

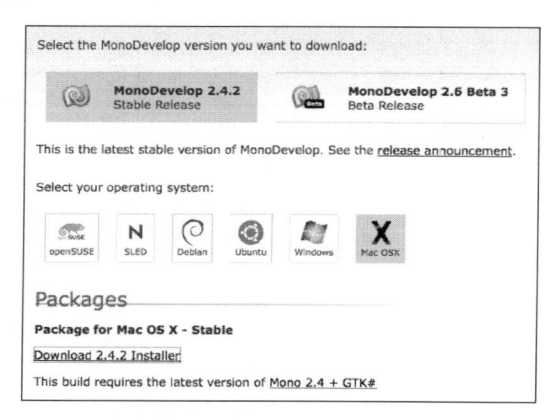

Figure 28-2: MonoDevelop download page

4. Once downloaded, run the installer and drag the icon labeled MonoDevelop to the Applications directory as indicated in Figure 28-3.

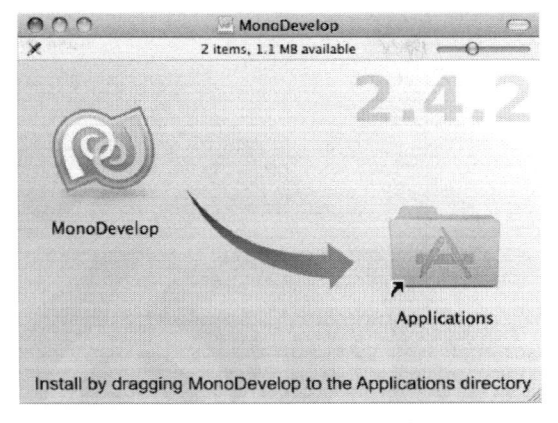

Figure 28-3: MonoDevelop Installation

5. Finally, download and install the MonoTouch SDK. This can be obtained at http://xamarin.com/trial for a trial version or click the link at monotouch.net to buy the fully licensed version.

Figure 28-4: MonoTouch Installation Complete

6. After all the components are installed, open the MonoDevelop IDE and, if necessary, perform the required updates by following the onscreen prompt.

28.2 Example

In this section you will be creating a simple "Hello World" style app.

1. Open the MonoDevelop IDE and choose "Start a New Solution." Then choose "iPhone Window-based Project" and name it "HelloMonoTouch" after choosing the location where you want to save it.

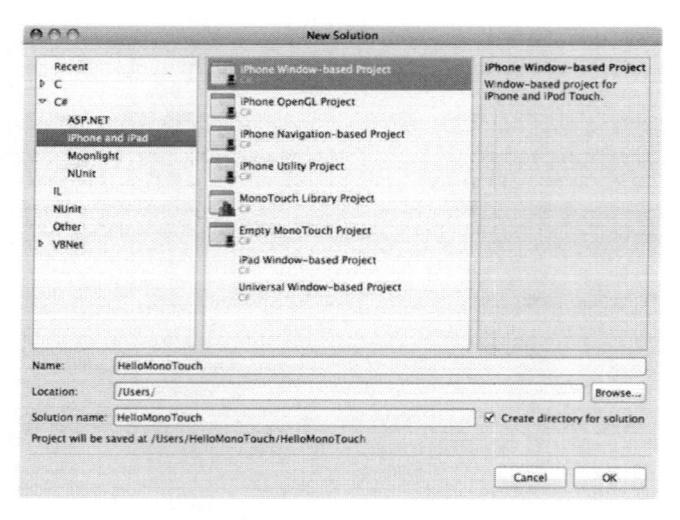

Figure 28-5: Creating new project.

Now that you have an empty project you can put some code into the appropriate file so that it will display the text "Hello World!" within the iPhone simulator.

2. In the Main.cs (Main Class) file, change the FinishedLaunching method to look as follows:

```
public override bool FinishedLaunching (UIApplication app, NSDictionary options)
{
    UIApplication.SharedApplication.SetStatusBarHidden(true, false);
    UILabel label = new UILabel();
    label.Frame = new System.Drawing.RectangleF(110, 215, 100, 50);
    label.Text = "Hello World!";
    label.TextAlignment = UITextAlignment.Center;
    window.AddSubview(label);
    window.MakeKeyAndVisible ();
    return true;
}
```

First, the status bar is hidden so that you have more screen real estate to work with. Next a UILabel is created and then set up with a frame that centers it in the middle of the screen. The text of the label is changed to display "Hello World!" and is aligned to the center of the label. Finally the label is added to the window with the AddSubview method. If the app were to be run, it would look similar to Figure 28-6.

Figure 28-6: iPhone Simulator

3. To add some more functionality to the app you will add two buttons that when pressed will change the text of the label. Before doing this, delete the code that you just added so that you can use Interface Builder to set up the label instead of using code to do so.

4. In order to add a button to the XIB file you need to use Interface Builder to open the MainWindow.xib file. It is important that this file isn't opened in Xcode 4 since Interface Builder is integrated into it, and it does not work with MonoDevelop. Open Interface Builder, then select "Open Existing Document" when it prompts you to select a template, navigate to your project workspace, then select and open the MainWindow.xib file.

5. Once the MainWindow.xib file is open, double-click the Window object to display the window.

Figure 28-7: Interface Builder window

Chapter 28: MonoTouch   297

6. Once it is open, drag a button from the library and place it in the window near the top. Drag another button from the library, this time placing it in the window near the bottom. Finally, drag a label from the library into the window and place it towards the center. Afterwards the window should look similar to Figure 28-8.

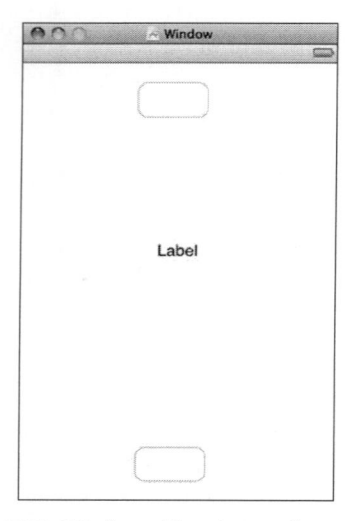

Figure 28-8: Window object in Interface Builder

7. Now that there are two buttons and a label in the window, an outlet needs to be set up for each of them so that they can be manipulated through code. This is done through the Library. Select the Classes tab and select the AppDelegate object. Click the drop down selector that says "Inheritance," at the bottom of the window and change it to "Outlets." The library window should now look similar to Figure 28-9.

8. Now you will add an outlet named "button1" with a type of UIButton. To do this click on the "+" sign and then type "button1" for the name, then click on the type where it says "id" and change it to UIButton. Create a second outlet named "button2" just like you did with the first outlet. Finally create an outlet for the label named "label" with a type of UILabel. Afterwards the library window should look like Figure 28-10.

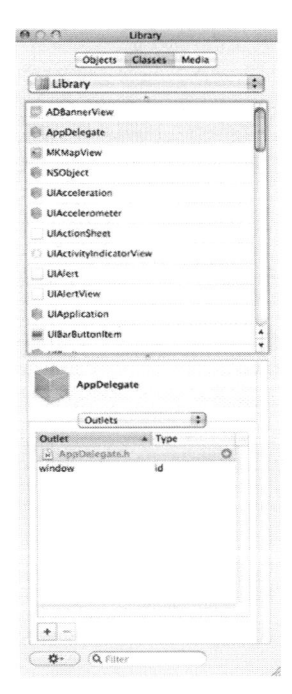

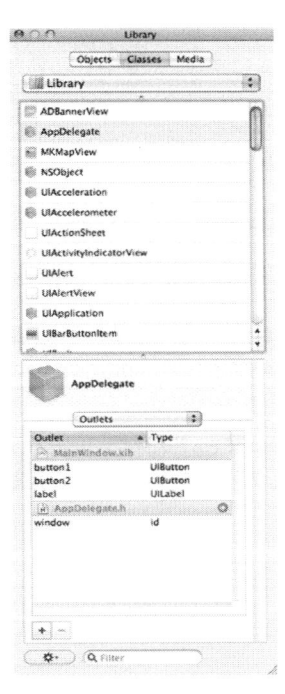

Figure 28-9: No Outlets   Figure 28-10: With Outlets

9. Now that there are three outlets, you will connect them to the two buttons and the label, where button1 corresponds to the top button and button2 to the bottom. The connections are created by right clicking on the App Delegate Object in the Interface Builder window, which will bring up the available connections as seen in Figure 28-11 which you can then click and drag onto the proper buttons.

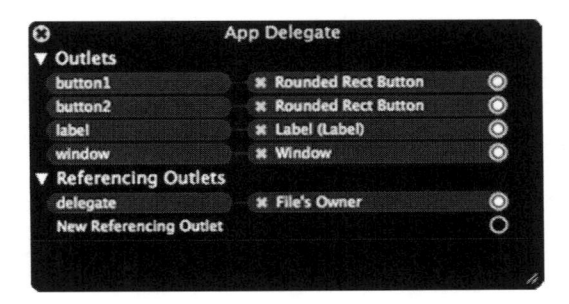

Figure 28-11: Connections for App Delegate

10. After the outlets are connected for the three objects, you will be able to manipulate them in code. However, in order for a button to trigger an event, an action needs to be configured for it. This is done in one of two ways. The first way it can be done through Interface Builder similar to how the outlets were set up. The second option is creating and assigning a method to be called every time the event is triggered through an event handler. Both of these options are covered in the following sections.

Chapter 28: MonoTouch   299

11. First, wire up the action through interface builder for your "button1" object. To do this, go to the library window and select the App Delegate object like before. However, this time select "Actions" instead of the "Outlets" label from the drop down menu. Click the "+" icon to add a new action and name it "buttonOnePressed." Now right-click on the App Delegate object and connect the new action to the top button and choose "Touch Up Inside" for the event as seen in Figures 28-12 and 28-13.

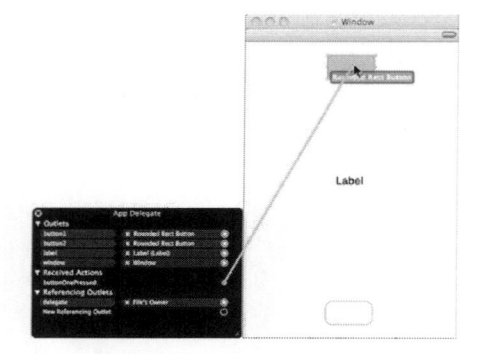

Figure 28-12: Attaching the action to Button1

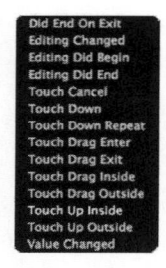

Figure 28-13: Select Event

12. Back inside the Main.cs file, implement the buttonOnePressed method in the UIApplicationDelegate class as follows:

```
partial void buttonOnePressed (UIButton sender)
{
        this.label.Text = "Hello World!";
}
```

This makes it so that when button1 is pressed, "Hello World!" is displayed in the text of the label.

13. Now that the first button is wired up, you need to wire up the second button's event handler inside of the FinishedLaunching method, which should look similar to the following:

```
this.button2.TouchUpInside += buttonTwoPressed;
```

Typing this should automatically create the method buttonTwoPressed just beneath the FinishedLaunching method.

14. Inside of the buttonTwoPressed method, add code to change the text of the label to "Hello Again!" The buttonTwoPressed method should look similar to below.

    ```
    void buttonTwoPressed (object sender, EventArgs e)
    {
            this.label.Text = "Hello Again!";
    }
    ```

15. Finally you need to set some text for the buttons so that they are easily distinguishable. This is accomplished by adding the following lines of code to the FinishedLaunching method.

    ```
    this.button1.SetTitle("Button 1", UIControlState.Normal);
    this.button2.SetTitle("Button 2", UIControlState.Normal);
    ```

This sets the title of the buttons to "Button 1" and "Button 2" respectively for the normal state of the control.

16. Run the program and test out the two buttons individually and you should see the label change as shown in Figure 28-14.

Figure 28-14: Final Output

28.3 Building for Distribution

Now that the app is completed it can be built for distribution through either Ad-Hoc distribution or the App Store. Before you can do this a few steps need to be taken to set up a configuration for a distribution build.

1. First, a new configuration needs to be added to MonoDevelop for the distribution build. To do this go to Project->Solution Options and a window will appear. Select "Configurations" in the window and you should see a list of the current configurations like in Figure 28-15.

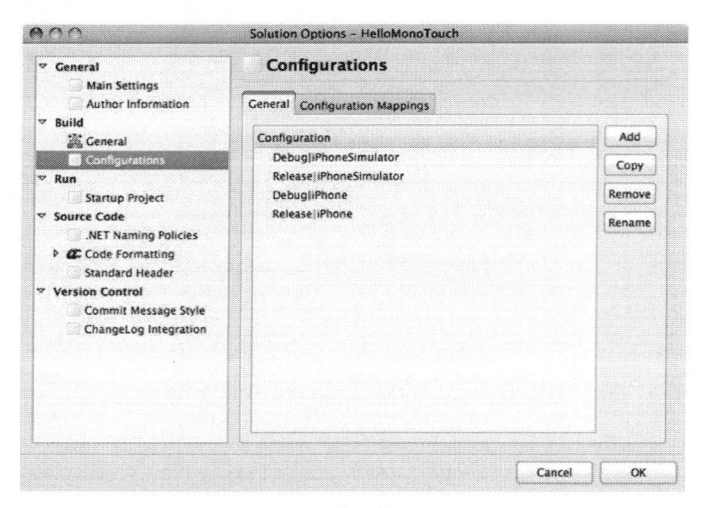

Figure 28-15: Configurations

2. Click on the "Add" button and in the new window create a name for the configuration and make sure that the platform is set to "iPhone" like Figure 28-16 and then click "OK."

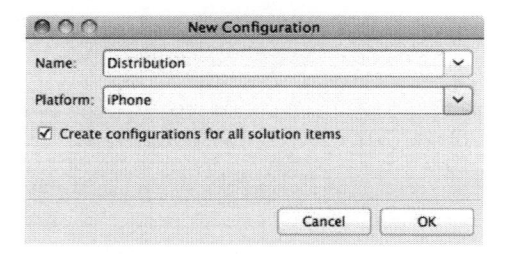

Figure 28-16: Set up a new Configuration

3. Next open the Project Options window by going to Project->ProjectName Options, where ProjectName is HelloMonoTouch in this case. Next select the "iPhone Bundle Signing" option at the left of the window as in Figure 28-17. Using the drop down menus, select the configuration created in the previous step. Next select the identity for signing the app and the provisioning profile which are both obtained from Apple's iOS Provisioning Portal as discussed in Chapters 9 and 38.

Figure 28-17: iPhone Bundle Signing

4. After you are done with the "iPhone Bundle Signing" options click on the "iPhone Application" option. In here set up the desired display name and the version number, and make sure that the bundle identifier is correct. If your app is going on the App Store then make sure you have a 57x57 png icon (114x114 for retina display) that can be added easily by clicking the "Icons" tab. When done with this click the "OK" button.

Figure 28-18: iPhone Application Options

5. Now that the options are set up, switch the active build configuration to the new profile. Build the project and you can find the bundle in the project folder by navigating to the bin/Distribution/iPhone directory. For instructions on how to submit your bundle to Apple, please refer to Section V.

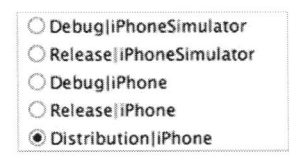

Figure 28-19: Select active build configuration

Chapter 28: MonoTouch   303

This concludes the chapter on MonoTouch. By now you should have a good idea about how the MonoTouch SDK works and can start developing your own unique apps.

Marmalade

Marmalade, previously named Airplay SDK, allows you create and build apps that run across several platforms, such as iOS, Android, Symbian, bada, BlackBerry PlayBook OS, webOS, Windows desktop (7, XP SP3), Mac OS X, and LG Smart TVs. Marmalade's code is written in C/C++, and its apps can be implemented using several IDEs like Microsoft Visual C++, Microsoft Visual Studio, and Xcode.

One of the major advantages of Marmalade is that unlike most other cross-platform development tools, you can achieve similar performance that you would expect to achieve from an app written in each platform's native language.

Another major advantage of Marmalade is that it helps you port iOS apps to Android. So instead of having to port your Objective-C code to Java you can just port it to C++, which is much easier. Once you have the code ported to C++ you can use that single codebase and deploy it to Android and iOS very easily. Future updates only need to be written once and can be deployed simultaneously.

The evaluation version of Marmalade, which is the version that will be used in this chapter, is free to download so that you can learn how to use it. However, if you want to publish and sell your app you will have to buy one of the paid versions. The Community version is $149 per seat/annum and only supports iOS and Android platforms. It comes with an In-app Marmalade Splash Screen that is displayed when the app loads. In order to get rid of this Splash Screen and get support for other platforms you will need to purchase the Indie or Professional versions. The Indie version is $499 per seat/annum and the Professional version requires you to contact them for purchase.

29.1 Installation and licensing for using Marmalade

The home page of Marmalade can be accessed at http://www.madewithmarmalade.com. You will find all sorts of detailed documentation and tools that can be used to create apps with Marmalade. In this chapter you will get a brief overview of how to install and use Marmalade for Windows and Mac. It will also guide you through the development of a "Hello World!" example that you can run using two IDEs (Visual Studio 2010 and Xcode).

29.1.1 Installation for Mac OS X
The following steps will guide you through the installation of Marmalade on Mac OS X:

1. Register on Marmalade's website by clicking on "Join" at the top right and fill in the necessary data.

Figure 29-1: Register with Marmalade

2. Download the latest version of Marmalade for your operating system. This can be done by going to the Downloads page, which is located at www.madewithmarmalade.com/downloads. Clicking the download button on Marmalade's Home page will also lead you there. Once there, find the link for the latest version for Mac, which will look similar to Figure 29-2. This link will prompt you to download a .pkg file.

Figure 29-2: Download Mac Release

3. To use Marmalade, you will need to either buy a license or get a free 90-day evaluation license. You can choose between these options on the page that appears when clicking on the "Buy" button on the Marmalade website.

4. Install Marmalade on your computer by double clicking on the .pkg file that you downloaded in step 2.

5. Activate the license when prompted during the installation process. You will need to enter the email address and password that was used to register for Marmalade and select your "Machine ID" (this should be selected by default). It should look similar to Figure 29-3.

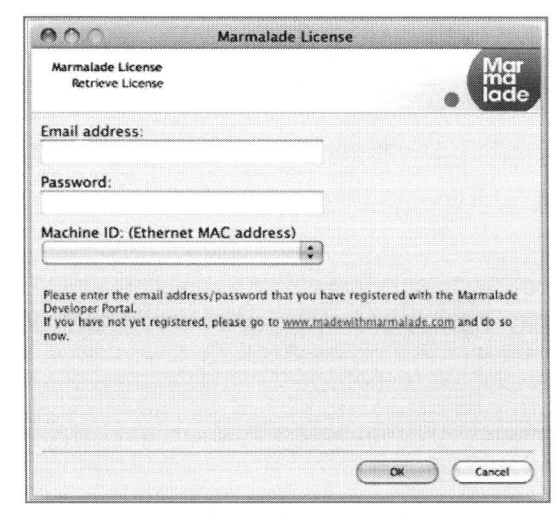

Figure 29-3: Activating license window on Mac

Now your installation of Marmalade for Mac OS X is complete.

29.1.2 Installation on Windows 7

The following steps will guide you through the installation of Marmalade on Windows 7:

1. Register on Marmalade's website by clicking on "Join" at the top right and filling in the necessary data.

Figure 29-4: Register with Marmalade

2. Download the latest version of Marmalade for your operating system. This can be accomplished from the Downloads page, which is located at www.madewithmarmalade.com/downloads. Clicking the download button on Marmalade's home page will also lead you there. Once there, find the link for the latest version for PC, which will look similar to Figure 29-5. This link will prompt you to download a .exe file. Install Marmalade on your computer by double clicking on the executable installer file.

Chapter 29: Marmalade   307

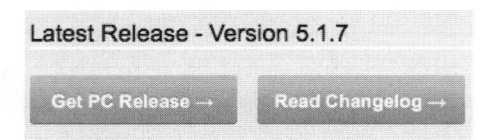

Figure 29-5: Download PC Release

3. To use Marmalade, you will need to either buy a license or get a free 90-day license. You can choose between these options on the page that appears when clicking on the "Buy" button on the Marmalade website.

4. Install Marmalade by opening the executable file that was downloaded in step 2 and then following the Setup Wizard.

5. During the Installation the Marmalade Configuration Utility will appear as shown in Figure 29-6. Click on the Activate License button to activate your license obtained from step 3.

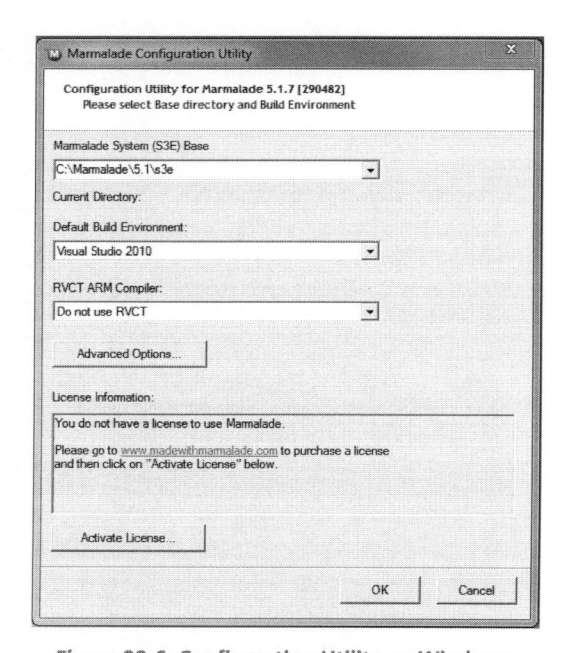

Figure 29-6: Configuration Utility on Windows

6. A window will appear requesting the email address and password that were used to register with on the Marmalade site, as well as the "Machine ID" (which should be selected by default). It will look like Figure 29-7.

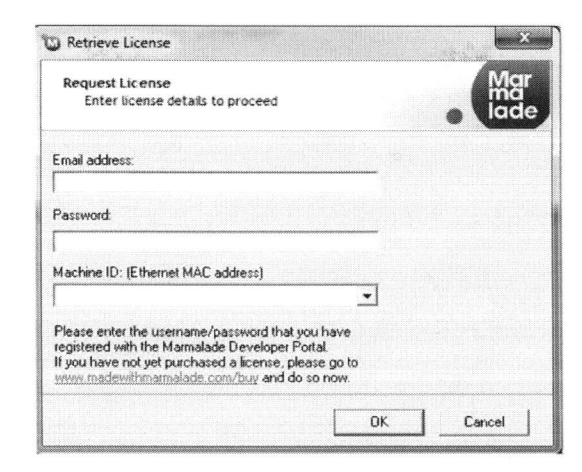

Figure 29-7: Activate Marmalade license

After activating the license on your machine, you will be ready to move on to the next section, where you will create your first app with Marmalade.

29.2 The "Hello Marmalade" Example

In this section the guide will walk you through creating your first project with Marmalade in Windows 7. The app will be a simple one where the string "Hello Marmalade" is displayed.

After installation, Marmalade LaunchPad should open. From LaunchPad you can access tools, documentation, examples, and launch new projects. To create a new app, follow these easy steps:

1. If LaunchPad is not open then go to Start Menu > All Programs > Marmalade > [Version #] > Marmalade LaunchPad to open it.

2. In LaunchPad, click on the Launch tab and choose Empty project as shown in Figure 29-8.

Chapter 29: Marmalade   309

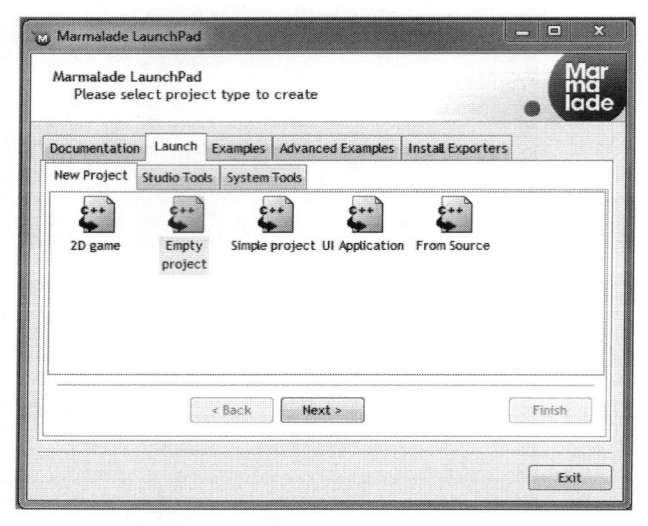

Figure 29-8: Create an Empty project

3. Name the project and choose a location for it to be saved as shown in Figure 29-9.

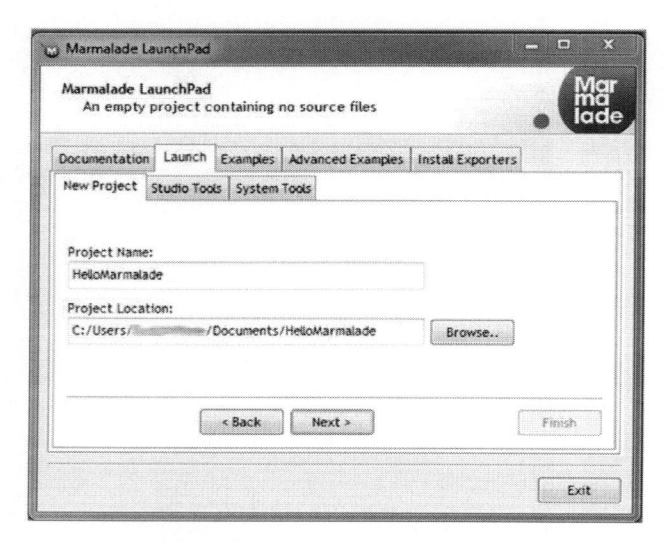

Figure 29-9: Choose Project Name and Location

4. Afterwards, *LaunchPad* will create the necessary files and it should load the project in the IDE that was specified during installation. In this case it will load Visual Studio 2010.
5. In Visual Studio, expand the project in the Solution Explorer so that it looks similar to Figure 29-10.

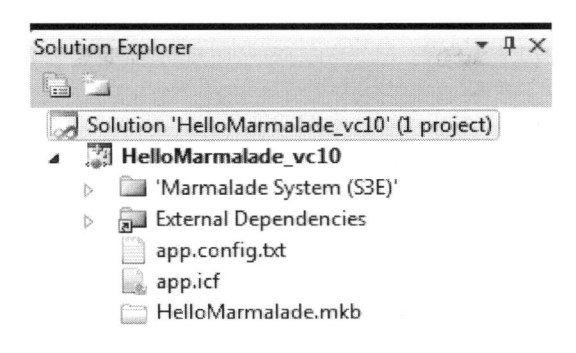

Figure 29-10: Solution Explorer

6. Right click on the project and choose Add > New Item then choose C++ file from the list of templates shown in Figure 29-11 and name it "main." This will be your main source file and will contain the `main()` function which is the standard entry point in Marmalade apps.

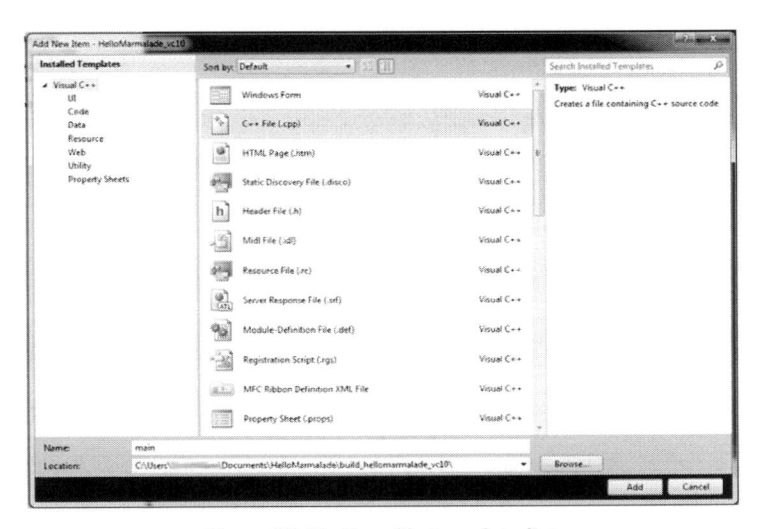

Figure 29-11: New file template list

7. Double click the *main.cpp* file that was created in the previous step and add to it the following code:

```
#include "s3e.h"
int main()
{
    while (!s3eDeviceCheckQuitRequest())
    {
    // Clears surface to color (r,g,b)
```

Chapter 29: Marmalade   311

```
            s3eSurfaceClear(0, 0, 0);
            // Prints the string "Hello Marmalade" at (120,150)
            s3eDebugPrint(120, 150, "Hello Marmalade", 0);
            // Displays the current surface to the device screen
            s3eSurfaceShow();
            // Yields device to OS
            s3eDeviceYield();
        }
        return 0;
    }
```

The first line of the above code is including the header file for S3E (Segundo Embedded Execution Environment), which is a C-style API that is used to control and provide access to core device services, events, and subdevice properties. This is the major component of Marmalade System, which is used to create apps that will work across multiple devices without the developer having to worry about device-specific issues.

The next line of code is the `main()` function. As mentioned previously, this is the entry point into your Marmalade app. Inside of this function is a loop, which repeatedly checks for any requests to quit the app. If a request is received it will break out of the loop and terminate the app when it reaches the `return 0;` statement.

Inside of the loop there are four S3E functions called:

- `s3eSurfaceClear()` - clears the surface to the color specified by the RGB value given, in this case the color black.
- `s3eDebugPrint()` - prints the specified string at the given location on the surface.
- `s3eSurfaceShow()` - displays the current surface to the device screen. -
- `s3eDeviceYield()` -makes the device yield to the OS for a specified period of time, in this case the minimum amount of time required by the OS to perform essential processing.

Now that you have your app put together and you know what everything does you can build and debug it. The project can be built for the x86 architecture or the ARM architecture. Building for the x86 architecture has the bonus of having all the features of the powerful IDE debugger available. Whereas building for the ARM architecture will use the ARM Emulator and debugger.

Follow these steps to build for the x86 architectures:

1. In Visual Studio 2010 go to Build > Configuration Manager.
2. A window will appear that looks similar to Figure 29-12. Select (x86) Debug as the Active solution configuration.

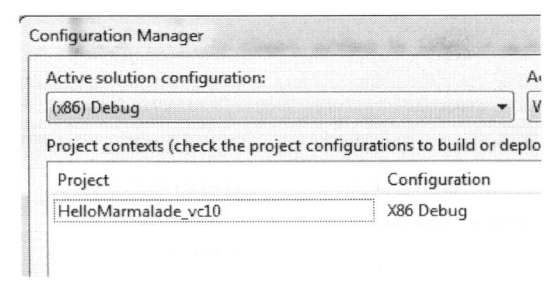

Figure 29-12: Configuration Manager

3. After selecting the proper configuration, build and run the project either through menu options or shortcuts. In Visual Studio 2010 theF7 key builds and the F5 key runs/debugs the project.

4. When you run the project for the x86 Debug Configuration the app will be launched on the Marmalade simulator and should appear similar to Figure 29-13.

Figure 29-13: Marmalade Simulator

Now that you can build for the x86 architecture, this can be used to debug your app with the IDE's (in this case Visual Studio 2010) powerful debugger.

There may come a point where you want to test your app on the ARM architecture and debug it or just get it ready to deploy to a device. To do this, first you have to build the project for the ARM architecture:

1. In Visual Studio 2010 go to Build > Configuration Manager and the Configuration Manager window will appear.

Chapter 29: Marmalade 313

2. Change the Active solution configuration to GCC (ARM) Debug so that it looks the same as Figure 29-14.

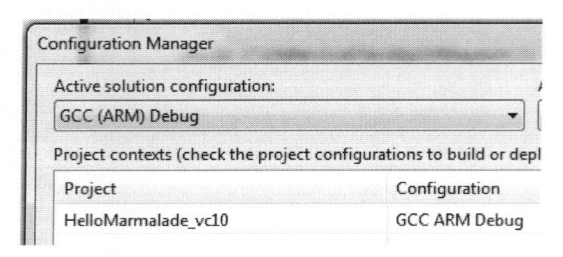

Figure 29-14: Active configuration set to ARM Debug

3. After selecting the proper configuration, build and run the project either through menu options or shortcuts. In Visual Studio 2010, "F7" builds and "F5" runs/debugs the project.

4. Several windows will pop up when you run/debug the project, one of them being the ARM Debugger, see Figure 29-15. In the ARM Debugger go to Control > Continue. This will close the debugger and load the app in the ARM Emulator depicted in Figure 29-16.

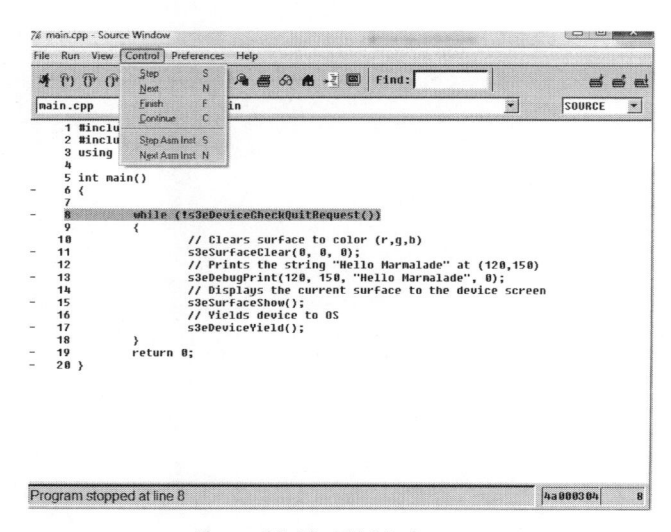

Figure 29-15: ARM Debugger

Figure 29-16: Marmalade ARM Emulator

Now with the app tested and working on the ARM Emulator, it is time to build the project for distribution. The process of building for distribution is similar to deploying the app to a device and will be detailed in the next section.

29.3 Building for Distribution and Deploying to a Device

Before building your project for distribution on the App Store or Android Market, or deploying it to an iOS device, there are certain requirements that must be met. For one, these platforms require additional software to be installed. The Marmalade documentation has guides for each of the platforms that the SDK supports, which can be found easily on the Marmalade website. For the purpose of this walkthrough you will only focus on the iOS platform.

The following are the requirements for deploying to an iOS device:

- Latest version of iTunes is installed
- An iPhone Developer Program Team Admin account with valid certificate for signing apps
- An iOS device (iPhone, iPad, iPod) set up with a provisioning profile to accept development apps

The latest version of iTunes can be downloaded from the www.apple.com/itunes website. The second requirement is explained in further detail in Chapters 6 and 9, which will walk you through the process of obtaining the account and certificate. The third requirement is covered in Chapter 9, which will walk you through setting up your device with a provisioning profile on a

Chapter 29: Marmalade   315

Mac. However for Windows PC development, Marmalade provides the iPhone Sign Request Tool that will generate keys and certificate requests.

29.3.1 Deploying to an iOS Device

Once the requirements are met the app can be deployed to a device by following these steps:

1. In Visual Studio 2010 go to Build > Configuration Manager, and the Configuration Manager window will appear.

2. Change the Active solution configuration to GCC (ARM) Release similar to how it was done in the previous steps.

3. After selecting the proper configuration, build and run the project either through menu options or shortcuts. In Visual Studio 2010, F7 builds and F5 runs/debugs the project.

4. Instead of the Debugger appearing, you should see the Marmalade System Deployment Tool, as seen in Figure 29-17. In the tool, select ARM GCC Release and then click on the button labeled "Next Stage."

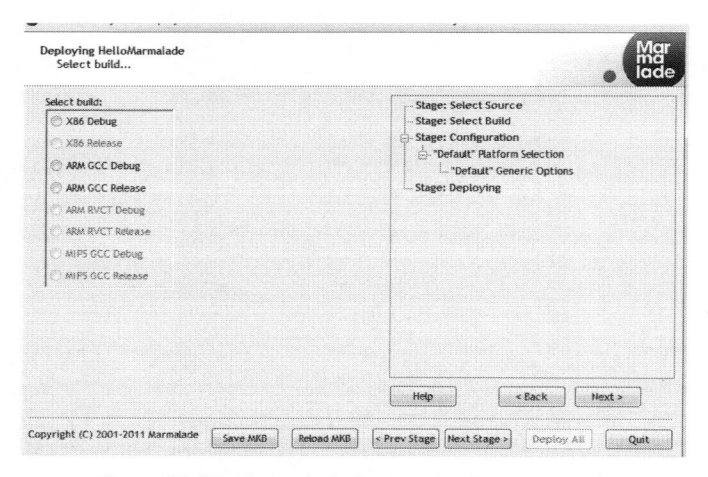

Figure 29-17: Marmalade System Deployment Tool

5. This will show a list of configurations. The currently selected "Default" should be fine for now. Click on the "Next Stage" button again.

6. This stage allows you to choose which Platforms that you want to deploy your app to. Select iOS as depicted in Figure 29-18 and click the "Next Stage" button again.

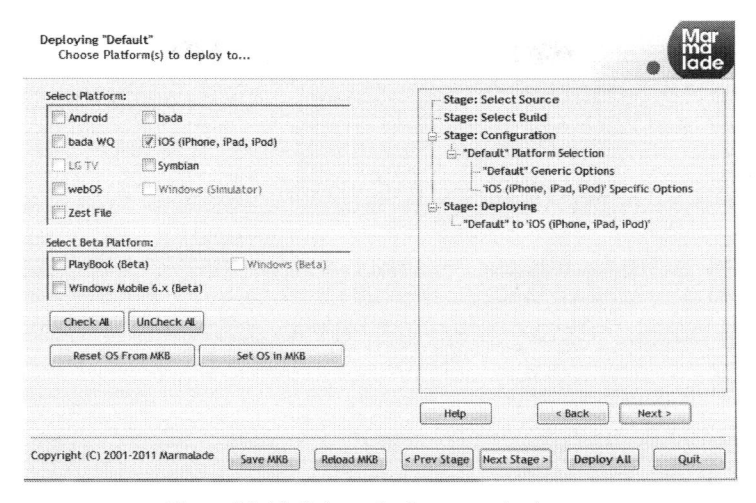

Figure 29-18: Select platforms to deploy to

7. The next screen is the Deployment Summary screen. You will be able to see the options for deployment for each of the platforms selected in the previous stage. Click the "Deploy All" button, which will create installer packages for the app for each of the platforms selected (in this case just iOS).

8. Click on the "Default" to "iOS" label under the "Stage: Deploying" heading to select the deployed app if you deployed to more than one platform.

9. Click the "Explore" button to bring up the folder, which contains the output ".APP" binary file (in this case .IPA file extension).

Once you have the IPA file you can put it on an iOS device through iTunes.

29.3.2 Building for Distribution

When you have completed testing your app, you can use the steps above to build your app for distribution; however, there are a few differences.

To build for distribution on the App Store the following need to be embedded in the deployment:
- Distribution Certificate
- Distribution Provisioning Profile

For more information on how to obtain a Distribution Certificate and Provisioning Profile please refer to Chapter 38: Preparing for Submission which covers this in detail. Besides the above requirements the only other difference is to make sure that Sign for Distribution is set to true and that a unique app ID is entered during the process above.

Once complete you will again have an IPA file that can then be submitted to the App Store with Application Loader. More information on how this is achieved can be obtained in Chapter 40: Submitting Apps Using Application Loader.

Adobe Flash Builder

Adobe Flash Builder is an Eclipse-based IDE for building mobile, web, and desktop applications. To create apps with Flash Builder, you will use ActionScript and Adobe's Flex framework. The Flex framework allows for creation of apps that are cross-platform. The supported platforms of the Flex framework are: Android, Blackberry Tablet OS, iPhone, and iPad. Web and desktop applications can be made using Flex as well, however some changes will have to be made to your existing mobile app code.

One of the advantages of using Adobe Flash Builder is that, for developers who are familiar with ActionScript and XML, it will be much easier to create apps using Flash Builder as opposed to learning Objective-C to create iOS apps. Another advantage is that the developer can use one code base and have an app across multiple platforms instead of having to create each one from scratch in different languages and environments.

A disadvantage to using Flash Builder is that in addition to any fees involved with publishing apps on their respective stores, you also have to purchase Flash Builder. The cost of Flash Builder is $249 for the Standard Edition or $699 for the Premium Edition. Depending on the size and type of app being created you may need the Premium Edition because it includes several professional testing tools.

30.1 Installation

Before you can start creating iOS apps with Adobe Flash Builder you need to download and install the latest version:

1. Click on the button labeled "Downloads" on the Adobe website shown in Figure 30-1 or go to www.adobe.com/downloads.

Figure 30-1: Downloads button

2. Once on the page select "Flash Builder 4.5" (or the latest version) from the drop down menu and click on the "Go" button as shown in Figure 30-2.

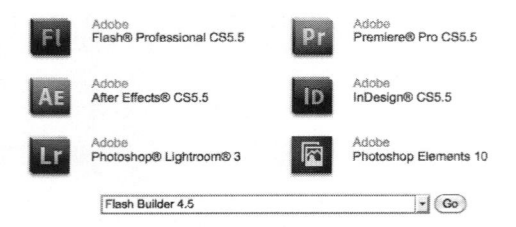

Figure 30-2: Flash Builder 4.5 download

3. Choose the version of Flash Builder that is appropriate for you. In this case the English version for Mac OS X. Then click the "Download now" button.

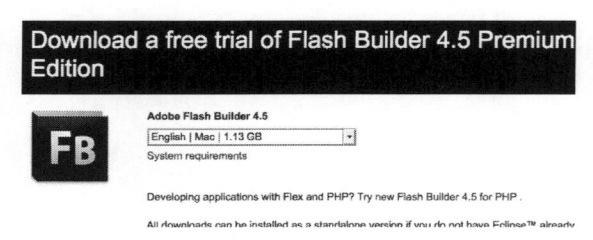

Figure 30-3: Flash Builder for Mac

4. Once downloaded install Adobe Flash Builder by opening the install file.

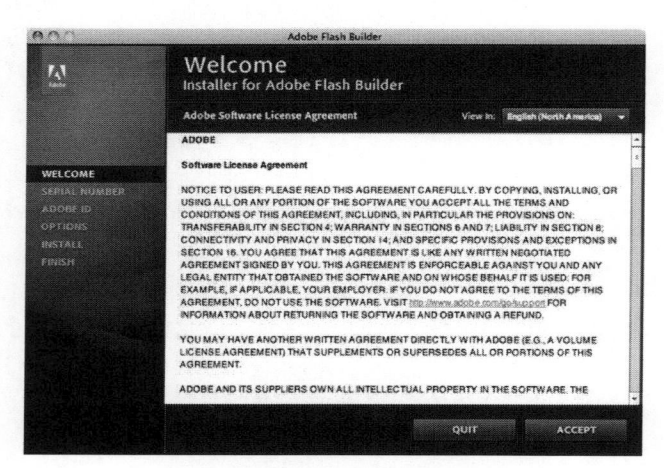

Figure 30-4: Flash Builder Installer

Once you have Adobe Flash Builder installed you will be ready to start building your first app.

30.2 Walkthrough

In this section, you will be walked through the creation of a simple "Hello World" style app.

1. Open Flex, then go to File > New > Flex Mobile Project to start a new project. Set the Project name to "HelloFlashBuilder" as seen in Figure 30-5 and choose a location for it to be saved then click the "Next >" button.

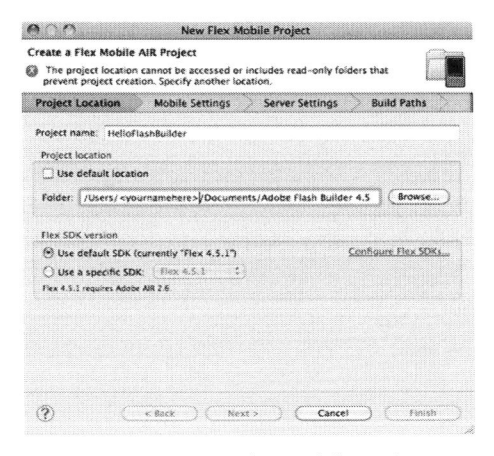

Figure 30-5: New Flex Mobile Project

2. Choose which platforms you wish to create the project for; in this case just choose Apple iOS, and then click Finish.

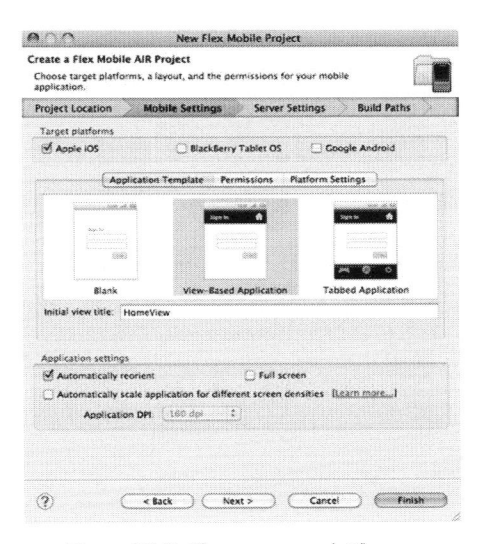

Figure 30-6: Choose your platforms

3. In the top-left of the Flash Builder window you will find the Project Explorer panel. Click on the HelloFlashBuilder Project. It will open in the editor and you will see an empty mxml file that is similar to Figure 30-7.

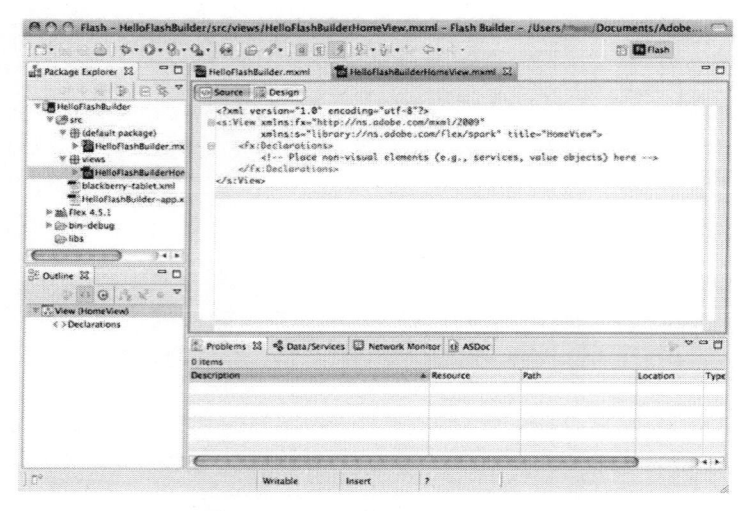

Figure 30-7: Open HomeView file

4. Add the following lines of code to the HomeView file before the closing `</s:View>` tag.

```
<s:VGroup width="100%" height="100%" verticalAlign="middle" horizontalAlign="center">
    <s:Label text="Hello Flash Builder!" id="lblOne"/>
    <s:Button label="Click Me" click="setLabel()"
    styleName="next" id="btnOne"/>
</s:VGroup>
```

This will add a label with the text "Hello Flash Builder!" and a button titled "Click Me" to the center of the screen. When the button is clicked it will trigger the `setLabel()` function that will be implemented in the next step.

5. Add the following function along with the `<fx:Script>` tag as shown below, to an area above the previously added code and the `<fx:Declarations>` tag.

```
<fx:Script>
<![CDATA[
    protected function setLabel():void
    {
        lblOne.text = "Good Job!";
    }
]]>
</fx:Script>
```

322   Chapter 30: Adobe Flash Builder

Now when the button is pressed the function `setLabel()` will change the text of the label to "Good Job!"

6. Now to test the app on the desktop hit the Run button on the main toolbar seen in Figure 30-8 as the second button from the left.

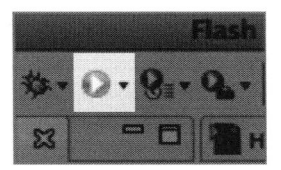

Figure 30-8: Run and Debug shortcuts

After clicking the Run button the Run Configurations window should appear as shown in Figure 30-9. This window allows you to choose what platform to build the app for, as well as what device you want to run it on and whether it is a simulator or an actual device.

7. Change the target platform to Apple iOS if it is not already chosen and then change the launch method to Apple iPhone 3GS or a different simulator if you so choose. The configurations should look similar to Figure 30-9.

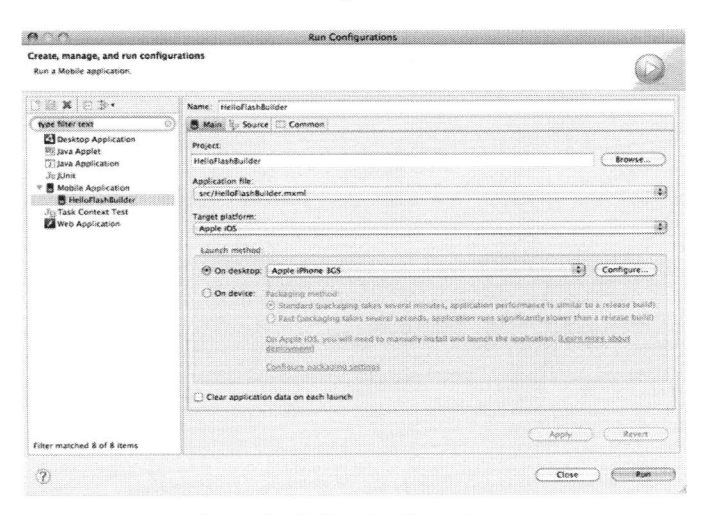

Figure 30-9: Run Configurations

8. Once the configurations are all set click on the "Run" button. After a short while you should see your app launch in the simulator you selected. If you click on the button the label should change as shown in Figure 30-10.

Chapter 30: Adobe Flash Builder   323

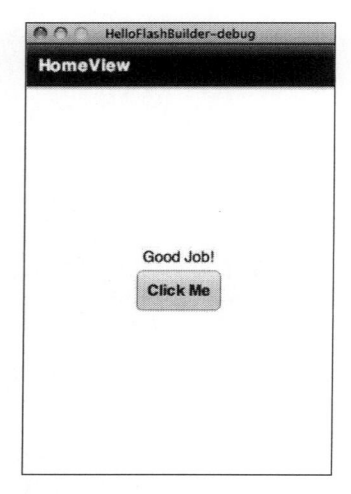

Figure 30-10: Output for Step 8

Now that you have a basic functioning app you can add a little extra to it to make it interesting.

9. First, add another button below the first one with the following code.

```
<s:Button label="Now Click Me" click="navigator.pushView(MyView)"
styleName="next" id="btnTwo"/>
```

This code makes it so that when this button is pressed a new View named MyView will be pushed on top of the navigation stack which results in it being displayed as the current screen. Since MyView does not yet exist it must be created in order for this button to work.

10. Go to File > New > MXML Component and a window will appear. Change the name to "MyView" and make sure the other options are configured as shown in Figure 30-11. After doing so click on the "Finish" button.

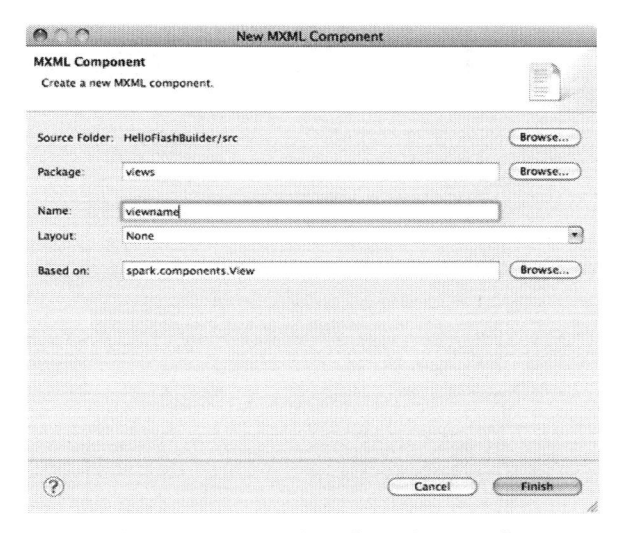

Figure 30-11: Creation of MyView.mxml

Now the `MyView.mxml` file should be added inside the views package shown in the Package Explorer.

11. Open the newly created file and add the following lines of code to it above the </s View> tag.

```
<s:VGroup width="100%" height="100%" verticalAlign="middle" horizontalAlign="center">
    <s:Label text="Congratulations!"/>
    <s:Button label="Start Over" click="navigator.popView()"
    styleName="back" />
</s:VGroup>
```

This will add a label with the text "Congratulations!" and a button labeled "Start Over" which will pop the current view from the navigation stack resulting in the program going back to the last view (screen).

12. Run the app and you will see two buttons now instead of one. The second button will take you to the new view shown in Figure 30-12.

Chapter 30: Adobe Flash Builder   325

Figure 30-12: Output for Step 12

Now you have two views in your app. Now you will change the code in the first view to make things a little more interesting.

13. Add the `visible` property to the button with the id `btnTwo` as shown below.

```
<s:Button label="Now Click Me" click="navigator.pushView(MyView)"
styleName="next" id="btnTwo" visible="false"/>
```

This makes the button invisible when the app first launches. Next, add some code to the `setLabel()` function so that when the first button is clicked it will make the second button visible.

14. Change the `setLabel()` function to look as follows.

```
protected function setLabel():void
{
  if (lblOne.text == "Hello Flash Builder!")
  {
    lblOne.text = "Good Job!";
    btnOne.label = "Click Me Again";
  } else if (lblOne.text == "Good Job!") {
    btnTwo.visible = true;
  }
}
```

326   Chapter 30: Adobe Flash Builder

This code still changes the text of the label to "Good Job!" and also changes the label of btnOne to "Click Me Again," but only when the label's text is "Hello Flash Builder!" In addition to this, if the button is clicked again once the label is changed to display "Good Job!" and it will make btnTwo visible.

15. Run the app again and you should see the label and button similar to what you saw in step 8.
16. Click the button and this time you will see its label change to "Click Me Again."
17. Click it again and a second button will appear with the label "Now Click Me" as shown in Figure 30-13.

Figure 30-13: Output for Step 17

18. Click this button and the view will change as it did in Step 12 depicted in Figure 30-12.

Congratulations you have completed your first app for iOS using Adobe Flash Builder. In the next section, you will learn how to build the app for distribution.

30.3 Building for Distribution

Once you have a completed app you can create a build for distribution purposes. Before doing so you will need an iOS developer account along with a developer certificate and provisioning profile. Information on how to do this can be found in Chapters 6-9. The following steps will walk you through the process of doing so.

1. Go to Project > Export Release Build and then the window shown in Figure 30-14 will pop up.
2. Make sure that the project is set to HelloFlashBuilder and the Application is set to HelloFlashBuilder.mxml. Select Apple iOS for the target platform and select "Signed packages for each target platform" for the "Export as:" option as shown in Figure 30-14.

Chapter 30: Adobe Flash Builder   327

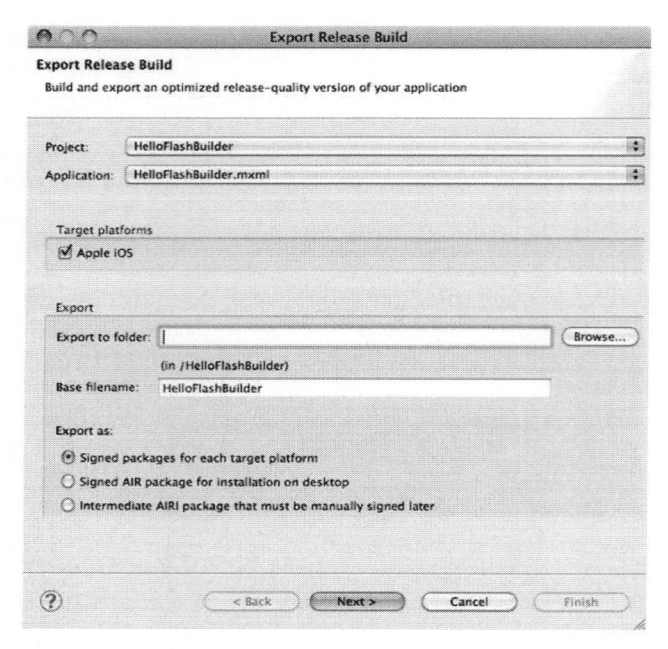

Figure 30-14: Export Release Build

3. Click on the "Next >" button to start the build process.

    A window will appear asking for a certificate, password, a provisioning file, and the package type as shown in Figure 30-15.

4. Select final release package for the package type and choose your certificate and provisioning profile.

    It is worth noting that the Apple iOS developer certificate needs to be in P12 format. If you need help with this Adobe has a guide on their site at the following URL: www.adobe.com/devnet/air/articles/packaging-air-apps-ios.html

5. When finished click on the "Finish" button to build the IPA file.

    Now with the IPA file you can upload it to the App Store following the procedures detailed in Chapters 38 and 40.

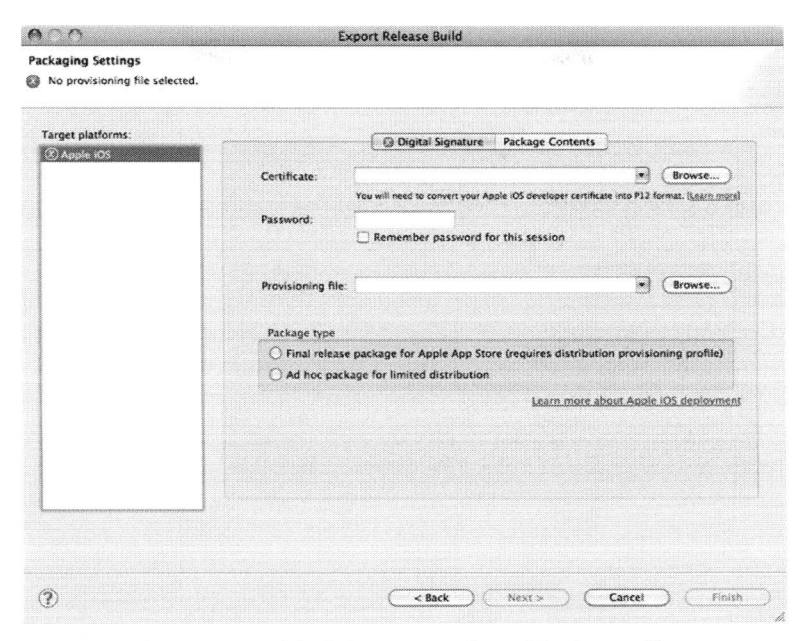

Figure 30-15: Digital Signature and Provisioning Profile

Now that you can build a simple app and can build for distribution, you are ready to make more advanced apps and sell them for profit on the various stores like the App Store and Android Marketplace.

Chapter 30: Adobe Flash Builder   329

31

Adobe Flash Professional

Adobe Flash Professional now offers the ability to publish your Adobe Flash and ActionScript 3 apps to iOS devices, which is great news for seasoned Flash developers who do not have the time or desire to learn Objective-C.

31.1 Adobe Flash Installation

Download and install the latest version of Adobe Flash Professional. Adobe Flash Professional CS6 is used as the example in this chapter.

1. First go to Adobe's website www.adobe.com to download Flash Professional. From here go to the Downloads page which can be located at the top of the screen as shown in Figure 31-1.

Figure 31-1: Downloads page

2. Once at the Downloads page click on the icon that says Adobe Flash Professional CS6 pictured in Figure 31-2, which will take you to the page to download it.

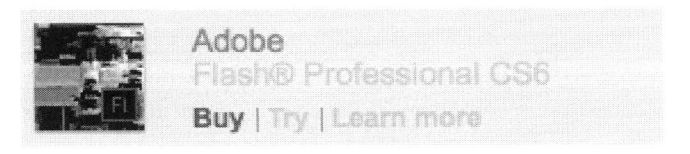

Figure 31-2: Adobe Flash Professional CS6

3. Once on the download page choose your language and then click on the button labeled "Download Now" shown in Figure 31-3.

Figure 31-3: Download Button

Chapter 31: Adobe Flash Professional   331

4. If you do not have Adobe's Download Assistant then it will be downloaded for you. After installing that, you will need to sign up for an Adobe ID if you do not currently have one.

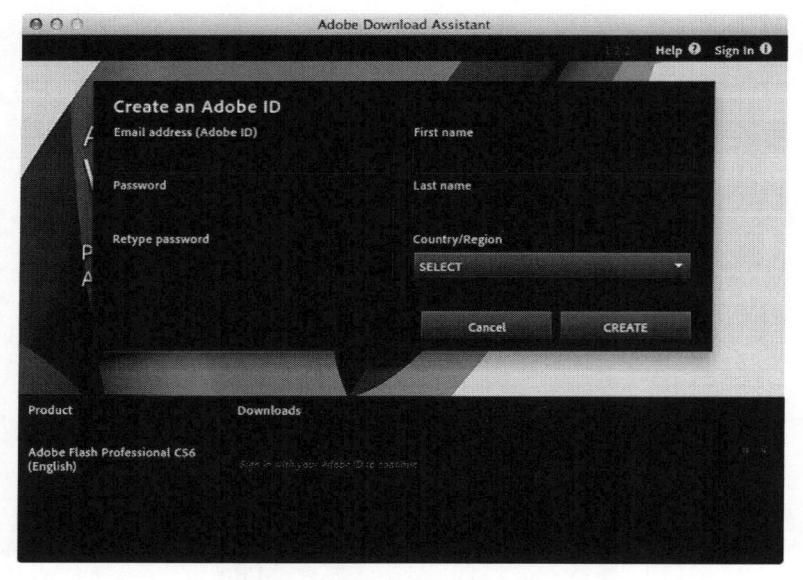

Figure 31-4: Adobe Download Assistant

5. After signing in with an Adobe ID the download should start and begin the installation process.
6. Accept the License Agreement shown in Figure 31-5.

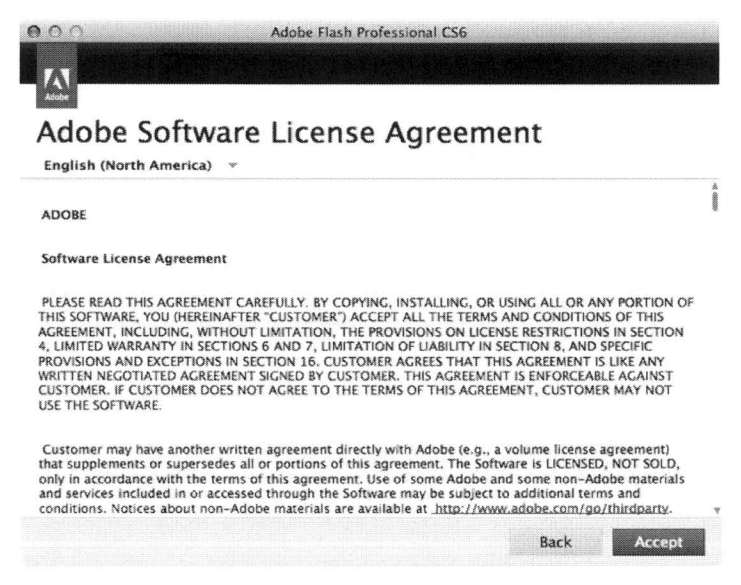

Figure 31-5: License Agreement for Flash Professional

7. After accepting the agreement, either enter a serial number from the purchased product or Install it as a trial. In the Install Options screen, as seen in Figure 31-6, make sure the option for iOS support is checked and click "Install" to begin installation.

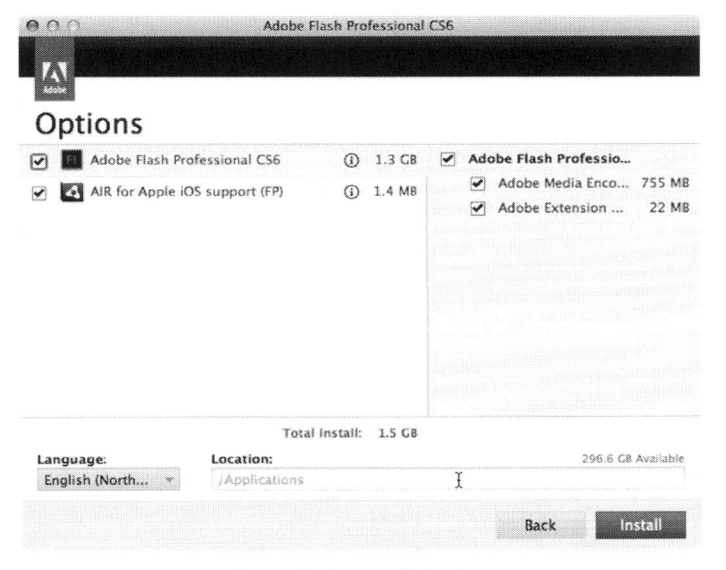

Figure 31-6: Install Options

8. The installation should take several minutes and once complete you will be ready to start using Adobe Flash.

31.2 Adobe Flash Walkthrough

When you first start flash, a dialog box will appear where you can either enter a serial number or click "Continue to use as a trial" to continue using it without a serial number - as shown in Figure 31-7. Click "Start Trial" to begin using Flash without entering a serial number. Afterwards you can sign in with your Adobe ID or just skip that step and go right into getting started with Flash.

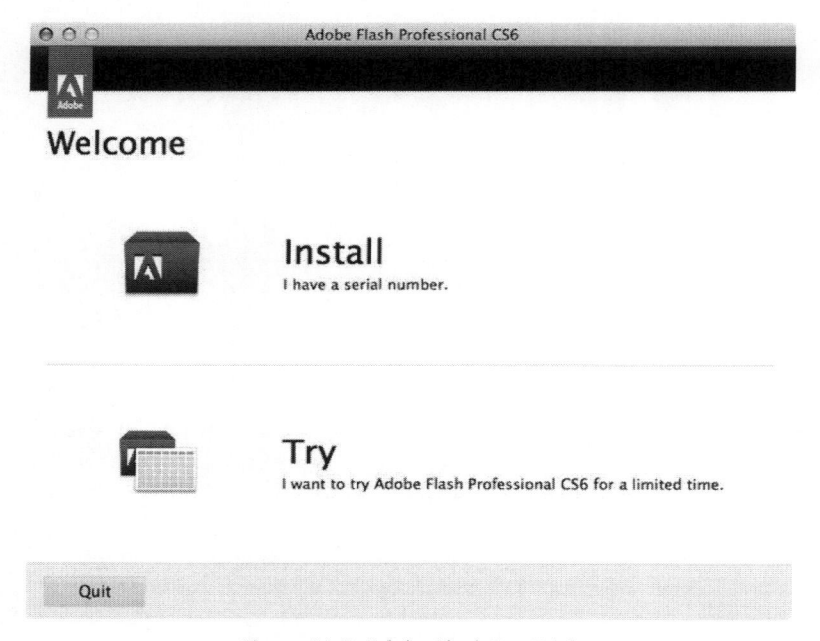

Figure 31-7: Adobe Flash Pro Trial

The next dialog box that pops up will show you the default templates that are available to help you start building your project. Click "AIR for iOS" as seen in Figure 31-8.

Figure 31-8: Create new AIR for iOS app

Flash will automatically size the stage to 320x480 pixels. This is the resolution for all iOS devices prior to the iPhone 4 and iPad. The iPhone4 device's retina display (640x960) and the iPad device's 768x1,024 display are both supported; however, for the sake of simplicity, only worry the basic 320x480 resolution will be covered for the purpose of this walkthrough.

If the Timeline is not displayed, click the "Timeline" tab at the bottom of the Flash window, then click the new layer button (bottom left) three times to add three new layers. Click the layer names to edit them and name them "ActionScript," "Labels," "Text," and "Button" so that your layers look like those in Figure 31-9. Unlike ActionScript 2.0, where you could apply code to individual objects, ActionScript 3.0 requires that all of the code be in one central location, which makes it easier to manage.

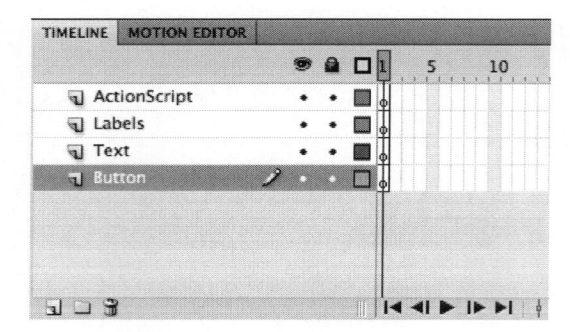

Figure 31-9: Timeline Layers

With the "Button" layer selected, select the rectangle tool in the toolbar (or use the shortcut R) and use it to draw a rectangle shape towards the bottom of the stage area, as seen in Figure 31-10.

Figure 31-10: Rectangle drawn for Button layer

Click on the Selection tool in the toolbar, which looks like a black mouse pointer (or use the shortcut V). Now right-click on the rectangle and select "Convert to symbol…" In the resulting dialog box, name the button "my_button" and make sure the Type is set to "Button," then click "Ok." By doing this, you have just made a button *type* named "my_button."

336   Chapter 31: Adobe Flash Professional

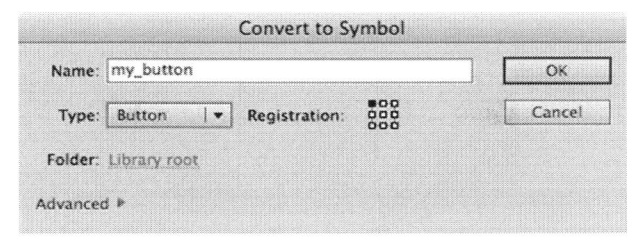

Figure 31-11: Create a new button

In order to have Flash recognize this *particular* button, you need to give it an instance name, so with the button selected, locate the "Properties" panel and where it currently says "<Instance Name>," type "Btn_1"

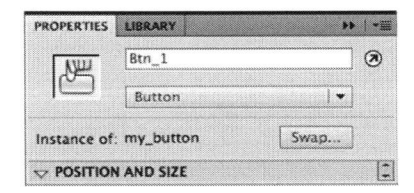

Figure 31-12: Change the instance name of the button

Next, click on frame 30 of the "ActionScript" layer, then, while holding "Shift," click on frame 30 of the "Button" layer to select all 4 layers. Right-click on the selection and then select "Insert Frame." The timeline should now look similar to the one shown here in Figure 31-13.

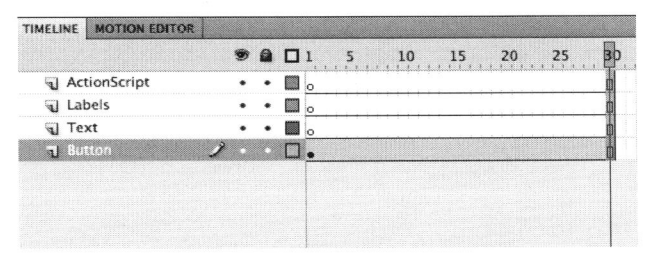

Figure 31-13: Selecting frames

Next create some keyframes, which will mark important spots on the timeline where you need to place code, labels and objects. Right-click on frame 15 of the "Text" layer and select "Convert to Keyframes."

With this frame still selected, use the Text tool (shortcut is T) to type "Hello" in a large font towards the top of the stage, which should now look similar to Figure 31-14.

Figure 31-14: Enter the text "HELLO"

Right-click and convert frame 30 of both the "ActionScript" and "Text" layers to keyframes as you did previously for frame 15 of the "Text" layer.

Your timeline should now resemble Figure 31-15.

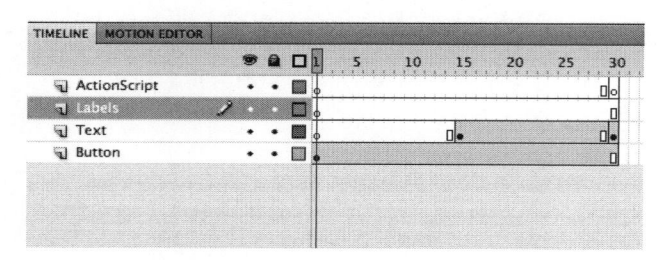

Figure 31-15: Updated Timeline

Now select frame 30 of the "Text" layer, and use the text tool to make another textbox below "Hello." This time, make it contain "World!" so when frame 30 is selected, your stage now resembles Figure 31-16.

338 Chapter 31: Adobe Flash Professional

Figure 31-16: "Hello World!" text for frame 30

Now, select frame 1 of the "Labels" layer, and in the "Label" section of the Properties panel, type "helloWorld," so that it resembles Figure 31-17.

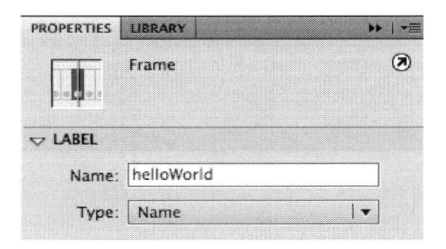

Figure 31-17: Set the "helloWorld" label

This applies the "helloWorld" label to frame 1 so that you can reference it in the code, which comes next.

Now open the Actions window by selecting Window>Actions from the main menu. Select frame 1 of the "ActionScript" layer and put the following code into the Actions window:

```
stop();
Btn_1.addEventListener(MouseEvent.MOUSE_DOWN, playMovie);
function playMovie(event:MouseEvent):void {
    gotoAndPlay("helloWorld");
}
```

The `stop();` function makes sure the app (which is essentially a movie with a button) does not automatically start playing when you run it. The next line adds an Event Listener to the instance of the button you created and tells it to execute the `playMovie` function when the button is clicked (`MOUSE_DOWN`). The `MOUSE_DOWN` event tells Flash to start playing the movie from the frame that you labeled "helloWorld."

Select frame 30 of the "ActionScript" layer and place the following code inside the Actions window:

```
stop();
```

One last important step is to turn off the "Runtime shared library" in the ActionScript settings. To do this, go to the "Publish" section under "Properties" and click the wrench icon next to the "Script:" drop down as pictured in Figure 31-18.

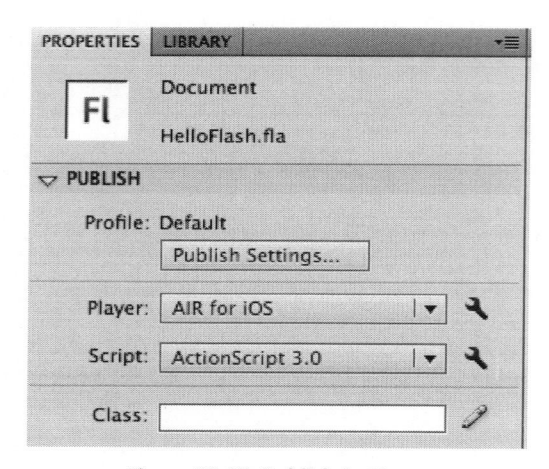

Figure 31-18: Publish Settings

In the next window click on the "Library path" tab, then at the bottom of the window, under "Runtime Shared Library Settings," click on the selector box that says "Runtime shared library (RSL)" and change it to "Merged into code," as displayed in Figure 31-19, and click "OK." This prevents an error that occurs when using text boxes (as you did with "Hello" and "World!") in apps developed for iOS devices.

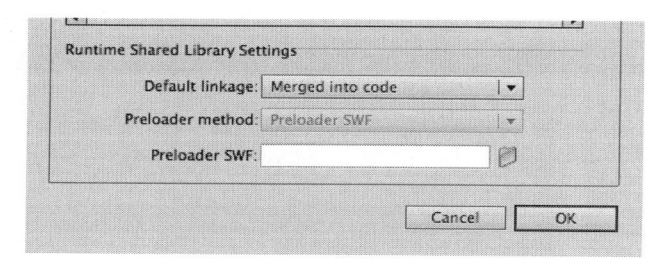

Figure 31-19: Runtime Shared Library Settings

To test the app, press the "command" + "return" buttons simultaneously (or Ctrl + Return if you are using Windows.) When the emulator appears, click the red button that you created. The app should run through the animation, displaying "Hello," then "World!" directly beneath it. Because you put the `stop();` function on frame 30, the app will stop playing at that point after displaying both words. Since the red button still has the `playMovie` function associated with it, if you click it again, it will reset the app to frame one, removing the text boxes and essentially resetting the app.

Congratulations on creating your first iOS app using Adobe Flash! Next, you will briefly review how to deploy your app to a device for testing.

31.3 Deploying to an iOS Device and Building for Distribution

You will need the Developer's Certificate and Provisioning Profile that you created in Chapter 9. In the Properties panel, click the wrench icon next to the "Player:" drop down menu. This will bring up a panel showing the General settings that you can adjust to get your app ready for deployment, and should look similar to Figure 31-20.

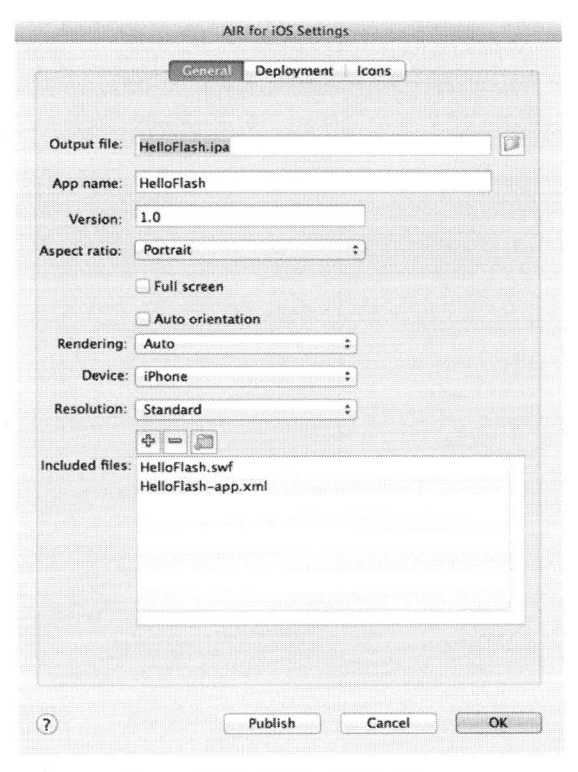

Figure 31-20: AIR for iOS Settings

For the demo app, leave these at their defaults, but take note of the options that are available to you when you are ready to publish your app to Apple's App Store such as Version number, Aspect Ratio, and Resolution.

The next tab, Deployment, is where you will upload the Developer's Certificate and Provisioning Profile that was mentioned earlier, as well as tell Flash what type of build you wish to create. The Developer's Certificate must be in a .p12 format which can be created in Keychain by navigating to the "Keys" category in the sidebar, selecting your private key, choosing "File" > "Export

Item...", then entering a new password. This password will be required when using the new certificate. Lastly, the Icons tab lets you associate appropriately sized icons to integrate into your IPA file.

When you have adjusted all of your settings appropriately, click "Publish" at the bottom of the window. Flash will build your app into an IPA file that can be deployed to an iOS device or uploaded to the App Store as discussed further in ***Section V: Submitting Your App***.

Cocos2d (via Xcode)

Unrelated to Cocoa from Apple, Cocos2d is a free, open source framework adapted for Xcode written in C and C++, wrapped into Objective-C, designed for building 2D games. It is considered "FOSS" (free and open source software under your Apple developer agreement), so be sure to verify any library or source framework source license in advance.

It uses OpenGL, supports 2D rendering and some 3D rendering. There is a large community behind Cocos2d. Whether it is on the official forum or in the documentation, support is easy to find on their site.

Cocos2d has many features designed to help you make your app quickly and efficiently. It offers scene management with transitions between scenes, an integrated physics engine, sprites and sprite sheets, and tile map support including orthogonal maps, isometric maps and hexagonal maps. Some other features are sound support, retina display support, high score server, text rendering and visual effects including but not limited to ripple, waves, lens and twirl. There are plenty more to be discovered if you go to the Cocos2d website: http://www.cocos2d-iphone.org/.

32.1 Installation

There is no installation necessary as Cocos2d is a source library. However you can install Cocos2d templates into Xcode for starting a Cocos2d-based project quickly.

To install the Cocos2d Xcode Templates:

1. Go to https://github.com/cocos2d/cocos2d-iphone
2. Click the "Downloads" link on the right of the page
3. Download as a .zip or tar.gz file
4. Extract the contents in your Downloads folder
5. Open Applications > Utilities > Terminal

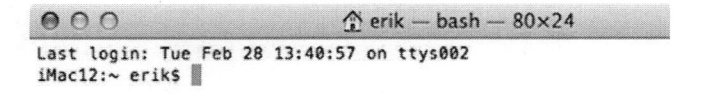

Figure 32-1: Terminal

6. In Terminal, type: cd Downloads
7. Press enter

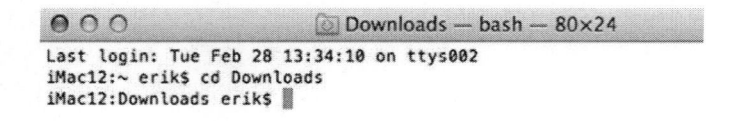

Figure 32-2: Change directory in Terminal

8. In Terminal, type: cd cocos2d-iphone-1.0.1
9. If you have a different version of Cocos2d or the folder you extracted is named something else replace "cocos2d-iphone-1.0.1" (or current version) with the name of the folder you created when extracting the download file.
10. After double-checking the folder name, press enter.

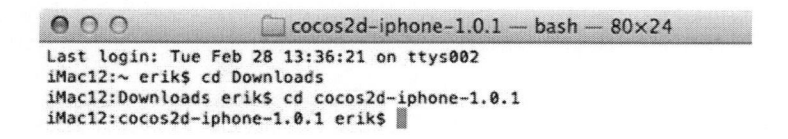

Figure 32-3: Change directory to Cocos2d folder

11. Type: ls (The first character is a lower case L)
12. Press enter. You will see a list of the contents of the folder you are in.

```
●○○                 cocos2d-iphone-1.0.1 — bash — 80×24
Last login: Tue Feb 28 13:40:57 on ttys002
iMac12:~ erik$ cd Downloads
iMac12:Downloads erik$ cd cocos2d-iphone-1.0.1
iMac12:cocos2d-iphone-1.0.1 erik$ ls
AUTHORS                         Resources-iPad
CHANGELOG                       cocos2d
CocosDenshion                   cocos2d-ios.xcodeproj
DONORS                          cocos2d-mac.xcodeproj
LICENSE_Box2D.txt               cocos2d.xcworkspace
LICENSE_Chipmunk.txt            cocoslive
LICENSE_FontLabel.txt           doxygen.config
LICENSE_TouchJSON.txt           doxygen.footer
LICENSE_artwork.txt             experimental
LICENSE_cocos2d.txt             external
LICENSE_cocosdenshion.txt       extras
LICENSE_libpng.txt              install-templates.sh
README.mdown                    templates
RELEASE_NOTES                   tests
Resources                       tools
Resources-Mac
iMac12:cocos2d-iphone-1.0.1 erik$
```

Figure 32-4: Run Cocos2d

13. Locate "install-templates.sh." This is a shell script that will install the Cocos2d templates for us. A shell is a software program that grants access to an operating system's main components. When run, a shell script executes a series of commands.

14. Run the script by typing: `sudo sh install-templates.sh`

15. Sudo is a command allowing Terminal admin access to your computer. Be careful when using the sudo command with unfamiliar commands, unfamiliar scripts, or both. Sudo is needed for this script because it needs write access to your computer in order to place the templates in an Xcode folder. For sudo to work, the user's password must be entered. Type the password then and press Enter.

Figure 32-5: Install Cocos2d templates

16. Allow the shell script to run. This may take several seconds.

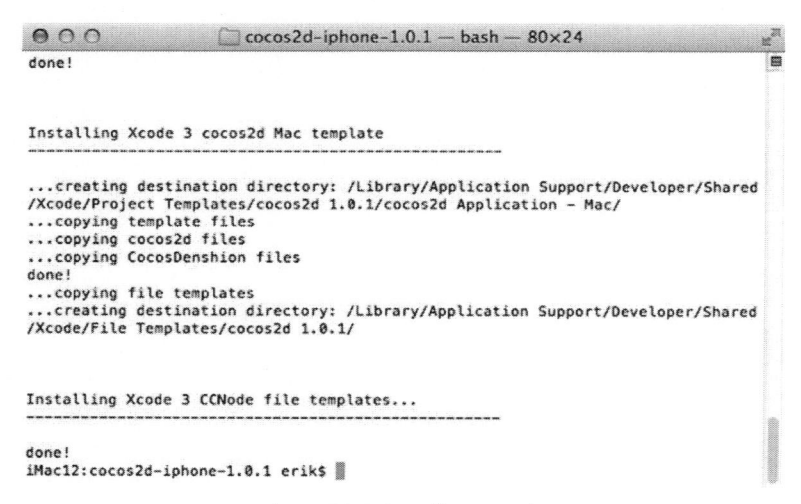

Figure 32-6: Install Cocos2d

17. Once the script has ended you will get a message saying "done!" signifying that all the templates are installed into Xcode.

32.2 Hello World!

Now that you have the Cocos2d templates installed you can get started on creating your first app. This section will be a walkthrough on making your first "Hello World!" app. It will be a simple interactive app that utilizes Cocos2d's sprites and actions objects.

This will consist of the following steps:
- Create a New Project
- Create a Sprite Sheet
- Create an Animated Sprite
- Handle Touch Events
- Make a Main Loop

32.2.1 Create a New Project

1. Open Xcode and start a new project.

Figure 32-7: Create a new Xcode project

2. Select "cocos2d" under the iOS menu then select the "cocos2d" template as depicted in Figure 32-8.

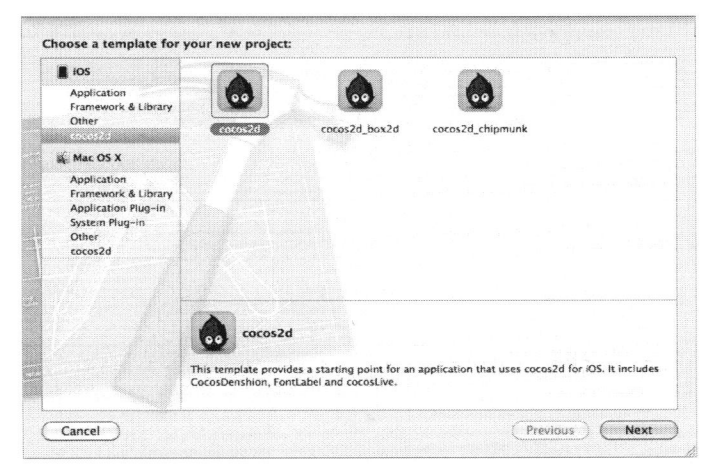

Figure 32-8: Choose a template for your new project

3. Name the project "HelloWorld."

Chapter 32: Cocos2d (via Xcode)   347

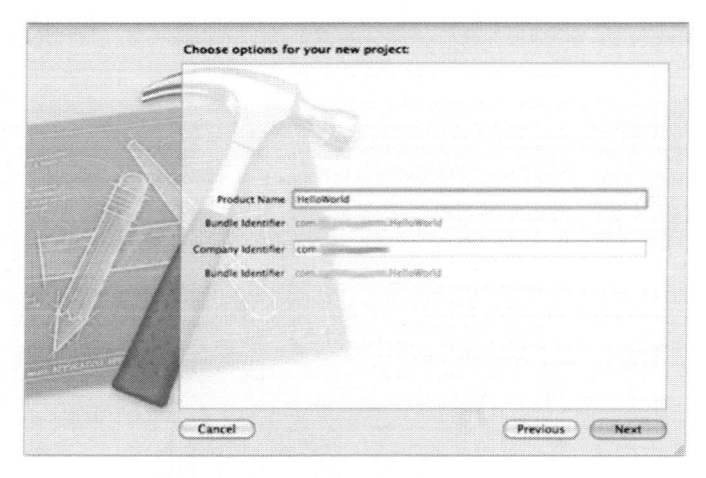

Figure 32-9: Choose options for your new project

4. Build the project and run it on the iPhone simulator.

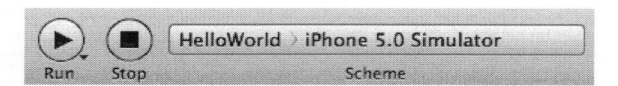

Figure 32-10: Xcode Runtime Controls and Scheme Settings

Figure 32-11: Cocos2d Hello World Project

In the iOS simulator you should see the words "Hello World" in white text on a black background as shown in Figure 32-11. The number at the bottom left of the screen is the app's current frame rate.

This project template contains three important files: AppDelegate, RootViewController, and HelloWorldLayer. The AppDelegate decides which scene should be run when the app starts. The RootViewController handles such things as the screen orientation and auto rotation. HelloWorldLayer is a CCLayer, a 2d frame that can contain such things as text, images, and objects.

1. Navigate to `HelloWorldLayer.h` then open it.

Figure 32-12: Xcode Project Hierarchy

2. Modify the contents of HelloWorldLayer.h to be the following:

```
#import "cocos2d.h"

@interface HelloWorldLayer : CCLayerColor
{
  BOOL isWalking;
  BOOL animationIsRunning;
  BOOL isMovingRight;
  BOOL hasBeenFlipped;
  CCRepeatForever *loopedAction;
  CCSprite *bearSprite;
  float speed;
  CGPoint destination;
```

```
}
+(CCScene *) scene;
-(void)mainLoop:(float)delta;
-(BOOL)point:(CGPoint)a isWithin:(float)distance toPoint:(CGPoint)b;
@end
```

The HelloWorldLayer originally extended CCLayer but now extends CCLayerColor. The CCLayerColor class allows for the layer to be assigned a color. The CCLayer class does not and will always be black. The individual use of these variables and methods will be shown and explained as they are used throughout the rest of this walkthrough.

32.2.2 Creating a Sprite Sheet

To make a sprite you need an image. To animate a sprite you need at least two images. A sprite sheet is a good way to store multiple images.

What is a sprite sheet?

A sprite sheet is a compilation of several images into one large image.

Why use a sprite sheet?

Sprite sheets save space and loading time. Instead of an app needing to load and store several images into memory it only needs to load one.

How to get started?

To make the sprite sheet you will need a program that can create and modify images. The program used in this walkthrough is GIMP, the GNU Image Manipulation Program, which is a free high-end graphics editor. It is available on most operating systems such as Microsoft Windows, Apple Mac OS X, and Linux. The official site for GIMP is http://www.gimp.org.

Figure 32-13: Sprite Sheet Example

The sprite sheet shown in Figure 32-13 is the one that will be used to create an animated sprite in this tutorial. The three images depict a polar bear walking. The image at the top will be used to convey the polar bear standing still. It is important for each image in an animation to be the same size so they line up correctly. These images are 64 pixels by 32 pixels, combined they are 64 pixels by 96 pixels.

32.2.3 Creating an Animated Sprite

To make use of the sprite sheet that was created, it needs to be added to the project. The image can be added using File > Add Files to "HelloWorld," or by dragging the image file directly into the hierarchy itself. Once the image is in the project your hierarchy should be similar to Figure 32-14.

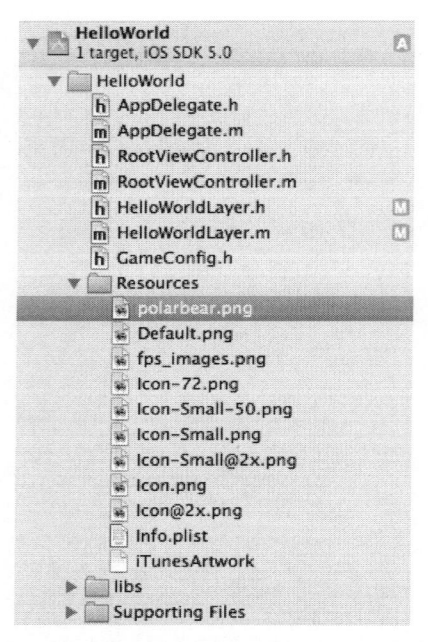

Figure 32-14: Xcode Project Hierarchy after adding image

The sprite sheet can now be implemented with a `CCSpriteBatchNode` inside the `init` method of the HelloWorldLayer.m file. Also, modify the call to the super class as shown in order to have a gray background. The method ccc4(r,g,b,a) creates a color with the given values for red, green, blue, and alpha (transparency).

```
-(id) init
{
  if( (self=[super initWithColor:ccc4(127, 127, 127, 255)])) {
    CCSpriteBatchNode *batch = [CCSpriteBatchNode
              batchNodeWithFile:@"polarbear.png"];
  }
  return self;
}
```

Think of `CCSpriteBatchNode` as a batch of cookies on a cookie sheet and each image is a cookie. Next, initialize `bearSprite` and the `CCSprite` object, with the following code immediately after the `CCSpriteBatchNode` creation:

```
bearSprite = [CCSprite spriteWithTexture:batch.texture
              rect:CGRectMake(0, 0, 64, 32)];
```

The argument `rect` asks for a `CGRect` that determines which section of the batch's texture is to be used. The top left of the sprite sheet is (0,0) also known as the origin; therefore, bearSprite will be the image at the top of the sprite sheet.

In order for the sprite to be displayed it must be added to the CCLayer. To add an object to the CCLayer you call its addChild: method shown below.

    [self addChild: bearSprite];

Place the above code after bearSprite's initialization. Take a look at Figure 32-15 to understand the coordinate system of a CCLayer on an iPhone when in landscape orientation. Notice that the coordinates are vertically reversed when compared to a CCSpriteBatchNode.

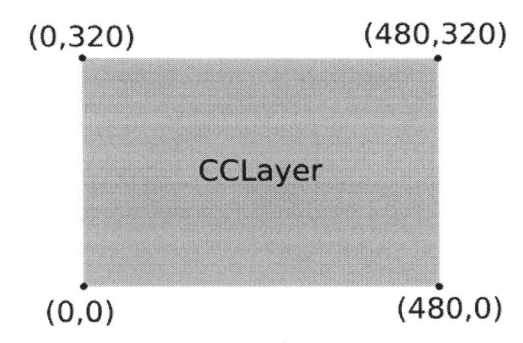

Figure 32-15: CCLayer Axes Coordinates for iPhone Landscape Orientation

When setting the position of a CCSprite the sprite's anchor point plays an important role. As seen in Figure 32-16, the default anchor point is the exact center of the CCSprite. If the anchor point is set to (0,0) then the anchor point is the very bottom left.

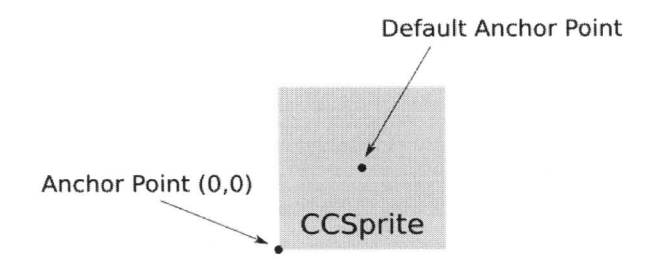

Figure 32-16: CCSprite Anchor Point

The anchor point defines the point on the CCSprite that is set to the position given to that same CCSprite. Cocos2d's function ccp(x,y) is a quick way to make a CGPoint. The values of an anchor point can range between 0 and 1. The default values of an anchor point are (0.5, 0.5). The bottom left of a CCSprite is anchor point (0,0) and the top-right is (1,1). The following is an example of setting a sprite's anchor point:

    [spriteName setAnchorPoint:ccp(0,0)];

Here is an example of setting a sprite's position:

```
[spriteName setPosition:ccp(0,0)];
```

Figures 32-17 and 32-18 illustrate the different locations of a sprite when the anchor point is changed but the position is the same.

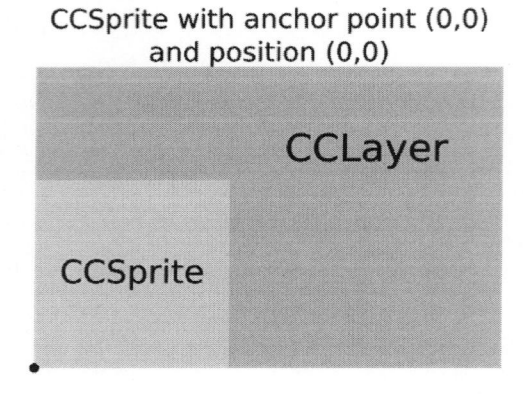

Figure 32-17: CCSprite Position Example 1

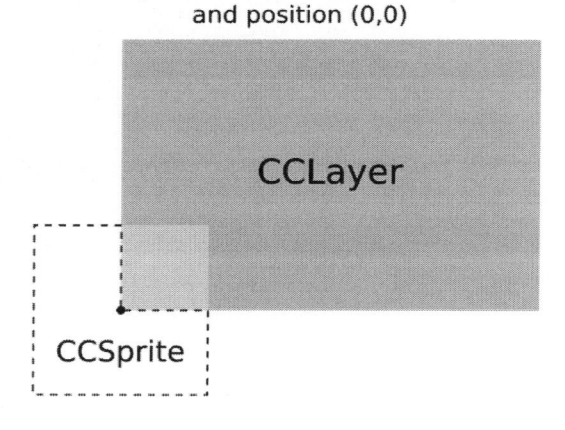

Figure 32-18: CCSprite Position Example 2

Keep the anchor point of the sprite at its default location. Set the position of the sprite to be the center of the CCLayer, which is half the width and half the height.

```
bearSprite.position = ccp(480/2,320/2);
```

In the interface of HelloWorldLayer, CGPoint destination was declared. This object will be the point on the CCLayer that the sprite will travel to. Set the value of destination to be the same as the sprite's position.

```
destination = ccp(480/2,320/2);
```

Run the project and the result should be the same as Figure 32-19.

Figure 32-19: Polar Bear Sprite at Runtime

To create an animation for the sprite you need to make an array filled with all the frames for the animation.

```
NSMutableArray* arrayOfFrames = [[NSMutableArray alloc]init];

[arrayOfFrames addObject:[CCSpriteFrame frameWithTexture:batch.texture
rect:CGRectMake(0, 0, 64, 32)]];

[arrayOfFrames addObject:[CCSpriteFrame frameWithTexture:batch.texture
rect:CGRectMake(0, 32, 64, 32)]];

[arrayOfFrames addObject:[CCSpriteFrame frameWithTexture:batch.texture
rect:CGRectMake(0, 0, 64, 32)]];

[arrayOfFrames addObject:[CCSpriteFrame frameWithTexture:batch.texture
rect:CGRectMake(0, 64, 64, 32)]];
```

When these four frames run it will appear as the bear is walking. Next the array is used to create a `CCAnimation` object.

```
CCAnimation *walkingAnimation = [CCAnimation animationWithFrames:arrayOfFrames
delay:1.0f/8.0f];
```

The `CCAnimation` object is then used to create a `CCAnimate` object. If the `CCAnimate` action were run on the sprite, the animation would run through once and disappear.

```
CCAnimate *walkingAction = [CCAnimate actionWithAnimation:walkingAnimation];
```

The `CCAnimate` object is used to create a `CCRepeatForever` object then that can be used to

Chapter 32: Cocos2d (via Xcode) 355

repeat the animation indefinitely.

```
loopedAction = [[CCRepeatForever actionWithAction:walkingAction]retain];
```

The CCRepeatForever loopedAction is declared in the HelloWorldLayer's interface. Retain is used to let the app know to keep the object in memory. If the CCRepeatForever was not retained the app would not know what loopedAction, because it would be essentially thrown away out of memory. Now assign values to the Booleans you declared and set the value to the float variable speed. These Booleans will be used to determine the CCSprite's animation and if the CCSprite's image itself should be mirrored. The float, speed, will determine how fast the CCSprite moves toward its destination.

```
isWalking = false;

animationIsRunning = false;

hasBeenFlipped = false;

isMovingRight=false;

speed=10.0f;
```

The next thing to do is to schedule the method that will be the main loop of the app. Calling CCLayerColor's method schedule will create a timer with the given interval.

```
[self schedule:@selector(mainLoop:) interval:1.0f/60.0f];
```

The previous block of code should be at the end of the HelloWorldLayer's init method. The following should be how the entire init method looks:

```
-(id) init
{
  if( (self=[super initWithColor:ccc4(127, 127, 127, 255)])){
    self.isTouchEnabled = true;
    CCSpriteBatchNode *batch = [CCSpriteBatchNode
                    batchNodeWithFile:@"polarbear.png"];
    bearSprite = [[CCSprite spriteWithTexture:batch.texture
                        rect:CGRectMake(0, 0, 64, 32)]retain];
    [self addChild:bearSprite];
    bearSprite.position = ccp(480/2,320/2);
    destination = ccp(480/2,320/2);
    NSMutableArray* arrayOfFrames = [[NSMutableArray alloc]init];
    [arrayOfFrames addObject:[CCSpriteFrame
                    frameWithTexture:batch.texture
                  rect:CGRectMake(0, 0, 64, 32)]];
    [arrayOfFrames addObject:[CCSpriteFrame
                    frameWithTexture:batch.texture
                  rect:CGRectMake(0, 32, 64, 32)]];
```

```objc
    [arrayOfFrames addObject:[CCSpriteFrame
                    frameWithTexture:batch.texture
            rect:CGRectMake(0, 0, 64, 32)]];
    [arrayOfFrames addObject:[CCSpriteFrame
                    frameWithTexture:batch.texture
            rect:CGRectMake(0, 64, 64, 32)]];
    CCAnimation *walkingAnimation = [CCAnimation
        animationWithFrames:arrayOfFrames delay:1.0f/8.0f];
    CCAnimate *walkingAction = [CCAnimate
            actionWithAnimation:walkingAnimation];
    loopedAction = [[CCRepeatForever
                actionWithAction:walkingAction]retain];
    isWalking = false;
    animationIsRunning = false;
    hasBeenFlipped = false;
    isMovingRight=false;
    speed=10.0f;
    [self schedule:@selector(mainLoop:) interval:1.0f/60.0f];
  }
  return self;
}
```

32.2.4 Responding to Touches

By default, touch interaction is disabled on the `CCLayer`. Placing the code below into the `HelloWorldLayer`'s `init` method will allow touches to be recognized.

```objc
    self.isTouchEnabled = true;
```

Now you need to implement Cocos2d's touch handling methods. Place the code below somewhere in HelloWorldLayer.m inside the implementation of HelloWorldLayer and outside of any method.

```objc
UITouch* touch;
CGPoint touchLocation;
-(void)ccTouchesBegan:(NSSet *)touches withEvent:(UIEvent *)event{
  touch = [touches anyObject];
  touchLocation = [[CCDirector sharedDirector] convertToUI:[touch locationInView:[touch view]]];
  destination=touchLocation;
}
```

Chapter 32: Cocos2d (via Xcode)   357

```
-(void)ccTouchesMoved:(NSSet *)touches withEvent:(UIEvent *)event{
  touch = [touches anyObject];
  touchLocation = [[CCDirector sharedDirector] convertToUI:[touch locationInView:[touch view]]];
  destination=touchLocation;
}
-(void)ccTouchesEnded:(NSSet *)touches withEvent:(UIEvent *)event{
  touch = [touches anyObject];
  touchLocation = [[CCDirector sharedDirector] convertToUI:[touch locationInView:[touch view]]];
  destination=touchLocation;
}
```

The moment a finger contacts the device's screen, ccTouchesBegan: is fired. When that finger moves while still contacting the screen, ccTouchsMoved: is fired. And when the finger loses contact with the screen, `ccTouchesEnded:` is fired. The same block of code is used in each method. The touch is stored in the `UITouch` object named `touch`. Then the location is converted using the CCDirector's method `convertToUI`. This conversion is necessary because the touch has coordinates respective to the device and not the Cocos2d user interface.

32.2.5 Make a Main Loop

Now that the app handles touch events, create a main loop to tell the `CCSprite` what to do with the touches. Start by creating the following method in the implementation of HelloWorldLayer:

```
-(void)mainLoop:(float)delta{
}
```

Before populating the contents of the main loop there is a method that needs to be written. This method will be used in the main loop to determine if the CCSprite is within a certain distance of the touch. The purpose of this distance check is to stop the CCSprite from moving if it is close enough to the touch.

```
-(BOOL)point:(CGPoint)a isWithin:(float)distance toPoint:(CGPoint)b
{
  float d = sqrtf( ((b.x - a.x)*(b.x - a.x)) + ((b.y - a.y)*(b.y - a.y)) );
  if (d <= distance){
    return true;
  }
  return false;
}
```

The method shown uses a math equation known as the distance formula to determine the distance between two points. That distance is then compared to the input distance. If the distance is less than or equal in the input value the method returns true, otherwise the method returns false.

In the main loop add the following statement:

```
isWalking=false;
```

The Boolean, `isWalking`, will be set to true if the CCSprite is going to be walking but for now you will assume it is false in the event that the bear will not move. Next will be the movement value of the CCSprite, which will be stored in a `CGPoint`.

```
CGPoint movement = ccp(0,0);
```

Now test the CCSprite's position to determine if it is too close to the touch.

```
if (![self point:bearSprite.position isWithin:2.5f toPoint:destination]){
}
```

The exclamation point negates the value following it. This means if the method to check the distance returns true, or if the CCSprite's position is within the specified distance, then the true that is returned will be turned into a false. Therefore this If Statement will only fire if the distance between the touch and the CCSprite's position exceeds the argument of a distance of two and a half.

Inside the `if` Statement, add the following code:

```
if (bearSprite.position.x < destination.x)
{
   isWalking=true;
        movement.x = 1;
   isMovingRight = true;
}
else if (bearSprite.position.x > destination.x)
{
   isWalking=true;
        movement.x = -1;
   isMovingRight = false;
}
if (bearSprite.position.y < destination.y)
{
   isWalking=true;
        movement.y = 1;
```

```
        }
        else if (bearSprite.position.y > destination.y)
        {
           isWalking=true;
                movement.y = -1;
        }
        bearSprite.position = ccpAdd(bearSprite.position, movement);
```

This code tests if the `bearSprite` is to the left of the touch, if it is then the bear will be set to walking, move a value of positive one on the x-axis, and set the bear to be moving right. The `isMovingRight boolean` will be used to flip the sprite horizontally. If the bear is to the right of the touch, then the bear will move a value of negative one on the x-axis. A similar test occurs but on the y-axis to determine if the bear should move vertically.

Outside of the If Block the position distance test will be this If Statement:

```
        if (isWalking){
           if (!animationIsRunning){
              [bearSprite runAction:loopedAction];
              animationIsRunning=true;
           }
        }
        else{
           if (animationIsRunning){
              [bearSprite stopAction:loopedAction];
              animationIsRunning=false;
           }
        }
```

The previous nested `if` block will determine whether or not the sprite should run the animation of the bear walking. The next nested `if` block will determine is the sprite itself should be mirrored. This is being done because the bear's sprite is facing left. If the bear moves to the right then the bear would be walking backwards. This is fixed by simply flipping the sprite whenever it moves to the right.

```
        if(isMovingRight){
           if(!hasBeenFlipped){
              bearSprite.flipX=true;
              hasBeenFlipped=true;
           }
```

```
      }
      else{
        if(hasBeenFlipped){
          bearSprite.flipX=false;
          hasBeenFlipped=false;
        }
      }
```

Here is what the complete main loop should look like:
```
    -(void)mainLoop:(float)delta
    {
      isWalking=false;
      CGPoint movement = ccp(0,0);
      if (![self point:bearSprite.position isWithin:2.5f toPoint:destination]){
        if (bearSprite.position.x < destination.x){
          isWalking=true;
          movement.x = 1;
          isMovingRight = true;
        }
        else if (bearSprite.position.x > destination.x){
          isWalking=true;
          movement.x = -1;
          isMovingRight = false;
        }
        if (bearSprite.position.y < destination.y){
          isWalking=true;
          movement.y = 1;
        }
        else if (bearSprite.position.y > destination.y){
          isWalking=true;
          movement.y = -1;
        }
        bearSprite.position = ccpAdd(bearSprite.position, movement);
```

```
        }
        if (isWalking){
          if (!animationIsRunning){
            [bearSprite runAction:loopedAction];
            animationIsRunning=true;
          }
        }
        else{
          if (animationIsRunning){
            [bearSprite stopAction:loopedAction];
            animationIsRunning=false;
          }
        }
        if(isMovingRight){
          if(!hasBeenFlipped){
            bearSprite.flipX=true;
            hasBeenFlipped=true;
          }
        }
        else{
          if(hasBeenFlipped){
            bearSprite.flipX=false;
            hasBeenFlipped=false;
          }
        }
    }
```

Run the project. The bear will now walk to any point you touch.

32.2.6 Building for iOS Distribution

Just like any Xcode project, you can build your project for distribution by archiving the project then compiling it into an IPA file. For instructions on how to do this please refer to Chapters 38 and 39.

32.3 Cocos2d Summary

Cocos2d is a free, open source iOS framework designed for 2d games. There is a large community behind Cocos2d. Whether it is on the official forum or in the documentation, support is easy to find on their site. Cocos2d has many features designed to help you make your app quickly and efficiently.

Corona SDK

Corona SDK is a mobile development platform for iOS, Android, Kindle Fire, and Nook. The software is oriented towards mobile apps such as games and e-books but can be used to create virtually any app.

Corona uses OpenAL for sound and sports a physics engine called Box2D. With Facebook and Twitter integration an app can be quickly tied into the social network. Corona also supports business needs like SQLite database access, HTTP connectivity, and JSON libraries. Device features like GPS, multi-touch, gyroscope, compass, camera, and photo albums are natively supported. Understand your customers with integrated analytics by getting information on how users interact with your app. Third-party tools (http://www.anscamobile.com/corona/tools/) exist to increase project productivity. Some of these tools include a sprite sheet maker, physics editor, and a project manager

Corona also uses the simple and lightweight scripting language: Lua. It is similar to ActionScript but more user-friendly. What takes 300 lines in Objective-C to display an image with the OpenGL library takes only 1 line of code in Lua with the same library.

The following table, Figure 33-1, provides a collection of Lua syntax examples.

Conditional Structure	Function	Loop

`if condition then`
    `block`
`end`
	`function functionName()`
	    `block`
	`end`
		`for variable = beginning, end, step do`
		    `block`
		`end`

`if condition then`
    `block`
`else`
    `block`
`end`
	`MethodAsVariable = function()`
	    `block`
	`end`
		`for index, element in ipairs(table) do`
		    `block`
		`end`

`if condition then`
    `block`
`elseif condition then`
    `block`
`else`
    `block`
`end`
		`repeat`
		    `block`
		`until condition`

		`while condition do`
		    `block`
		`end`

Figure 33-1: Lua Syntax Examples

There are three types of subscriptions for Corona SDK; an iOS subscription that costs $199/year, an Android subscription that costs $199/year, and an all platform subscription that costs $349/year.

33.1 Installation

Go to http://www.anscamobile.com/corona/ and select "Try Corona" as seen in Figure 33-2.

Figure 33-2: Corona Website Menu

You will be prompted to register before you can download. Enter your email address and a password. Once the download is finished install Corona by opening the file and following the instructions. Before moving on be sure that you have a text editor. All project files in Corona are text files. If you do not have a text editor, Text Wrangler is a free text editor for Mac OS X. Browse to http://www.barebones.com/products/textwrangler/ to get Text Wrangler.

33.2 Hello World

Once Corona is installed you will be ready to create your first app with Corona SDK. The following steps will walk you through the process.

1. Open Corona SDK and click on "New Project."

2. Select the Blank template and set the App Name to "Balloon" as shown in Figure 33-3.

Figure 33-3: Create New App

3. Select a location to save the project.

Chapter 33: Corona SDK   367

Figure 33-4: Project Completed Successfully

4. After the project has successfully completed select the button labeled "Show in Finder…" as seen in Figure 33-4.

33.2.1 Understanding the Workspace

In the directory of the project there are three text files: build.settings, config.lua, and main.lua shown in Figure 33-5.

Figure 33-5: New Project Contents

The build.settings file specifies the settings of your build. Things such as the default and supported device orientations of your program are defined in this file. The plist is also defined in this file. The plist determines the name of your icon files, which icon file sizes are supported by your app, and whether or not the status bar is hidden when your app is being run.

The default contents of the build.settings file:

```
settings = {
    orientation = {
        default = "portrait",
        supported = { "portrait", }
    },
```

```
        iphone = {
            plist = {
                UIStatusBarHidden = false,
                UIPrerenderedIcon = true, -- set to false for "shine"
    overlay
                UIApplicationExitsOnSuspend = true,
            }
        },

        --[[ For Android:

        androidPermissions = {
            "android.permission.INTERNET",
        },
        ]]--
    }
```

The config.lua file configures certain parts of your application.

```
        application = {
            content = {
                width = 320,
                height = 480,
                scale = "letterBox",
                fps = 30,

                --[[
                imageSuffix = {
                    ["@2x"] = 2,
                }
                --]]
            }
        }
```

Chapter 33: Corona SDK   369

The main.lua file contains the code that makes your app run. In the blank template this file is empty.

33.2.2 Writing the Code

Now it is time to write the code for main.lua, the main section of the program. You will initialize, update, and draw objects in this file.

1. First, add the physics engine and start it.

   ```
   local physics = require("physics")
   physics.start()
   ```

2. In order for objects to fall you need to set the gravity.

   ```
   physics.setGravity(0, 9.8)
   ```

3. Corona SDK defaults to allowing only one touch but you want multitouch. Turn on multitouch.

   ```
   system.activate("multitouch")
   ```

You already made an image called background.png, which is in the balloon folder on the desktop. The image is 320 pixels wide by 480 pixels high.

4. Create a local object called background, load the image into it and display it.

   ```
   local background = display.newImage("background.png")
   ```

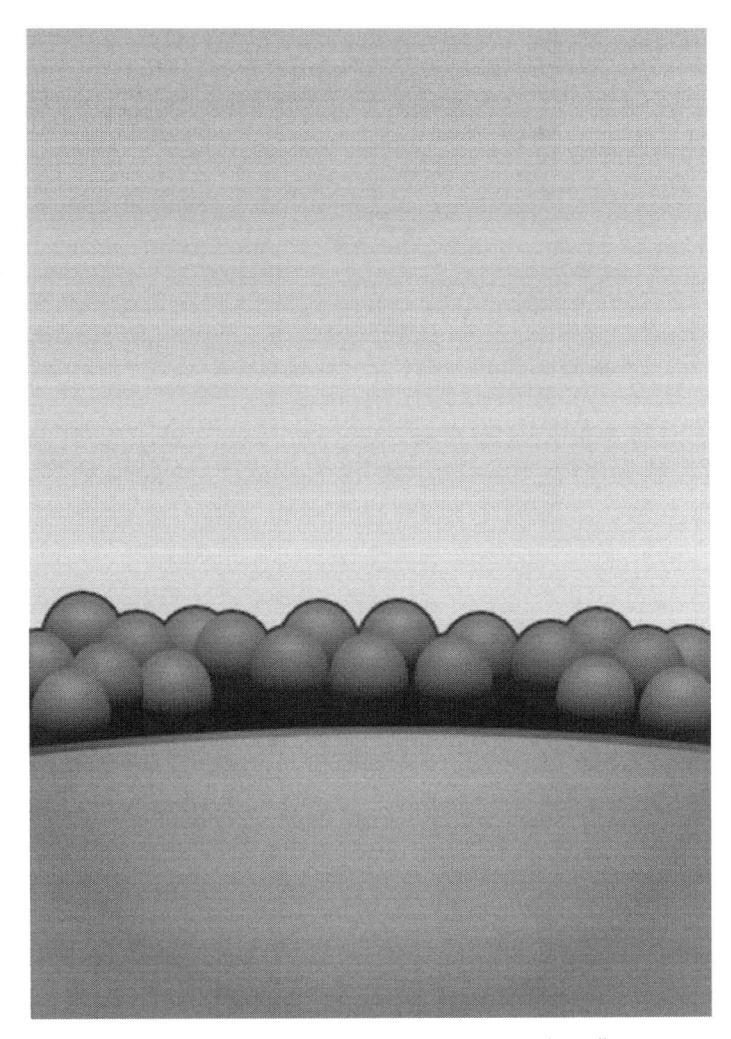

Figure 33-6: The Background, "background.png"

5. Next you will create a shadow that will stay on the ground under the balloon. The shadow is 100 pixels wide by 25 pixels tall.

   ```
   local shadow = display.newImage("shadow.png")
   ```

Figure 33-7: The Balloon's Shadow, "shadow.png"

6. Next, add the balloon image. The balloon.png image is 100 pixels wide by 125 pixels tall.

   ```
   local balloon = display.newImage("balloon.png")
   ```

Figure 33-8: The Balloon, "balloon.png"

7. Position the image to be in the middle of the screen on the X-axis.

   ```
   balloon.x = display.contentWidth/2
   ```

8. To position the shadow under the balloon, create a function that will be called by a timer to update the shadow's position using the balloon's position.

   ```
   function alignShadow()
       shadow.x = balloon.x
       shadow.y = 436
   end
   tmr = timer.performWithDelay(10, alignShadow, -1)
   ```

9. In order for the balloon to be physics enabled you need to add it to the physics engine as a body.

   ```
   physics.addBody(balloon, {bounce = 0.5, friction = 1.0, radius=50})
   ```

   The parameters above define the physical properties of the balloon. The bounce parameter determines how much bounce this object will have when colliding with other objects as well as how much bounce other objects have when they collide with this object.

   The friction parameter determines if the object receives a resistant force when colliding with or sliding along other objects

   The radius parameter makes the object a circular object with the radius of the value given. If the radius were not added to the parameter list then the hit-box of the image would be the edges of the picture itself.

10. Now you are going to add the image that will represent the grass. The grass will not be the object colliding with the balloon. Instead, the grass will be an overlay so it looks as though the balloon sinks into the grass. Create the image object.

    ```
    local grass = display.newImage("grass.png")
    ```

11. Now you will position the grass to be rendered at the bottom of the screen.

    ```
    grass.y = display.contentHeight - grass.contentHeight/2
    ```

12. To ensure that the balloon will not go off screen when being pushed by the player, you will add walls on the sides and a ceiling to act as static physics objects.

    ```
    local leftWall = display.newRect(0,0,1,display.contentHeight)
    local rightWall = display.newRect(display.contentWidth-1,0,1,display.contentHeight)
    local ceiling = display.newRect(0,0,display.contentWidth,1)
    local floor = display.newRect(0,440,display.contentWidth,1)
    ```

13. Now you will add the rectangle objects to the physics engine.

    ```
    physics.addBody(leftWall, "static", {bounce = 0.1, friction = 1.0})
    physics.addBody(rightWall, "static", {bounce = 0.1, friction = 1.0})
    physics.addBody(ceiling, "static", {bounce = 0.1, friction = 1.0})
    physics.addBody(floor, "static", {bounce = 0.1, friction = 1.0})
    ```

14. The default display of the rectangles is to draw them as being white. I want these walls to be invisible. To change the rectangle, you will use the transition.to method with a time of zero, so it will be instantaneous, and change the alpha to zero, making them invisible.

    ```
    transition.to(leftWall, {time = 0, alpha = 0})
    transition.to(rightWall, {time = 0, alpha = 0})
    transition.to(ceiling, {time = 0, alpha = 0})
    transition.to(floor, {time = 0, alpha = 0})
    ```

15. Now make the status bar hidden. To do this add this line:

    ```
    display.setStatusBar(display.HiddenStatusBar)
    ```

16. You are now going to write the first function for your game.

    ```
    function moveBalloon(event)
        local balloon = event.target
        x=0.2
        y=0.2
        if event.x > balloon.x then
            x = -x
        end
        if event.y > balloon.y then
    ```

```
            y = -y
        end
        balloon:applyLinearImpulse(x, y, event.x, event.y )
    end
```

This function gets the object event and does not return anything. It moves the balloon using linear impulses. The event object is from touch events handled in the next line of code.

17. You will now add an event listener to your object, the balloon.

    ```
    balloon:addEventListener("touch",moveBalloon)
    ```

    On any "touch" the moveBalloon method is called. This is only done for the object balloon.

18. You now will add one more method. This method will change the direction of gravity based on the tilt of the device. This is made possible using the accelerometer.

    ```
    local function onTilt( event )
        physics.setGravity( 10 * event.xGravity, -10 * event.yGravity )
    end
    ```

19. Finally, add an event listener to the Runtime that handles the accelerometer and passes parameters to the method onTilt that you defined above.

    ```
    Runtime:addEventListener( "accelerometer", onTilt )
    ```

20. Save the file.

33.2.3 Running the Simulator

You can run the Corona SDK simulator one of two ways.

1. Without console output:
   Applications > CoronaSDK > Corona Simulator
2. With console output:
   Applications > CoronaSDK > Corona Terminal

Either way will open the simulator but only the latter will open Terminal linked with the Corona Simulator. The Terminal is needed to display any output that you put in your code using:

```
print("Hello Console")
```

1. With the Corona Simulator active, select
   File > Open
2. Select your main.lua file in your balloon folder on your desktop.
3. Before you open the file, notice that you can select a device to simulate on. You can choose iPhone, iPhone4, iPad and more.
4. With the device of your choice selected, open your file.

Now you have a complete working app that can be published on the App Store.

33.3 Building for iOS Distribution

33.3.1 Compiling

To compile into an app:

1. Open Corona Simulator
2. Open your main.lua file
3. Select *File>Build>iOS*
4. Enter an app name
5. Enter a version number
6. Make sure the "Build for" option has Device selected
7. Choose which devices you want to support - choosing Universal is recommended
8. Select your code signing identity in the drop down menu
9. Choose a directory as the Save to Folder
10. Select Build

33.3.2 Adding to a Device

Using iTunes:

1. Open iTunes
2. Add the app to your Library
3. Connect your device
4. Sync the app to your device

Using Xcode:

1. Open Xcode
2. Select Window > Organizer
3. Select the Devices tab
4. Drag the app file to your device under DEVICES on the left hand menu

33.3.3 Publishing

After you have built and tested your app as instructed in the previous steps, publish your app to the App Store by following the steps detailed within ***Section V: Submitting Your App***.

33.4 Advertising

Corona SDK supports advertisements from the services InMobi and inneractive. The Corona SDK reference pages below have instructions on signing up for each service and how to use them in your apps.

InMobi Official Site

http://www.inmobi.com/

Corona SDK InMobi Reference

http://developer.anscamobile.com/reference/banner-ads/

inneractive Official Site

http://www.inner-active.com/

Corona SDK inneractive Reference

http://developer.anscamobile.com/reference/ads-inneractive/

GameSalad

GameSalad offers the ability to start creating games without any prior programming knowledge. It's a drag and drop based game builder that offers a great introduction into game design. The basic software is free, and allows you to build and publish your games to either the web or iTunes, which requires you to have an Apple Developer account, which is covered in Chapter 6. Upgrading to the "Professional" version costs $299 a year and gives you access to several advanced options. These options include the ability to take advantage of Apple's iAd mobile advertising platform, In-App Purchasing, and Game Center functionalities in your app. These are Apple's mobile advertising, In-App purchase, and score-tracking services. Also included with the Professional version is the ability to publish Android and Windows 8 games.

While GameSalad offers a user-friendly interface to create and publish iOS, Android, Windows 8, web, and Mac OS X games, it's not without its limitations. Currently, GameSalad does not have the ability to make 3D or multiplayer games. It also does not provide access to all of the iPhone's functionalities such as its GPS, compass, and gyroscopes.

It should also be noted that you never receive the actual source code for your project. When you complete your game, you submit it to GameSalad; they process it and send you back an application (.app) file. You then submit it to Apple's App Store using the methods explained in Chapter 40: Submitting Apps Using Application Loader.

34.1 Installation

1. Go to http://gamesalad.com and download the latest version of the GameSalad Creator.
2. Open the installer and drag the GameSalad icon into your "Applications" folder.
3. It is now installed and ready to use.

34.2 Hello, World!

For the demonstration app, you will create a simple game containing a shooter, a bullet, and a brick. A tall rectangle will act as the shooter, and will be positioned on the left side of the screen. The shooter will fire a bullet at a brick, located on the other side of the screen, whenever the user taps the screen. When the bullet hits the brick, the text "Hello, World!" will be displayed in front of it.

1. When you open GameSalad, you will be presented with several options on the left side of the window.

   - *Home:* Read about recent GameSalad-related success stories and events
   - *News:* A link to the GameSalad Blog
   - *Start:* Guides, documentation and tutorials to get you started with GameSalad
   - *Profile:* Login, check your GameSalad messages or update your personal settings
   - *New:* Start creating your own GameSalad game, or edit and preview the included demos
   - *Recent:* Quickly locate and load your recent GameSalad projects
   - *Portfolio:* A collection of all your GameSalad creations

2. Click the plus sign labeled "New" to start building a new app.

Figure 34-1: GameSalad's main screen

3. Select the first icon labeled "My Great Project," and then click the button labeled "Edit in GameSalad Creator" in the bottom right of the window. This will open up a blank project for you to work with.

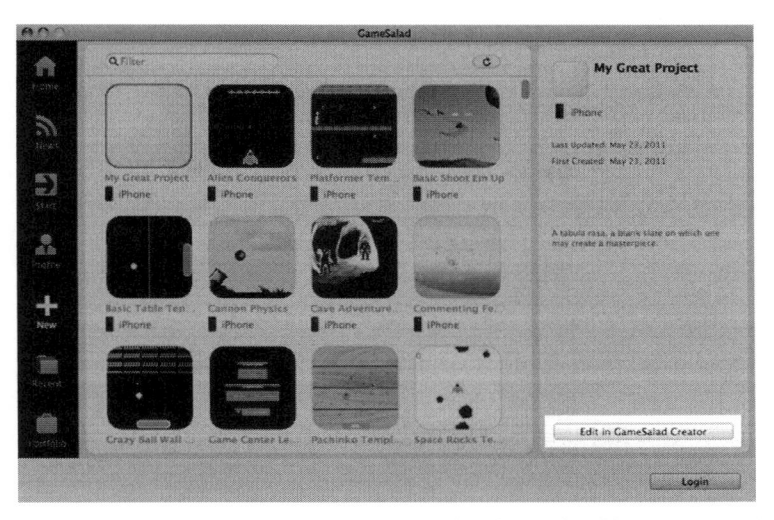

Figure 34-2: Click "Edit in GameSalad Creator" to begin building your project

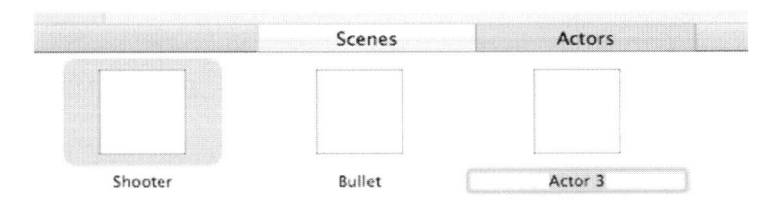

4. In between the top and bottom halves of the window, you should see two tabs, labeled "Scenes" and "Actors." Click the Actors tab. Actors are the elements of your game, you place them in a Scene and tell them how to behave when they interact with other actors or inputs such as a mouse click or tapping on the screen. Locate the ⊞ button beneath the section that says "No items yet. Click + to add items," and click it 3 times, you should now have three actors. Click on their names and change them to "Shooter," "Bullet," and "Brick."

Figure 34-3: Highlight and change the Actor's names

5. If you want to include custom graphics in your game, GameSalad makes it easy to do so by dragging images into the Actors' attribute window. However, since this is just a basic tutorial, you are just going to change the size of your actors just to differentiate them from one another. Double click on "Shooter" to open up the attributes for that actor. This is where you can set the properties of the actors, such as their size, color, and position. Click the arrow next to the "Size" attribute and change the width to "50."

Chapter 34: GameSalad 379

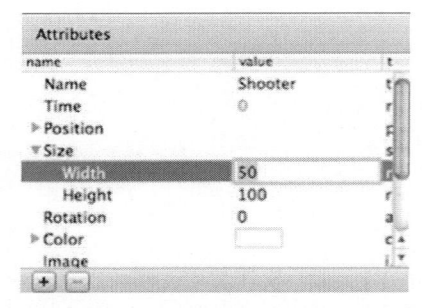

Figure 34-4: under the Size attribute, highlight and change the Width to 50

6. "Rules" tell your actors how to behave when they interact with each other.

Here, you are going to make a rule to tell the actor you named "Shooter" to perform an action when the user clicks their mouse pointer anywhere inside the app. Click the "Create Rule" button towards the top right of the window. The default values are fine for now.

7. In the bottom left of the GameSalad window, there is a panel labeled "Library" where you can select the behaviors, images and sounds associated with the selected actor. Scroll down until you see the behavior called "Spawn Actor," then click and drag it into the section labeled "Drag your behaviors here" *inside* the Rule box you just created. In the resulting "Spawn Actor" window, change the "Actor:" section to "Bullet." Now, when you click the mouse, the "Shooter" will create, or "spawn" the "Bullet". You will assign attributes to the "Bullet" actor in the next step.

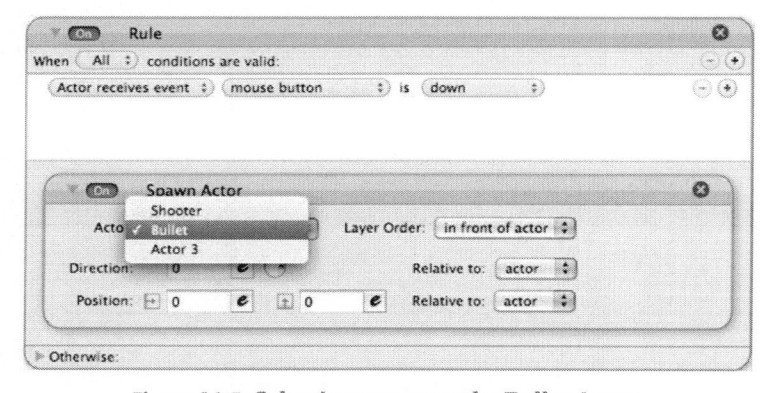

Figure 34-5: Selecting to spawn the Bullet Actor

8. Click the Back arrow in the top left, and then double-click on the "Bullet" actor. Change its size attributes to Width = 10 and Height = 10. While still in the "Attributes" panel, scroll down until you see the "Motion" option, click the arrow to show more options, and under "Linear Velocity," change "X" to 300. Now you have a relatively small, 10x10 "Bullet" that, when spawned, will move positively along the x-axis (meaning from left to right).

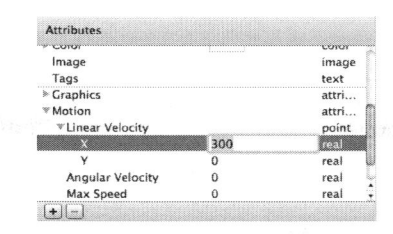

Figure 34-6: Change the Bullet's linear velocity along the X-axis to 300

9. Click the "Back" arrow at the top left of the screen to get back to the screen that shows the three actors you've created. Double-click the "Brick" actor to open its panel. Click "Create Rule," then click the selector that is currently set to "mouse button," and change it to display the "overlaps or collides" option. Then change the last selector to "Bullet," so that it matches Figure 34-7.

Figure 34-7: Setting a rule for the Brick actor

10. Now, when the "Brick" actor collides with the "Bullet" actor, it knows that something needs to happen. Normally, you might tell it to disappear, show an explosion, add some points to your score, or all three. For this demonstration, you will just have it display some text. From the "Behaviors" panel, drag and drop a "Display Text" behavior *inside* the behaviors panel of the Rule you just created. The text property inside the "Display Text" panel you just added should be set to "Hello world!" by default. Change the color to a shade of red to make it more visible by clicking the box next to "Color" and using the color picker.

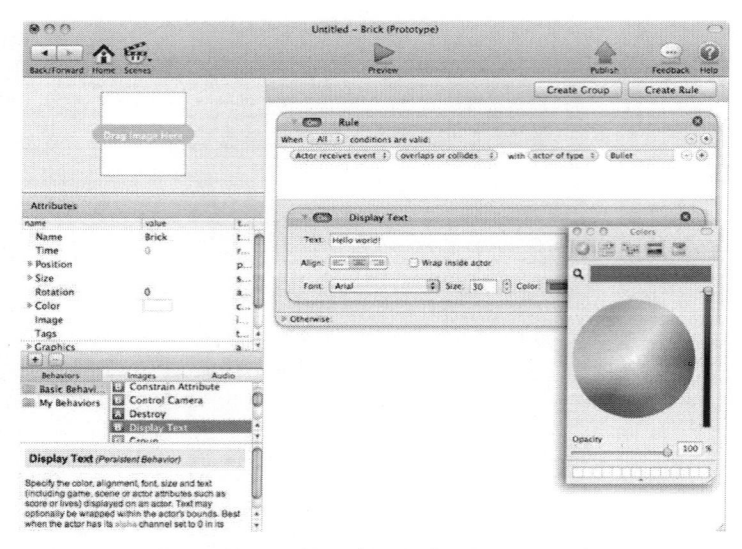

Figure 34-8: Setting the Display Text options

11. Click the "Scenes" icon in the top left corner of the window and select "Initial scene." This is where you will place all of the actors you just created, providing them a stage where they can interact. Each scene that you create represents a different stage, level, or menu screen for your game. Drag and drop the "Shooter" actor to the left side of the blank canvas, then place the "Brick" actor to the right side, so they line up horizontally. You do not need to place the "Bullet" actor because the "Shooter" actor will spawn it when clicked.

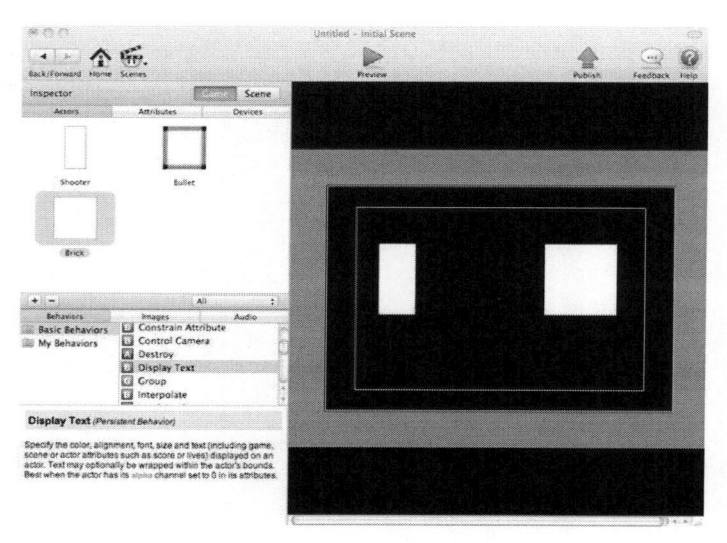

Figure 34-9: Placing your actors into the Initial Scene

382   Chapter 34: GameSalad

12. Save your file, and then touch the green "Preview" arrow at the top of the window to test your first GameSalad game. When you click the mouse button while the pointer is inside the scene, the "Shooter" actor will launch a "Bullet" actor that, when it collides with the "Brick" actor, displays the text "Hello world!" briefly.

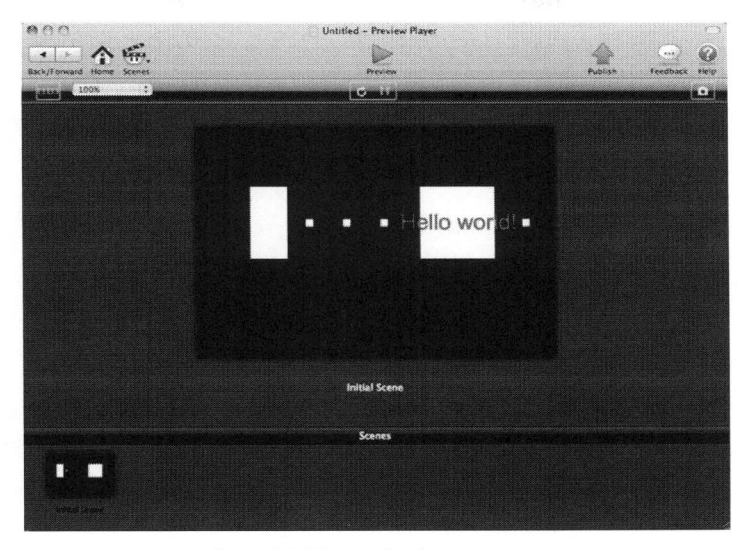

Figure 34-10: Previewing your app

Congratulations! You've just created a (very simple) game using the GameSalad Creator!

34.3 Testing on an iOS Device and Distribution

GameSalad's iOS simulator is a useful tool for prototyping and aiding in design, but it's no substitute for testing on an actual device, which will be covered next. If you have not signed up for Apple's iPhone Developer Program, you will need to do that first, which is covered in Chapter 6. If you have all of that ready to go, you are ready to start testing your app on an actual iOS device.

GameSalad makes it relatively easy to get your app up and running on an iOS device, but they are going to ask you to fill out some basic information during the publishing process. For this reason, it's good to have a few things prepared beforehand, such as an app description, a 1024x1024 PNG image to represent the app on the App Store, keywords, screenshots and so forth. Because you are still testing your app, it's not vital that all of these are filled out completely, but it will definitely make it easier if you have everything all ready to go before you start.

One thing that is required is a screenshot of your app, which, fortunately, GameSalad makes really easy to do. While previewing your game, click the little camera icon ( ) in the top right corner of the window, which will bring up a window showing a series of screenshots. Click the button below the middle window labeled "Set project screenshot" to populate that section and then click "Done." Following is a screenshot associated with the app.

Figure 34-11: The screenshot management window

1. In order to put your new app onto an actual device, the project you generated in GameSalad needs to be converted into an application file - with the .app extension. Start by clicking the orange arrow labeled "Publish" which is to the right of the green "Preview" arrow in GameSalad. This tells GameSalad you are ready to upload your project to their website, have them convert it to a .app file and send it back to you for testing on an iOS device. In order to test your app on a device or publish it to the App Store, you need to register with GameSalad. Registration is free and the only information that is required is an email address, Username, Password, and acknowledgement that you have read and agreed to their Terms of Use and Privacy Policy. Create an account and log in to continue. The next window that pops up will ask you which device you wish to publish to. Currently, the options are web, iPhone, iPad, and Mac. Select "iPhone" to continue.

2. The GameSalad Publishing Manager will pop up. Click the "Create New Game" button in the bottom right-hand corner to continue. This is where you will set all of your categories, screenshots, descriptions and so forth. Change the title to "Hello World," then drag and drop the 1024x1024 image that you want to use to represent the game on iTunes into the Icon box in the lower left corner of the window. Normally, here is where you would set different options for your app such as the primary, secondary, and sub-categories, as well as its title, description, and copyright info. GameSalad will not let you continue without typing in at least a few keywords, so type in some generic ones for now, then click "Next."

Figure 34-12: GameSalad Publishing Manager

3. The next screen will ask you for the provisioning profile you wish to use, which, hopefully you have already set up through Apples Developer Portal. The "Advanced Options" button allows you to set options such as which orientations to support (landscape, portrait, etc.) and your bundle identifier, which should uniquely identify your company and app. The general format of the bundle identifier is usually "com.yourdomainname.yourappname." The default values are fine for now since you do not intend to actually publish this app to Apple's App Store.

Figure 34-13: The Advanced Platform Settings Window

4. Next, GameSalad will ask you for a YouTube link to put beside your game. This is a nice option if you want to show people a video of your game being played in addition to the screenshots on iTunes. It isn't a requirement, though, so for this project, just leave it blank and click "Next."

5. This screen is where you upload your screenshots, of which you are required to have at least one. The screenshot of the app that created earlier by clicking the camera button

should show up in the bottom bar here. Drag it into the first box and click "Next."

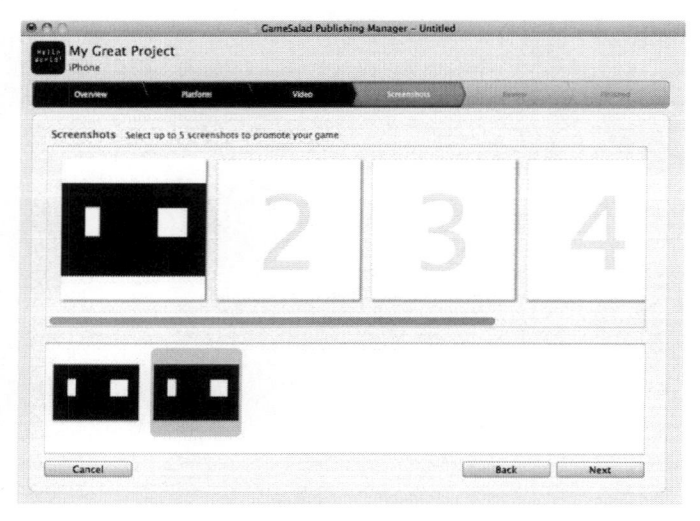

Figure 34-14: Selecting a screenshot for your app

6. The next tab will show you a summary of all the information just supplied. Review everything to make sure it's correct. If you find something that needs to be corrected, click the "edit" button above that section and make the necessary corrections. Scroll down to find the "Publish" button, and click it when you are ready to publish your app. GameSalad's Submission Terms and Agreement will pop up, after you have read and understood it, click "Agree and Upload."

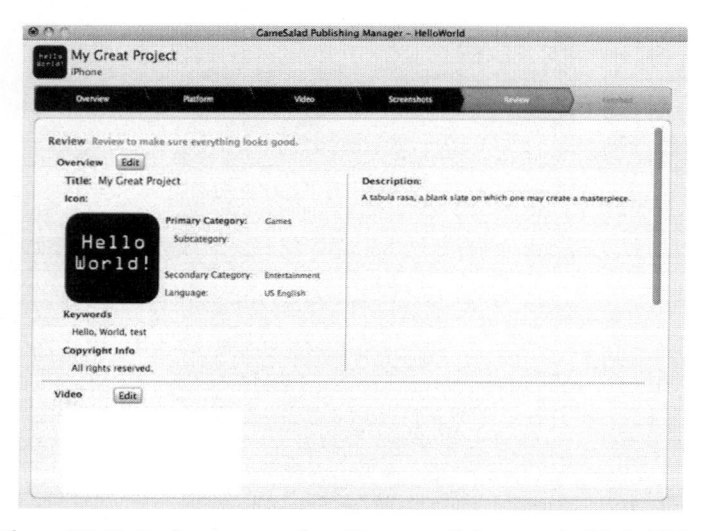

Figure 34-15: Review your app's options, scroll down and click Publish

7. Your game will be sent to GameSalad's site and converted into a packaged .app file. Then, a window will pop up asking you where you want to save the file. Select a location that you will remember and click "Choose."

Figure 34-16: GameSalad will receive, convert, sign and return your app

8. Provided everything went as planned, GameSalad will offer you Congratulations.

Figure 34-17: A successful upload prompts several options on how to proceed

9. You'll be presented with several options on how you wish to proceed with your app, including testing it on a device or showing it in the Finder to copy or archive it. Make sure you have a registered device plugged into your computer and click "Test." This will open iTunes and sync the program you just created to your testing device. On your iOS device, locate your app and open it. It should function as it did when you previewed it, except here, the shooter reacts to a finger tap as opposed to a mouse click.

10. Now that you have your app file, you are ready to upload it to Apple's App Store using Apple's Application Loader or directly through Xcode 4, as described in ***Section V: Submitting Your App***.

Chapter 34: GameSalad   387

Titanium Studio

Titanium Studio allows you to write apps using JavaScript, HTML, and CSS. The main advantage being that using it grants you access to native iOS user interface elements, allowing you to make your app truly look like it was developed for iOS using Xcode and Objective-C.

One disadvantage is that it adds another layer of programming conversion to your app instead of writing it in native Objective-C code. Because of this, developing in Titanium Studio will tend to produce larger and slower apps than developing in Xcode and Objective-C.

35.1 Installation & Setup

Before you get started developing the demonstration app, you will need an Apple Developer account. You will also need the iOS SDK and Xcode installed on your machine in order to test your app on a device, and eventually upload it to iTunes. For help with completing these steps, please refer to Chapter16.

1. To get started, visit http://appcelerator.com to download and install Titanium Studio. It is a free download, but you will need to create an account, verify it through email, and then login to reach the download section.

Figure 35-1: Titanium Studio is available for a variety of platforms

2. The first time you start Titanium Studio, it will ask you to login. Sign in with the account you created previously to continue.

3. It will then ask you where you want to locate your workspace. This will be the directory where all of your project's files will be located on your local machine. After selecting your workspace, Titanium Studio will write the necessary files and open to the Dashboard view.

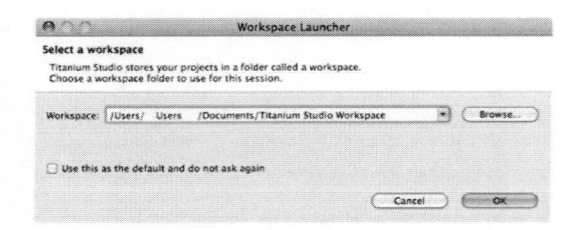

Figure 35-2: Selecting your workspace location

35.2 Hello World!

When you begin creating an app with Titanium Studio, it creates a default project in the designated workspace. In this tutorial, you will slightly modify the default project, a simple app demonstrating tab switching, so that you can learn how to navigate Titanium Studio's interface.

1. After Titanium Studio starts, you will be brought to the Dashboard view. On the left are two buttons, "Create Project" and "Import Project." Once you begin developing your app, this will be the "Project Explorer" panel, which will show all associated directories and files associated with your app. Alternatively, switching tabs in this panel changes it to the "App Explorer" panel, which allows you to easily navigate between other projects you've created with Titanium Studio. The large window on the right is the main window - where you will be able to edit the contents of the files selected in the Project Explorer to build your app.

Below the main window is a panel containing the Console, Terminal and Error panels, which are selected using the tabs at the top of the panel. To begin, click the "Create Project" button in the left panel.

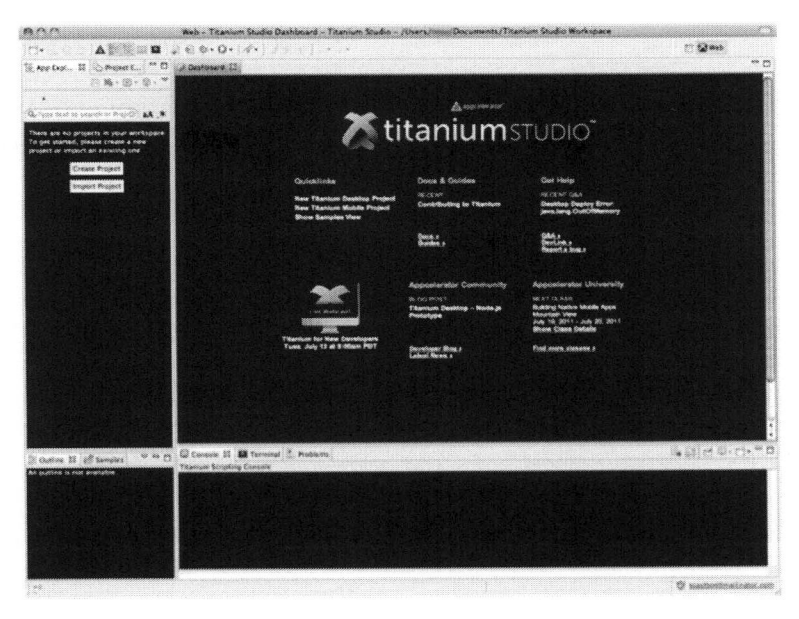

Figure 35-3: Titanium Studio's main screen

2. A window will pop up asking you to select the wizard to use to develop your app. Since you are developing for iPhone app, select "Titanium Mobile Project" and then click the "Next >" button.

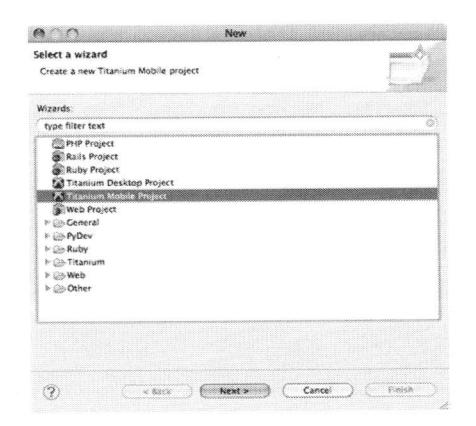

Figure 35-4: Wizard selection window

3. The next screen will ask you to fill in some details regarding the app that you are about to create. Name the project "Hello World," enter an appropriate app id (traditionally *com.yourdomainname.yourappname*) and the website you wish to associate with your app (your personal or company domain) in the provided fields, then click "Finish."

Chapter 35: Titanium Studio   391

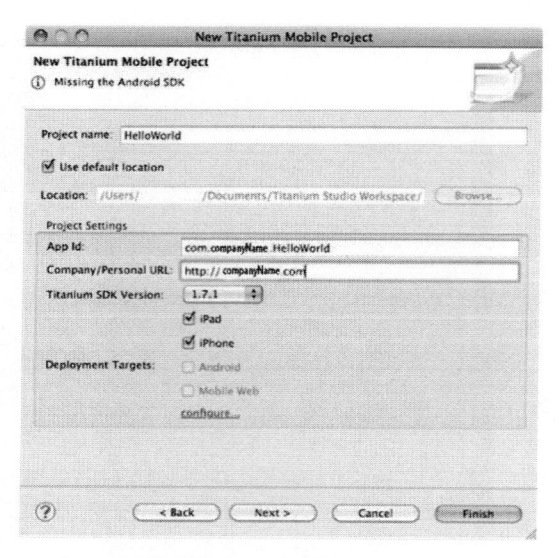

Figure 35-5: Titanium mobile project settings

4. Titanium Studio's main screen will open and show you the details of your app that you just entered. There will also be some additional information that you may wish to modify at this time if you wish, such as the version number, publisher, copyright details, and the description. Because this is just a simple introductory app, the default values are fine for right now.

Figure 35-6: The configuration options for the app

392   Chapter 35: Titanium Studio

5. In the left panel, you will see several files and folders related to your app. Click the arrow next to the "Resources" folder to expand it, then double click on the "app.js" file to open it directly in Titanium Studio. You'll see the JavaScript code for a very simple app that Titanium includes when you start a new project. Change the text attributes of the variables `label1` and `label2` to "Hello World!" and "Hello again!" respectively, as it appears in the following example:

   ```
   var label1 = Titanium.UI.createLabel({
       color:'#999',
       text:'Hello, World!',
       font:{fontSize:20,fontFamily:'Helvetica Neue'},
       textAlign:'center',
       width:'auto'
   });
   ...
   var label2 = Titanium.UI.createLabel({
       color:'#999',
       text:'Hello again!',
       font:{fontSize:20,fontFamily:'Helvetica Neue'},
       textAlign:'center',
       width:'auto'
   });
   ```

6. Save the file, then run the app in the simulator by selecting Run > Run from the main menu. It might take 10-20 seconds for the simulator to start, but this generally only happens the first time you run a test, subsequent load times will be considerably shorter.

Figure 35-7: The demo app as it appears in the iOS Simulator

When the simulator does pop up, you can see and interact with the small demo app. There are tabs at the bottom. When each is clicked, you can see the labels that you renamed. This completes the Titanium Studio demo.

If you want a more comprehensive overview of the UI elements that are available through Titanium Studio, Appcelerator has included a sample app called "Kitchen Sink" for just this purpose.

In the bottom left corner of the Titanium Studio window, there is a tab called "Samples." Click the arrow next to the "Titanium Mobile" item under that tab to display the samples included in this release. Right click on the "Kitchen Sink" listing and select "Import sample as project…"

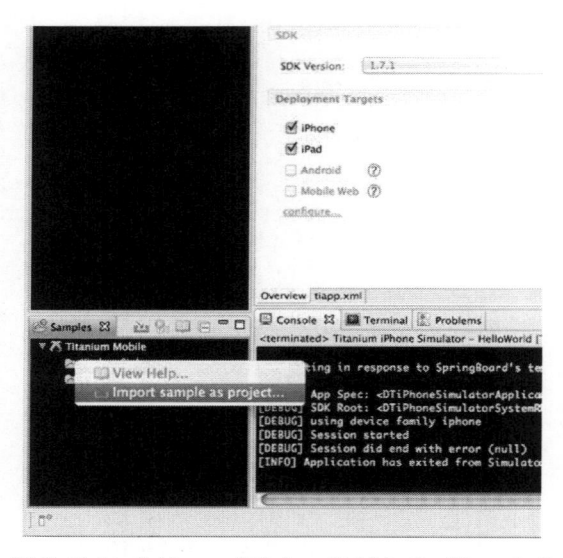

Figure 35-8: Right-clicking on "Kitchen Sink" in the "Samples" panel

Click "Finish" in the window that pops up to have Titanium Studio download the Kitchen Sink Git repository, making it available to Titanium Studio on your local machine. To start it in the simulator, make sure it's selected in the "Project Explorer" panel (it should be located just below the HelloWorld app you just made) then select Run > Run from the main menu. You will be prompted with a window asking you which simulator or emulator you wish to use. In this case, you would select "iPhone Simulator." Kitchen Sink is meant to function as a fairly comprehensive collection of all of the iOS native UI elements that are available when developing with Titanium Studio.

35.3 Testing on a Device and Building for Distribution

To test your app on a device, from either the App Explorer or Project Explorer panel, click the run icon, which looks like an arrow pointing right, then select "iOS Device."

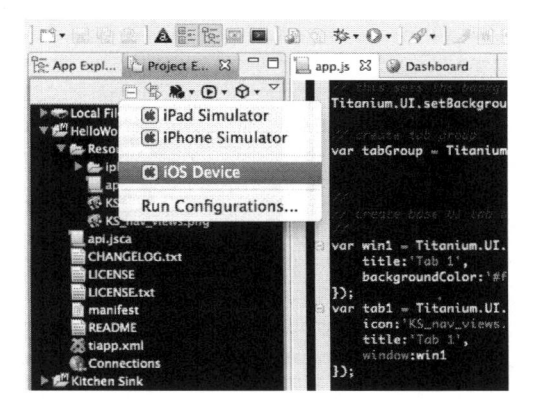

Figure 35-9: Click and Hold the run icon, then select "iOS Device."

A window will appear, showing you the requirements that you need to fulfill before you can run a test on a device. These include registering your device with Apple, obtaining various development certificates, and provisioning profiles. Please refer to Chapter 9 for information on completing these steps. Once these are fulfilled, click "Finish" to have Titanium Studio deploy the app on your device for testing.

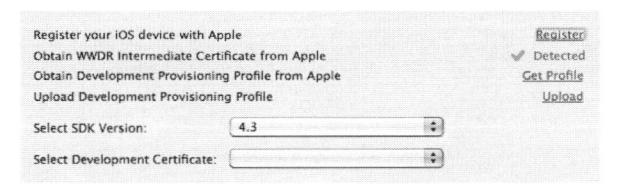

Figure 35-10: Requirements for device testing

The process of packaging your app for distribution is similar to that of testing it on a device. When you are ready for distribution, click the cube icon right next to the "Run" icon in the Project or App Explorer window, then select "Distribute – App Store."

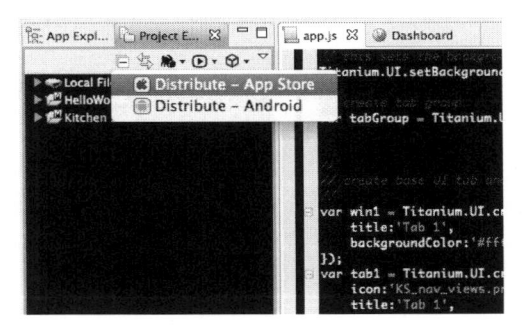

Figure 35-11: Packaging your app for distribution

Chapter 35: Titanium Studio   395

The next window that appears will have the requirements that need to be fulfilled to distribute your app on the App Store. It will initially look very similar to the requirements for testing on a device. The main difference: while only the "Development" provisioning profile is required for testing, here the "Distribution" profile and certificate are needed. For help on creating these certificates and provisioning profiles, please refer to Chapter 9.

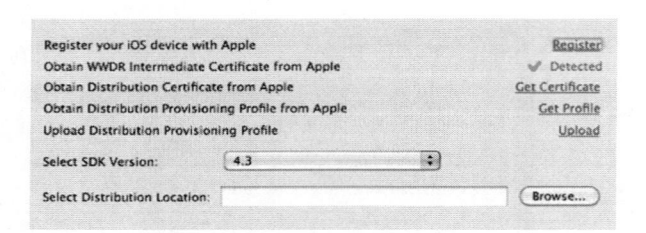

Figure 35-12: Requirements for app distribution

Once you fulfill the requirements, click "Finish." Titanium Studio will then package your app and provide you with an IPA file that will be installed in the Archives tab in Xcode. You can then submit it to the App Store as explained in Section V.

MoSync

MoSync is an SDK for developing mobile apps for multiple platforms simultaneously. Apps can be created with only HTML5 and JavaScript. In addition, MoSync has an Eclipse-based IDE that can be used for C/C++ app development, as well as, to create hybrid apps that combine both JavaScript and C/C++. MoSync supports a multitude of mobile operating systems, including Android, iOS, Windows Phone 7, Windows Mobile, Symbian S60, and Java ME, and Moblin/MeeGo, and Blackberry. MoSync supports three Development Models: the Classic Procedural Model, the Event-Driven, Object-Oriented Model, and the Full GUI-Based Model.

One of the major advantages of MoSync is that you can write and build an app for many different Mobile Platforms at the same time, thus eliminating the need to program in several different languages to support different devices. In addition to the many portable devices supported by MoSync, it can also be used to develop apps for Windows and OS X.

Another advantage is the Native UI support for Android and iOS platforms. This allows you to create apps that will look like native Android or iOS apps. However, this feature is currently *only* supported by the Android and iOS platforms.

One of the disadvantages of MoSync is that when developing for multiple devices you have to design your layout with all the potential screen sizes in mind. Another disadvantage is that the documentation isn't as organized and user friendly as some of the other alternative development options in this book.

36.1   Installation

Before installing the MoSync SDK you will need to install Xcode with the iOS SDK in order to build iOS apps. You will also need to join the Apple Developer Program. Help and information on how to do each can be found in Chapters 16 and 6, respectively.

First download and install MoSync SDK from their website at www.mosync.com.

Figure 36-1: Download MoSync

1. Download MoSync, click on the appropriate version according to your operating system. Once downloaded, open and click on the package to begin installation.

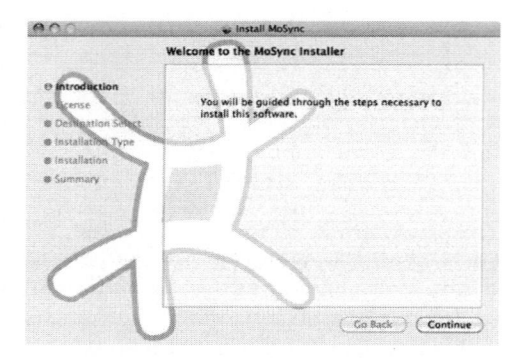

Figure 36-2: MoSync Installer

2. Proceed through the installation process, reading over and agreeing to the license, and choosing the install location. Once the installation is complete, you will be prompted to log out in order to finish installation.

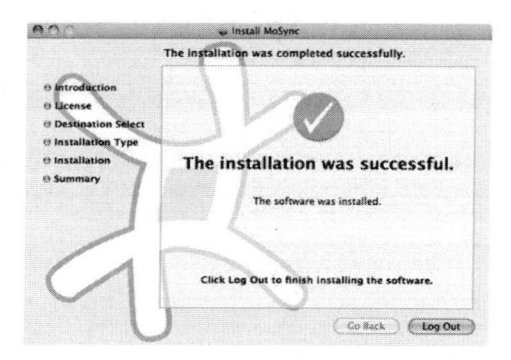

Figure 36-3: MoSync Installation Complete

3. Log back in, open MoSync and select a workspace location to save your projects.

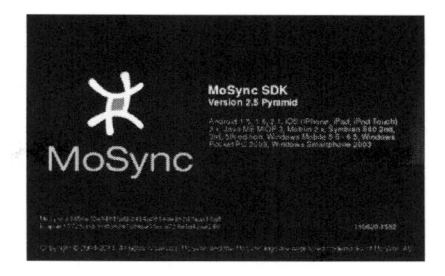

Figure 36-4: MoSync load screen

4. Register your copy of MoSync.

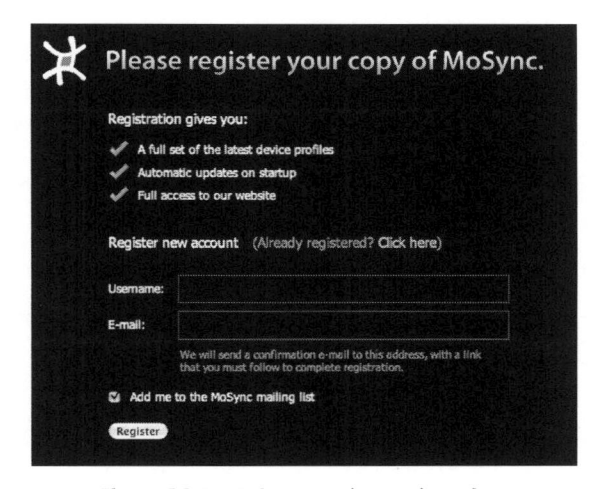

Figure 36-5: MoSync version registration

Register your copy of MoSync by following the on-screen instructions. You will need to enter a valid email address and create a username and password. Before continuing, check your email and click the activation link from MoSync Support to complete your registration.

Once registered, you will be ready to create apps using MoSync.

36.2 Example

In this section you will learn how to create your first iOS app in MoSync.

1. Create a new project by going to File>New>Project. Select MoSync> MoSync Project from the Wizard menu to create a new MoSync Project.

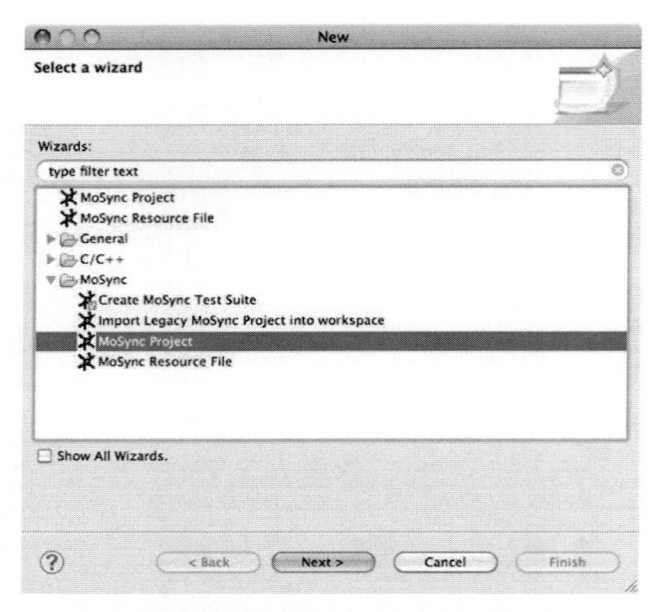

Figure 36-6: MoSync Project Creation Wizard

2. Name the Project HelloWorldMoSync and choose the save location.

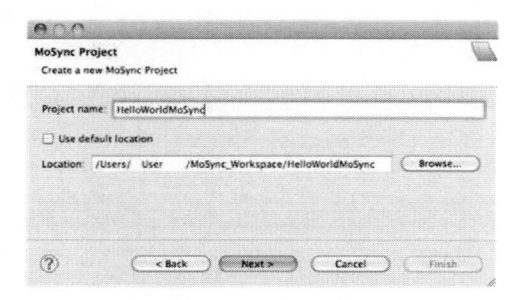

Figure 36-7: Create a new project

3. Choose the template for a MoSync Native UI Project, then click "Finish."

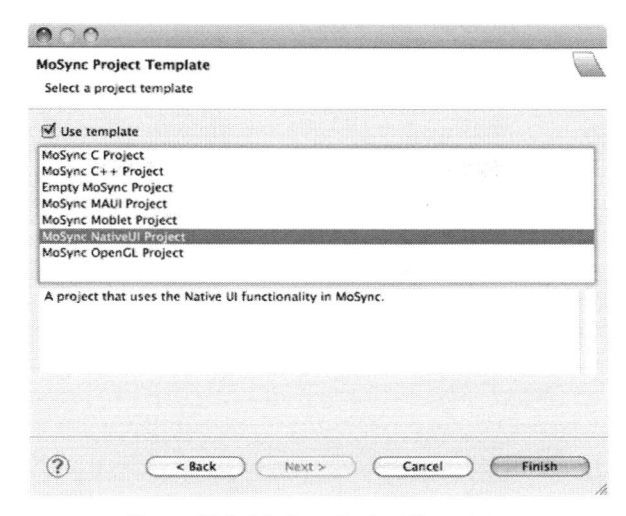

Figure 36-8: MoSync Project Template

A new project will be created for you with a button that when clicked, counts the number of times it was clicked and displays that number in the title of the button.

You are going to modify this so that there are two buttons and a label. The label will change its text when one of the buttons is clicked. To do this you must edit and add some code inside the `createUI()` method.

1. First you need to declare the buttons and label inside of the private section of the class.

```
private:
    MAHandle mButton1;
    MAHandle mButton2;
    MAHandle mLabel;
//  MAHandle mButton;
    int mButtonClickCount;
};
```

As you can see, the mButton variable is commented out because in the next step you will be modifying it.

2. First modify the code that is used to create a button.

```
// Create button1.
mButton1 = maWidgetCreate(MAW_BUTTON);

widgetSetPropertyInt(mButton1, MAW_WIDGET_WIDTH,
MAW_CONSTANT_FILL_AVAILABLE_SPACE);

widgetSetPropertyInt(mButton1, MAW_WIDGET_HEIGHT, MAW_CONSTANT_WRAP_CONTENT);
```

Chapter 36: MoSync   401

```
maWidgetSetProperty(mButton1, MAW_BUTTON_TEXT_VERTICAL_ALIGNMENT,
MAW_ALIGNMENT_CENTER);

maWidgetSetProperty(mButton1, MAW_BUTTON_TEXT_HORIZONTAL_ALIGNMENT,
MAW_ALIGNMENT_CENTER);

maWidgetSetProperty(mButton1, MAW_BUTTON_TEXT, "Button 1");

maWidgetSetProperty(mButton1, MAW_BUTTON_FONT_SIZE, "26");
```

This will create a button with the text "Button 1" and a font size of 26. The button is configured so that it will fill the available space horizontally so it will change with differences in screen size.

3. Next, create another button that will be almost exactly like the first but with the text "Button2."

```
// Create button2.

mButton2 = maWidgetCreate(MAW_BUTTON);

widgetSetPropertyInt(mButton2, MAW_WIDGET_WIDTH,
MAW_CONSTANT_FILL_AVAILABLE_SPACE);

widgetSetPropertyInt(mButton2, MAW_WIDGET_HEIGHT, MAW_CONSTANT_WRAP_CONTENT);

maWidgetSetProperty(mButton2, MAW_BUTTON_TEXT_VERTICAL_ALIGNMENT,
MAW_ALIGNMENT_CENTER);

maWidgetSetProperty(mButton2, MAW_BUTTON_TEXT_HORIZONTAL_ALIGNMENT,
MAW_ALIGNMENT_CENTER);

maWidgetSetProperty(mButton2, MAW_BUTTON_TEXT, "Button 2");

maWidgetSetProperty(mButton2, MAW_BUTTON_FONT_SIZE, "26");
```

4. Create a label by placing the below code after the code just created for the two buttons.

```
// Create a label.

mLabel = maWidgetCreate(MAW_LABEL);

widgetSetPropertyInt(mLabel, MAW_WIDGET_HEIGHT,
MAW_CONSTANT_FILL_AVAILABLE_SPACE);

widgetSetPropertyInt(mLabel, MAW_WIDGET_WIDTH, MAW_CONSTANT_FILL_AVAILABLE_SPACE);

maWidgetSetProperty(mLabel, MAW_LABEL_TEXT, "Hello World!");

maWidgetSetProperty(mLabel, MAW_LABEL_TEXT_HORIZONTAL_ALIGNMENT,
MAW_ALIGNMENT_CENTER);
```

You now have a label with the text "Hello World!" centered in the middle of the screen. By setting the height and width to fill available space it makes it so the label will use up any remaining space horizontally and vertically. So if this label is placed between the two buttons it will be centered in the screen.

5. Add the buttons and label to the layout, which was automatically created for you in the template. Now add the following code to create the three widgets.

    ```
    // Add the widgets to the layout.
    maWidgetAddChild(layout, mButton1);
    maWidgetAddChild(layout, mLabel);
    maWidgetAddChild(layout, mButton2);
    ```

    This will add the three widgets to the layout which controls where they are displayed on the screen.

6. Modify the `customEvent` method to configure functionality for the two buttons by modifying the code

    ```
    void customEvent(const MAEvent& event)
    {
      if (EVENT_TYPE_WIDGET == event.type)
      {
        // Get the widget event data structure.
        MAWidgetEventData* widgetEvent = (MAWidgetEventData*)
                                          event.data;

        // Has the button been clicked?
        if (MAW_EVENT_CLICKED == widgetEvent->eventType &&
                    mButton1 == widgetEvent->widgetHandle)
        {
          ++mButtonClickCount;
          char buffer[256];
          if (1 == mButtonClickCount)
          {
            sprintf(buffer, "You clicked me once!", mButtonClickCount);
          }
          else if (100 == mButtonClickCount)
          {
            sprintf(buffer, "Wow! 100 times!", mButtonClickCount);
          }
          else
          {
            sprintf(buffer, "You clicked me %i times!",
                                    mButtonClickCount);
          }
          maWidgetSetProperty(mLabel, "text", buffer);
        }
        else if (MAW_EVENT_CLICKED == widgetEvent->eventType &&
                    mButton2 == widgetEvent->widgetHandle)
        {
          maWidgetSetProperty(mLabel, MAW_LABEL_TEXT, "Hello Again!");
        }
      }
    }
    ```

7. To build your project for iOS, go to the Device Profiles tab on the right hand side of the IDE as shown in Figure 36-9. In the tab, right click on Apple > iPhone, then select "Set Target Phone." In the main menu bar, select Project > Build Project. MoSync will then build an Xcode project in your MoSync Workspace folder.

Figure 36-9: MoSync Device Profiles

8. Navigate to your MoSync Workspace folder and then go to HelloWorldMoSync > FinalOutput > Release > Apple > iPhone > package > xcode-proj and open the file named HelloWorldMoSync.xcodeproj to build the project and run on the iOS Simulator.

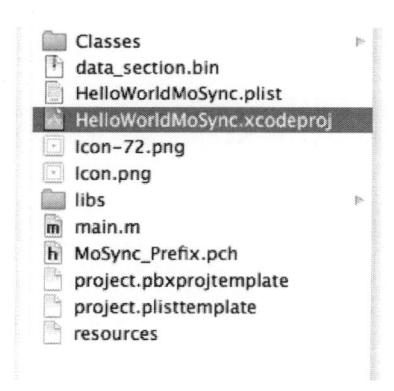

Figure 36-10: Xcode Project

9. Now that you have the Xcode project you can run the iOS Simulator. If you click "Button 2" the text of the label will change to "Hello Again!" and if you click "Button 1" three times you should see results similar to Figure 36-11.

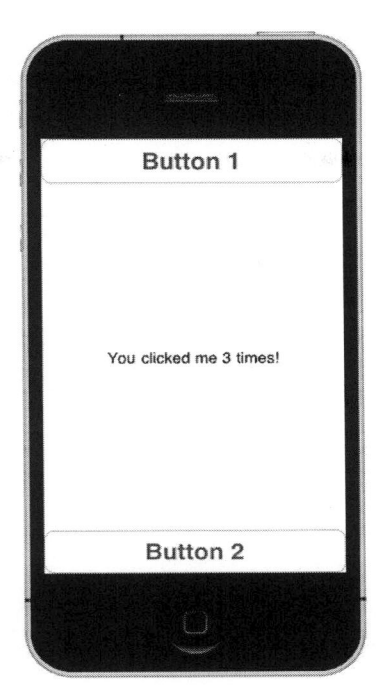

Figure 36-11: Output for HelloWorldMoSync project

36.3 Building for Distribution

Now that you have an Xcode project, you can follow the steps in Chapters 38 and 39 to build you app for distribution.

That concludes the walkthrough of the MoSync SDK. After completing this walkthrough you should be able to go on to create more advanced apps using the MoSync SDK. For more information on MoSync please refer to its documentation at www.mosync.com/documentation.

SECTION V

SUBMITTING YOUR APP

The following chapters in Section V will guide you through the process of uploading your app to Apple's App Store for review, while avoiding some of the common pitfalls.

Covered first is the means to upload your app through both the Application Loader and through the Xcode framework, and then focus on how to prepare your app for a successful upload to Apple for review.

Upon submission, your app may be instantly rejected and then the chapters of this section cover some of the top reasons as to why your good app may have gone bad, at least according to Apple. Additionally, this section has a focus on using submission tools to walk you through the potential minefield of messy metadata, image sizes, API calls, and assumed reality.

As an added bonus, this section concludes with a chapter containing a large collection of "Gotchas" that the members of the Unknown.com, Inc. team have encountered over the years of producing iPhone, iPod touch, and iPad apps, and working with Apple.

App Submission Overview

Submitting apps to Apple for distribution on the App Store can be a daunting process but becomes easier with practice. Before an app can be submitted, several steps must be taken to ensure not only a smooth delivery but to make sure that your app will pass Apple's review process.

First, the developer must login to the iOS Dev Center to create an App ID and Bundle ID through the iOS Provisioning Portal. Next, the app settings and metadata must be configured in iTunes Connect. Once the metadata for the app in iTunes Connect is correct, the app must be marked as "Ready to Upload Binary" in iTunes Connect. When this has been successfully completed, the app build settings must be configured for the project. Finally, the developer must ensure that the proper distribution certificate is used for code signing and referenced within the build settings before the app is built for distribution. These steps will be covered in more detail in the next chapter: Preparing for Upload.

Once the above steps have been completed, double checked and verified by the developer, the developer has the ability to upload the app using one of two different methods.: the standalone tool from Apple named "Application Loader," or use Xcode 4 itself. Both submission methods will be discussed at length in Chapters 39 and 40 and both require a Mac running OS X 10.6 (Snow Leopard) or above.

Although the required steps may seem overwhelming at first, after reading the next chapters and completing an initial upload, the process becomes easier. It is important to note that Application Loader can be used to upload projects created in Xcode 4; the developer is not limited to using Xcode 4's built-in submission tools. In most cases, Xcode 4 cannot upload binaries created by third-party development tools and Application Loader must be used.

Once Apple has received the app, it will be reviewed for accuracy, stability, and compliance with Apple's Human Interface Guidelines:
https://developer.apple.com/library/IOs/#documentation/UserExperience/Conceptual/MobileHIG/Introduction/Introduction.html

App Store Review Guidelines: https://developer.apple.com/appstore/guidelines.html

Also, review the Developer Program License Agreements:
https://developer.apple.com/membercenter/index.action#agreements.

The review process can take anywhere from a few days to a few weeks. Once complete, the developer will be notified by Apple of their app status via email. If the app is rejected, Apple will explain why.

There are many reasons why Apple may reject an app ranging from a lack of functionality, to inappropriate content, to an improper use of a button or other visual control. These "Gotchas" are covered in more detail in Chapter 41: Bonus Chapter of Gotchas!

Once Apple approves your app, it will become available on the App Store within 24 hours or on the date and time specified by the developer when creating the app page in iTunes Connect.

After the app is released, the developer can view analytics regarding sales and other statistics through Apple's iTunes Connect.

37.1 Xcode 4 App Submission Overview

It is possible to upload apps to Apple directly through Xcode 4. This is relatively new to the app submission system. It is useful for projects created with Xcode 4 or when an Xcode project is created by a third-party development environment or tool. For example, Unity 3D produces an Xcode project that can be uploaded using Xcode 4's embedded app submission features.

Xcode 4's app submission features are similar to Application Loader in the sense that it is required to log in with the team admin Apple ID and then choose the app desired for upload from a drop down list containing a list of apps that have been marked as "Ready to Upload Binary" from the iTunes Connect web portal. The actual submission process is done through the Archives section of the Organizer in Xcode 4 as seen in the following figure.

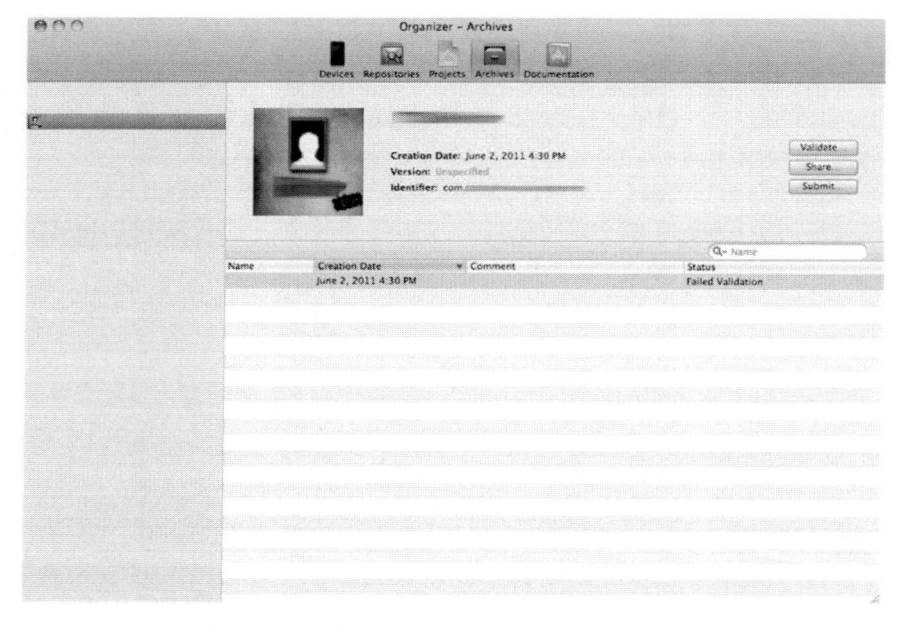

Figure 37-1: Xcode 4 Organizer – Archive Section

The Xcode 4 App Submission process will be described in more detail in Chapter 39.

37.2 Application Loader Overview

Application Loader is a standalone tool provided by Apple and comes bundled with Xcode on systems running Mac OS X 10.6 and above. Essentially, Application Loader allows the developer to login to iTunes Connect with their Apple ID and select the app they wish to upload that has been marked as "Ready to Upload Binary" in iTunes Connect from a drop down menu. The developer then chooses the corresponding binary for the app from the developer's local workstation and uploads the app to Apple.

It is important to note that the majority of the third-party development environments, capable of producing iOS apps and native Mac OS X apps, require Application Loader to upload apps for distribution on the App Store (or Mac App Store). Whenever a third-party development option builds a single package file, .app or binary, Application Loader must be used.

This process will be described in more detail in Chapter 40. Figure 37-2 displays a screenshot of the Application Loader welcome screen.

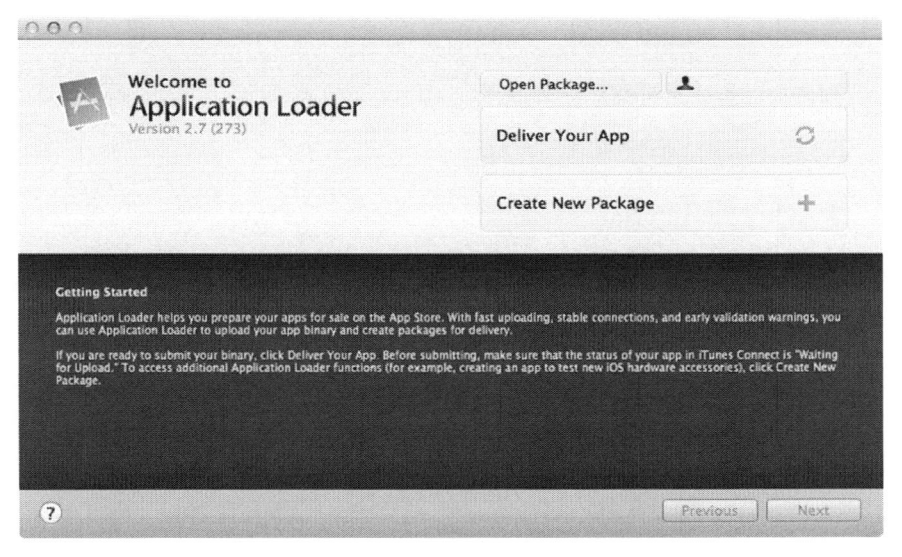

Figure 37-2: Application Loader Welcome Screen

37.3 Submission Methods Overview

Depending on what development environment was used, the developer will either upload their app using Xcode 4 or Application Loader. It is entirely possible to upload an Xcode 4 project using Application Loader however it is not possible in most cases to upload a binary created using a third-party development tool using Xcode 4. If a third-party development environment builds a single binary or .app file, Application Loader must be used. It is always a good idea to

Chapter 37: App Submission Overview   411

use the most recent version of Application Loader or Xcode when uploading an app to the App Store to ensure successful delivery.

During the upload, Apple performs validation checks on your binary to ensure that required assets are not missing, that the Bundle ID and version number correspond correctly to what is on file with iTunes Connect, and to ensure that the app was code signed with the appropriate distribution certificate. Creating, obtaining, and installing a distribution certificate required for code signing is covered in detail in Chapter 38.

Once the upload is complete, the developer will be notified, via email, that the app has been successfully submitted. iTunes Connect will show the app's state as "Waiting for Review" or "Upload Received." If there was an error with the upload, the developer will be notified during the upload process.

After Apple has reviewed the app, the developer will be notified of acceptance or rejection. If the app is rejected, the developer will be given a reason why and be directed to the Resolution Center for the app available through iTunes Connect. Possible rejection reasons and ways to avoid them are discussed in the following Chapter 38.

If the app is rejected, make the required changes and re-submit to Apple following the steps mentioned in the next chapters. If your app is accepted, congratulations! The app will become live on the App Store within 24 hours or on the date that was pre-defined on the app page in iTunes Connect.

38

Preparing For Submission

Once an app is complete and ready for distribution, there are steps required to prepare for its submission.

- Creating a distribution certificate
- Verifying all metadata is correct in iTunes Connect
- Ensuring that project build settings are configured for distribution
- Code signing the app for distribution

38.1   Avoiding Instant Rejection

Your app project will contain many values and settings regarding its assets, or content. Just a few examples include the app's detailed public description that customers see in the Apple iTunes App Store, its screen shots for marketing purposes, and even its little desktop icon (the graphic you touch on the device to launch the app itself). All of these little details are considered the app's *Metadata*.

Other than problems with the programming code itself, one of the biggest reasons for app rejection is due to some issue relating to its metadata.

Even if your app code compiles, works as designed without errors, and looks great on your test devices, it may still be rejected due to errors in its metadata. Understanding the possible reasons for rejection due to metadata - and where, exactly, it's stored - is critical to maintaining your sanity while resolving the issue.

38.1.1   Locating the Metadata

Your app metadata is primarily located in the following two areas:

The app itself: These values and settings, including app content such as the desktop icon, are specified in whatever development platform you used to create your app (e.g. Xcode, Unity3D, etc.), and saved within a special file *your_app_name.plist* as wrapped into the ".app" file itself.

iTunes Connect: These values and settings, including store content such as the written description of the app itself, are specified in the iTunes Connect website portal (e.g. your developer account), and saved or otherwise stored on Apple's servers.

The project's locally-stored critical metadata is primarily located in default filename "Info.plist" (in XML format). This .plist file contains very important information about the app project, such as its desktop icon filename and location, and its desktop name. It's hiding in Project Name -> Target Name -> Resources.

There are two ways to view the metadata located in this critical file. Normally, the data is viewed and edited within the GUI's Target View. However, in the event it is desired to over-ride the project data, and to have complete control for manual tweaking, then you can also directly view and edit the .plist file itself.

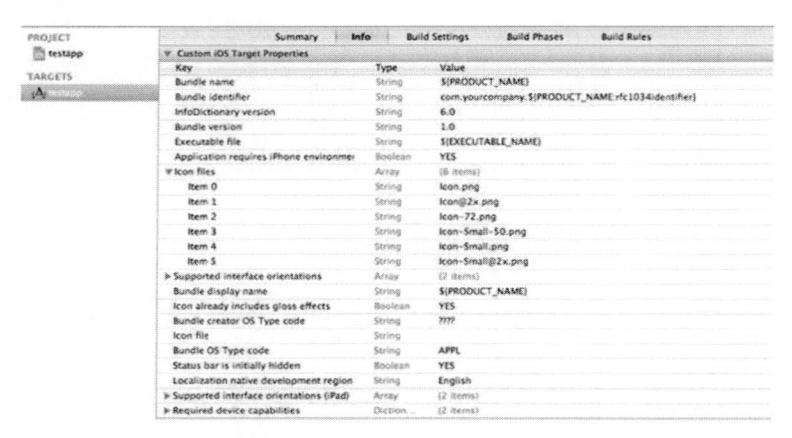

Figure 38-1: GUI Target View of Data Contained in INFO.PLIST

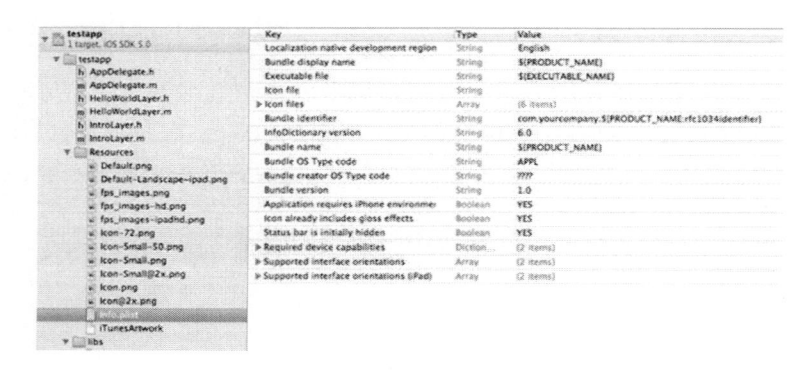

Figure 38-2: Directly Editing the Data Contained in INFO.PLIST

> **TIP:** *Avoid using the direct-edit method unless absolutely necessary. Not only is using the GUI easier, it's also safer.*

There are two major "Key" items of interest within this .plist file:

"Bundle display name": This Key is where you can tweak the device-name of your app shown under its desktop icon, exactly as it is displayed on the real devices' desktop. *Note: NOT "Bundle name" (!!) which is a similar-sounding yet different Key-name. CFBundleName is the short name of the bundle itself. CFBundleDisplayName is the name displayed for the bundle where ever necessary (e.g. app name on iPhone home desktop screen under its icon).*

It's very important to verify the app's assigned desktop name itself is not too long, to ensure its name (under its icon) isn't truncated with "..." notation on end-users devices. This should also be similar to the full name of the app as shown in iTunes Connect.

Unfortunately, there isn't an exact number of set characters to know if the app name will or will not be truncated. This is due to issues such issues as font kerning (spacing of the individual characters) used on the device's desktop. However, the general rule is 15 characters or less to be verified on a real test device. *For this reason, the app's desktop name is almost always a very short or otherwise abbreviated version of the "full" app name shown in iTunes Connect (and ultimately, the end-user iTunes App Store).*

"Icon files": Notice the "Icon files" Key-section in the screenshots above. Those XML entries specify the type and location of each variation of desktop icon, as displayed on the actual end-user device itself. Those images must match graphically, yet be in the required resolution for the particular device type (e.g. standard vs. retina). Details are outlined later in this chapter.

38.1.2 Potential Metadata Issues

Apple's review process has many automatic checks it performs to your app and its metadata immediately upon upload. This is done essentially to avoid wasting the Apple reviewers' time having to reject apps which have obvious and easy-to-check issues.

These "issues" can essentially be broken down into two categories, both of which are covered in detail by this book:

The first and obvious category is that the metadata might have a technical flaw, such as a typographical error pointing to a file (causing the file to not be found).

The other and perhaps more difficult category to identify, is that the metadata might have content which is potentially against Apple's policy (or maybe even against the law).

These verification checks for issues are a *good thing*, as they can help avoid a lengthy delay from being rejected after the manual review process – by quickly discovering an easily identifiable mistake *in advance* – and even has the potential for helping you to avoid legal conflict with third parties!

38.2 Top Technical Reasons for Rejection

The app's metadata is thoroughly checked by Apple's automated systems upon upload. If there's any *obvious* glitch with the metadata, the upload process will *immediately* fail its verification checks, and you will be automatically notified of the problem without having to wait for a manual review. This is a great time saver!

Most of the time, errors are a result of a simple - yet very easy to miss - oversight with a critical setting, such as the app's desktop icon having a different filename than specified in its project settings.

The most common bullet points that can result in an automated rejection are discussed below.

38.2.1   Incorrect image sizes used for device

Ensure both iTunes Connect, and the project's device-target itself, each contain the proper resolutions for every currently-supported device. For example, Apple is known for rejecting apps that fail to contain the latest retina-resolution (high-quality) images for those users who may have retina devices. Those high-quality images are then down-sized automatically as required automatically by Apple.

> ***TIP:*** *Don't try to "upscale" / "upconvert" low-resolution or old existing graphics. Otherwise, the image will become grainy, and if an Apple reviewer notices - it may be rejected.*

It is recommended to create and work with the higher resolution graphic, then downscale/downconvert the graphic to the various requirements in order to maintain the most razor-sharp image possible. The PNG format in 24bit color is recommended, although JPG is also permitted. Note the new Retina requirements for iTunes Connect / App Store as well.

As of this writing, the required full-screen pixel-resolutions for each image are listed below and sorted by device-type. Note that some image requirements are unique for each device's Landscape mode (holding the device horizontally right-left), versus its Portrait mode (holding the device vertically up-down).

ALL Launch/SplashScreens must be in full resolution. The screenshots for iTunes Connect must remove the status bar, if applicable.

> ***TIP:*** *For the app to be properly designed and accepted as a Universal App, meaning specially tagged in the App Store as fully compatible with and tuned for both the iPhone/iPod touch as well as the iPad, then both sets of images must be included in the app's project for upload to iTunes Connect.*

Quick Reference Guide: Resolutions

iPhone 2G/EDGE, 3G, 3GS & iPod touch 1, 2, 3:
- Full-resolution Landscape: 480 x 320.
- Minus Status Bar Landscape: 480 x 300.
- Full-resolution Portrait: 320 x 480.
- Minus Status Bar Portrait: 320 x 460.
- App Desktop Icon: 57 x 57.

iPhone 4, 4S & iPod touch 4th-Gen: *Retina*
- Full-resolution Landscape: 960 x 640.
- Minus Status Bar Landscape: 960 x 600.
- Full-resolution Portrait: 640 x 960.
- Minus Status Bar Portrait: 640 x 920.
- App Desktop Icon: 114 x 114.

iPhone 5 & iPod touch 5th-Gen: *Retina*
- Full-resolution Landscape: 1136 x 640.
- Minus Status Bar Landscape: 1136 x 600.

- Full-resolution Portrait: 640 x 1136.
- Minus Status Bar Portrait: 640 x 1096.
- App Desktop Icon: 114 x 114.

iPad 1, 2:
- Full-resolution Landscape: 1024 x 768.
- Minus Status Bar Landscape: 1024 x 748.
- Full-resolution Portrait: 768 x 1024.
- Minus Status Bar Portrait: 768 x 1004.
- App Desktop Icon: 72 x 72.

iPad 3rd-Gen: *Retina*
- Full-resolution Landscape: 2048 x 1536.
- Minus Status Bar Landscape: 2048 x 1496.
- Full-resolution Portrait: 1536 x 2048.
- Minus Status Bar Portrait: 1536 x 2008.
- App Desktop Icon: 114 x 114.

iTunes App Store (via iTunes Connect):
- ALL App Device Home Desktop Icons for App Store: 1024 x 1024.
- ALL Retina Screenshots should be cropped to remove the 40-pixel Status Bar (e.g. the Telco carrier's name, signal strength, etc.).
- ALL Standard Screenshots should be cropped to remove the 20-pixel Status Bar (e.g. the Telco carrier's name, signal strength, etc.).
- Note that these images must be crystal-clear at these resolutions, as they may be viewed by iTunes users on a regular full-sized computer. Again, if these images are grainy or otherwise of poor quality, the App may be rejected.

Banner Advertisements (3rd-Party):
- Traditional: Width x 50. (e.g. iPhone 3GS can be 320x50). *Double for Retina.*
- Many advertisers have unique requirements that must be accounted for early on in the design phase of the app itself, sometimes requiring additional libraries to be added to the app's project, etc.

Reminder: Be careful to account for the size of all banner ads in the app's overall design and graphics *before* getting started making images. To avoid major problems with graphical overlap, full-resolution images must be reduced by the *exact* size of the banner ad.

As of this writing, no developer account login is required to view most of Apple's iOS Developer Library (documentation):

http://developer.apple.com/library/ios/navigation/

TIP: *See the "Custom Icon and Image Creation Guidelines" contained within the official Apple "iOS Human Interface Guidelines" PDF, for additional information and full details for all image-type requirements.*

Note that the higher-resolutions for iTunes images are now required, but do not worry: Apple automatically down-sizes the images as required for lower-resolutions.

38.2.2 Incorrect, missing, or misplaced imbedded device desktop icon

One of the more frustrating mistakes to troubleshoot is a missing desktop icon, a desktop icon in the wrong resolution for the device type, or incorrect location information for said desktop icon as contained within the info.plist metadata file.

As of this writing, Apple now requires *both* image types to be included in *every* new project: one for the original resolution devices, and one for the new retina devices. Both of these icon file images are defined and specified within the Info.plist "Key" name "Icon files" described and shown in the screenshots above.

Do not smooth the edges or use any alpha transparency for the device desktop icon, as those processes are done automatically for icons.

38.2.3 Mismatch between iTunes Connect image and device's icon

The iTunes Connect marketing banner is required to be a very *close approximation* of the device's desktop icon. With the advent of the Retina screens and the release of iOS6, this iTunes Banner image size requirement has now increased (from the previous 512 x 512 pixel requirement) to the new 1024 x 1024 pixels image size requirement. Don't worry, Apple automatically down-sizes the images as required for the older devices and their resolutions.

> **TIP:** Note the wording "close approximation" of the iTunes Connect image based on the device's desktop icon, which is unfortunately somewhat subject to the individual Apple reviewer's opinion. It's very common for developers to want to have "enhanced" iTunes Connect versions of the much smaller desktop icon shown to end-users on the actual device. As a general rule, if the enlarged iTunes Connect image is overly modified from the small device icon image, to add items such as callout tags (e.g. "All New Version! Download Now!"), then the entire app may be rejected based on this mismatch.

Essentially, the only big concern to avoid rejection here is in ensuring that when the end-user downloads an app from the iTunes App Store, that it can be easily and instantly identified with the nearly-identical graphical image (and name) on the device desktop itself. Note that Apple reviewers may reject Apps with cheesy or outright obnoxious marketing tags (by way of example only, such as a big red diagonal "New Version!") blasted across the icons and/or screenshot images.

38.2.4 Feature or API used for incompatible device type

As a general rule, most developers compile with a target iOS of the lowest possible version to support the features they require. This is to maintain as much compatibility with as many older devices as possible. For example, the millions of original iPhones still in use today are forever limited to iOS version 3.1.3. This means if any feature used within an app requires iOS version 4.0 or higher, then those millions of original iPhone devices will not be able to run that app.

> **TIP:** *It might be "cool" to create the app to require the latest iOS version, but be prepared in advance to Say Goodbye to Millions of Potential Customers.*

An excellent resource for major device features, their required iOS versions, and additional nitty-gritty details, can all be found here:

> http://en.wikipedia.org/wiki/List_of_iOS_devices

It's for this very important reason that the iOS target version is carefully considered, and the lowest possible denominator be used whenever at all possible for maximum end-user coverage.

38.2.5 Mismatch between App Store and reality

Yes, it's true. The app must actually perform the actions outlined within the app's description and screenshots as entered in the metadata in iTunes Connect, which upon approval is shown to end-users in the App Store. The Apple reviewers even go so far as carefully reading the description, reviewing the screenshots, and compare it all to the app's graphics, features and usable functionality. If any portion of the iTunes Connect metadata is deemed by the Apple reviewer to be misleading or outright false advertising, then the app will be rejected.

38.3 Walkthrough on Creating a Distribution Certificate

This is necessary for submitting apps to Apple for distribution on the iTunes App Store. It is very similar to the process described in Chapter 9, Preparing for Development, in which a walkthrough of creating development certificates is discussed. A distribution certificate is different than a development certificate in the sense that its purpose is to code-sign the app for distribution on the App Store. While a development certificate's purpose is to code-sign apps for development and debugging on devices. Only the Team Admin can create distribution certificates through the iOS Provisioning Portal.

1. Log in to the iOS Developer Portal (http://developer.apple.com/ios) and browse to the "iOS Provisioning Portal" section.

Figure 38-3: iOS Developer Program

2. Select "Certificates" from the menu on the left hand side and then "Distribution" from the tabs across the top.

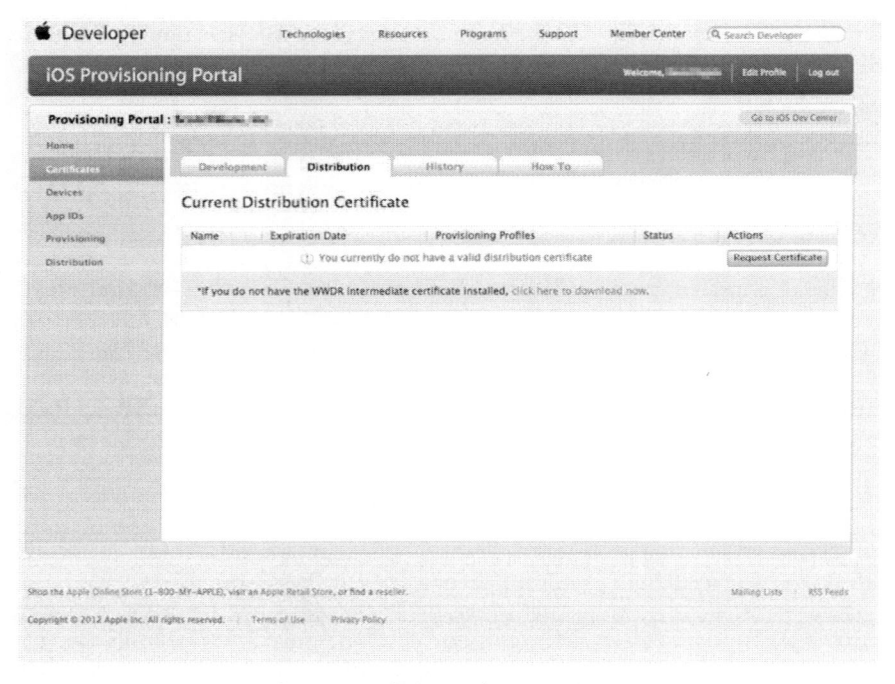

Figure 38-4: iOS Developer Program

3. Download the WWDR intermediate certificate by clicking the link and opening the certificate file, "AppleWWDRCA.cer," once it has finished downloading. This will open the "Keychain Access" application and display the prompt shown in the figure below. Click "Add" and the certificate will be added to the login keychain.

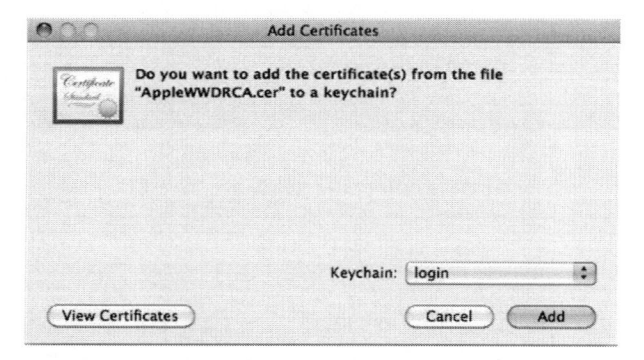

Figure 38-5: Adding the AppleWWDR Intermediate Certificate to Keychain

4. Within Keychain Access, go to the menu Keychain Access -> Certificate Assistant -> Request a Certificate from a Certificate Authority. The screen shown in Figure 38-6 will be displayed.

5. Enter the email address and company name, select save to disk, and click "Continue" as shown in the figure below.

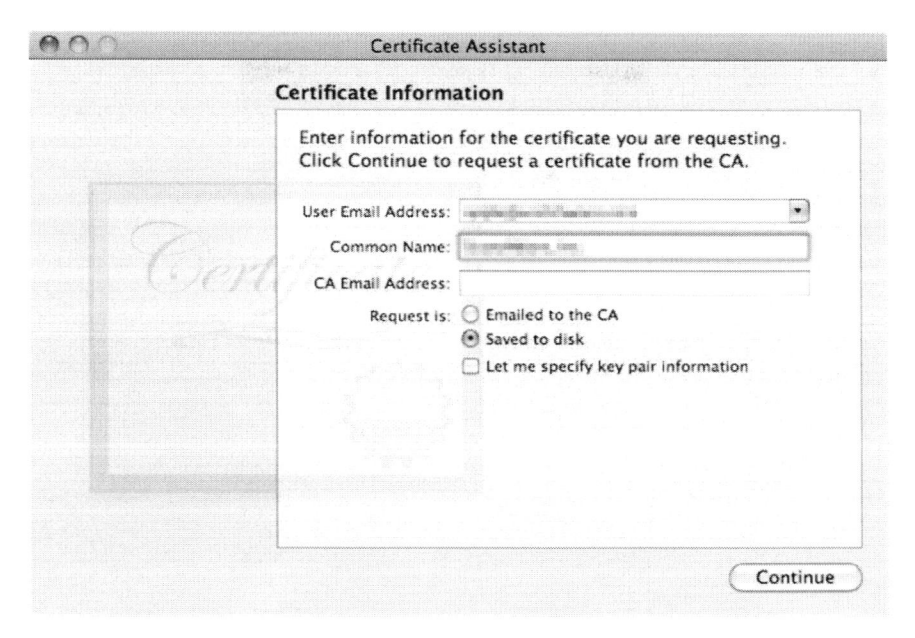

Figure 38-6: Requesting a Certificate

6. Save the certificate to the desktop.
7. Return to the certificates provisioning portal in the iOS Developer Portal. Select "Request Certificate."

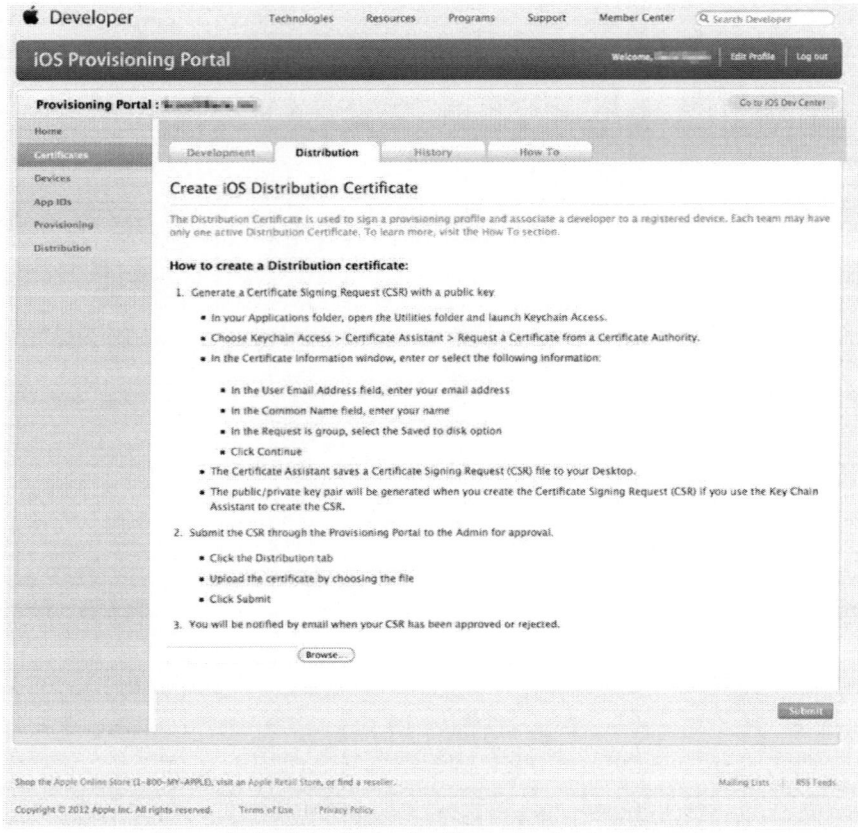

Figure 38-7: Uploading a Certificate for Apple Approval

8. Select "Browse..." and choose the certificate that was just created and saved to the desktop. Click "Submit."

9. After submitting the certificate for approval, you will be redirected to the Distribution Certificates section of the iOS Provisioning Portal. It may be necessary to refresh the page for the request to be processed. You will then be able to download the distribution certificate.

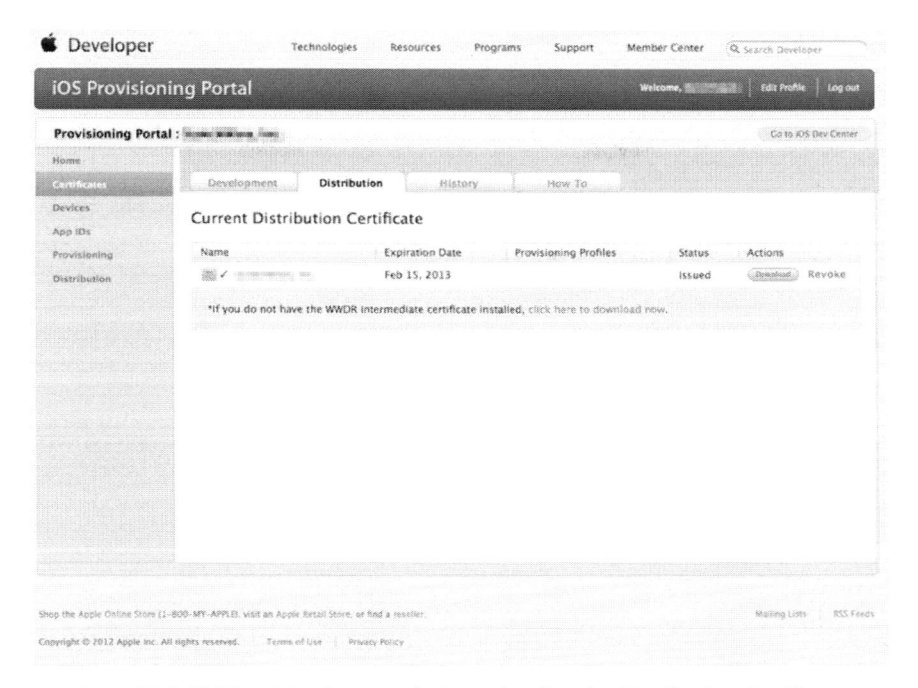

Figure 38-8: iOS Provisioning Portal – Downloading the Distribution Certificate

10. Once the distribution certificate is downloaded, open the distribution certificate file to import it into the keychain similar to the above step 3.

Now that the distribution certificate has been created, it can be used for code-signing apps for distribution on the App Store.

38.4 Generating and Importing Profiles into Xcode

In order to upload an app to Apple for distribution on the iTunes App Store, an app specific provisioning profile set for distribution must be included with the app upon submission. This provisioning profile essentially notifies Apple of the intended distribution method (App Store), the distribution certificate that will be used (which was created in the previous section), and the App ID in which the provisioning profile is intended for use.

1. Login to the iOS Developer Portal -http://developer.apple.com/ios.
2. Select "Provisioning" from the menu on the left hand side.
3. Select the "Distribution" tab located at the top of the module.

Chapter 38: Preparing For Submission   423

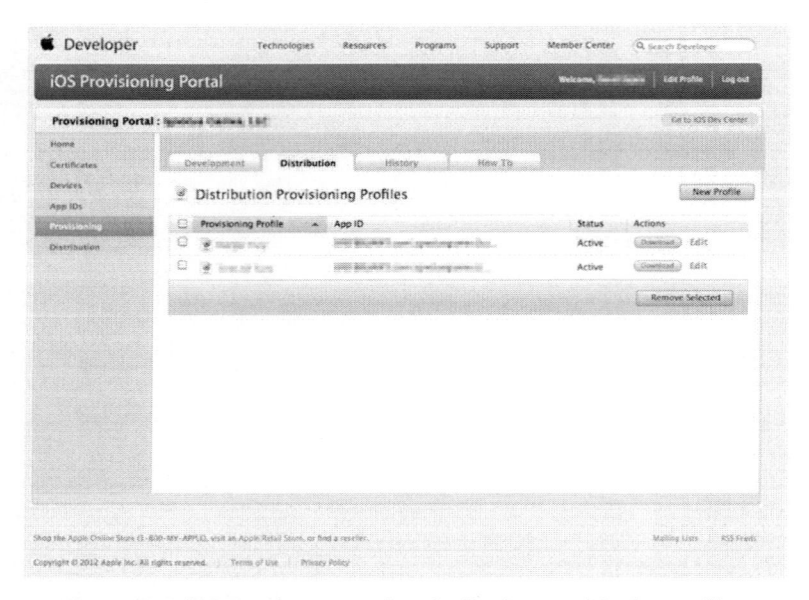

Figure 38-9: iOS Developer Portal – Distribution Provisioning Profiles

4. Select New Profile.

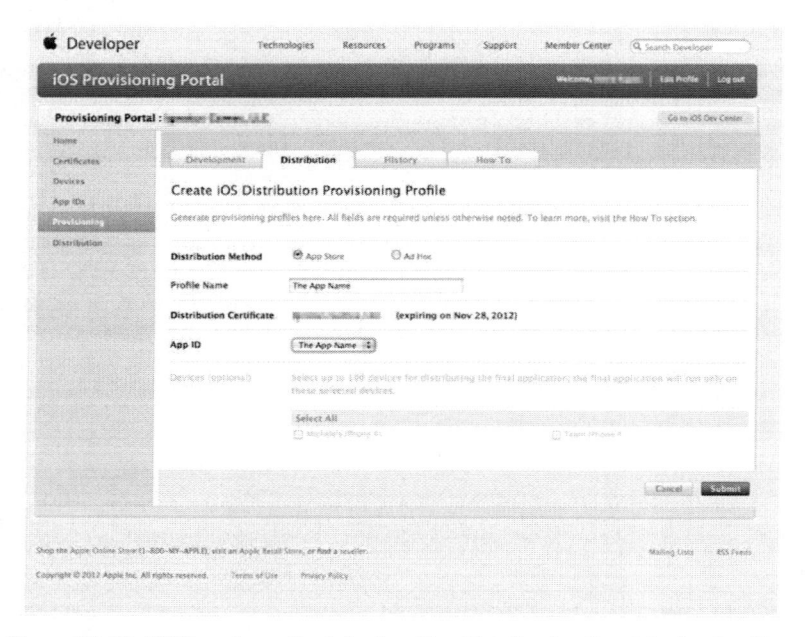

Figure 38-10: iOS Developer Portal – Creating Distribution Provisioning Profiles

5. For the Distribution Method, select "App Store".

6. In the "Profile Name" field enter a name for the profile - this is usually the name of the app.
7. The "Distribution Certificate" field should automatically display the distribution certificate that was created in the previous section.
8. Under "App ID, select the App ID corresponding to the app that is to be uploaded.
9. Select "Submit" located in the bottom right-hand corner of the module.
10. After clicking "Submit", you will then be redirected back to the distribution tab of the provisioning section in the iOS Developer Portal. The new provisioning profile for distribution will appear. If a "Download" button does not appear and you cannot download it or the status reads as "Pending", refresh the page.

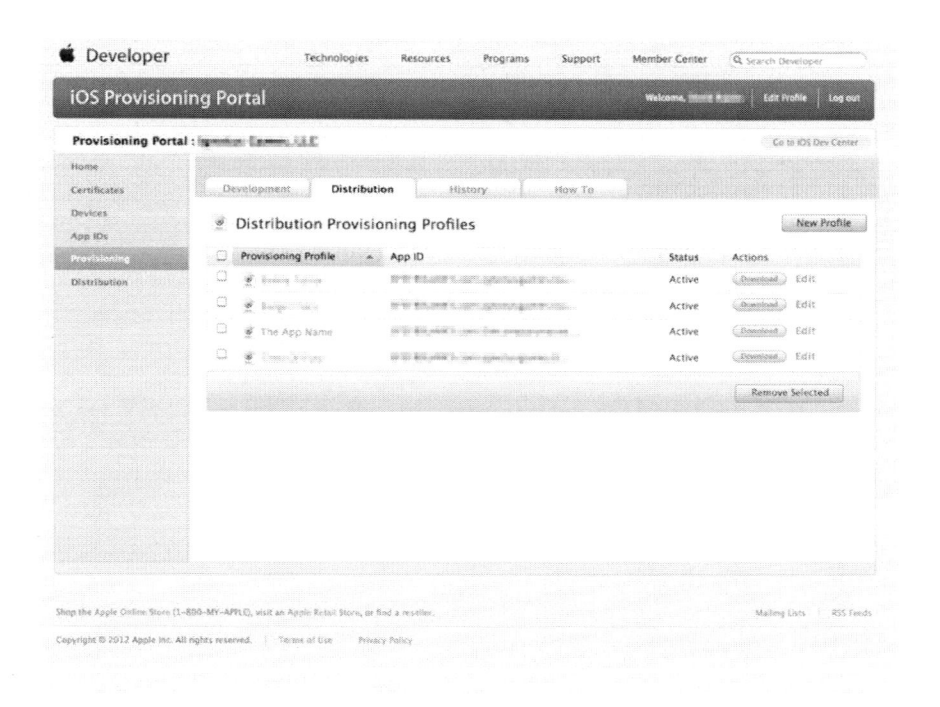

Figure 38-11: iOS Provisioning Portal – Obtaining Distribution Provisioning Profiles

11. Download the provisioning profile by selecting the "Download" button.
12. Open the downloaded file. This will import the provisioning profile into Xcode and copy the file to the necessary directory.

38.5 Verifying Items and Set as "Ready for Upload"

It is important to remember to set up the following before you submit your app for review. If you forget to do this then you must either reject the app before it is approved or do this in a version update. The following sections provide a collection of valuable checklists of things to do prior to submitting your app for review.

38.5.1 Set up In-App Purchases in iTunes Connect.

To set up In-App Purchases in iTunes Connect click on the "Manage In-App Purchases" button on the App Summary Page. Click "Create New" to create a new In-App Purchase. You must select from the following:
- Consumable: One-time use items such as currency in a game.
- Non-Consumable: Many-time use items such as new levels in a game.
- Auto-Renewable Subscription: Content purchased for a set duration which auto renews at the end of the duration such as magazine subscription.
- Free Subscription: Can only be offered in Newsstand-enabled apps for developers who wish to put free subscription content in Newsstand.
- Non-Renewing Subscription: Similar to Auto-Renewable Subscription without the auto-renewing part.

38.5.2 Each In-App Purchase needs to have the following metadata:
- Reference name.
- Product Id similar to *com.company.appname.productid*.
- Price tier.
- Screenshot of In-App purchase in action must have a resolution of at least 640x920 pixels and 72 dpi.
- Localized for at least one language with a Display name and Description to be shown in the App Store.
- Auto-renewable subscription needs a selected duration so that it knows when to renew. The options are: 7 days, 1 month, 2 months, 3 months, 6 months, or 1 year.

38.5.3 Activate Game Center in iTunes Connect

Before you can set up leaderboards or achievements in iTunes Connect you have to enable Game Center. To do so, select "Manage Game Center" from the App Summary Page and then choose to enable for a single game. Then you will be presented with a page to manage leaderboards and achievements.

38.5.4 Set up Leaderboards

Click on "Add Leaderboard" under the "Leaderboards" section to add a new leaderboard. Choose "Single Leaderboard" when presented with the option.

Each Leaderboard will need the following:
- Reference name
- Leaderboard ID similar to *com.company.appname.leaderboardID*
- Format Type: Integer; Fixed Point – to 1,2, or 3 Decimals; Elapsed Time – Minute, Second, or Hundredth of a Second; Money – Whole Number, or To 2 Decimals
- Sort Order (High to Low or Low to High)
- Localization (Language, Name, Score Format)

38.5.5 Set up Achievements

Click on "Add Achievement" under the "Achievements" section to add a new Achievement. Each Achievement will need the following:
- Achievement Reference Name
- Achievement ID similar to *com.company.appname.acheivementID*

- Point Value: 100 points maximum per achievement with total of 1000 points for all achievements.
- Hidden: Achievements can be marked hidden so they are not visible on Game Center until they have been achieved
- Achievable More than Once: If Yes users can accept Game Center challenges for achievements that they have earned.
- At least one localization setup for the achievement which requires the following: language, title, pre-earned description, earned description, and a 1024x1024 image.

38.5.6 Set up iAds in iTunes Connect.

Click on "iAd Network Settings" and choose whether primary audience is under 17 yrs. old and click "Enable iAds" (**CANNOT BE DISABLED ONCE SAVED**).

38.6 Finding, Verifying and Tuning Metadata

It is also VERY important to verify the metadata associated with the app before submitting the app to Apple for review. Metadata can be accessed through the "Manage Your Application" module of iTunes Connect. Once all metadata is verified, the app's status must be changed to "Ready for Upload." Once the status has been updated, Apple's submission system is ready to accept the app.

1. Login to iTunes Connect (http://itunesconnect.apple.com) and browse to the "Mange Your Applications" module.
2. Select the app being uploaded from the list of apps displayed. The app's detail page will be displayed.

Figure 38-12: iTunes Connect – view app details

3. Select "View Details" for the version being uploaded and verify the metadata associated with your app is correct.
4. Press the button labeled "Ready to Upload Binary." Once this is done, metadata cannot be updated.

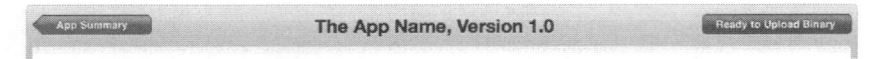

Figure 38-13: iTunes Connect – view app details: Ready to Upload Binary

38.7 Configuring App Build Settings

Before submitting an app to Apple for review and distribution, it is necessary to ensure that the project build settings are configured properly. This section may not apply if the app was developed using one of the third-party tools discussed in this book. Xcode 4 configures most of the build settings for you but it is important to verify they are correct.

1. From the project navigator, within Xcode, select the project itself. This will display a menu with the sections "Project" as well as "Targets." This is shown in Figure 38-14.

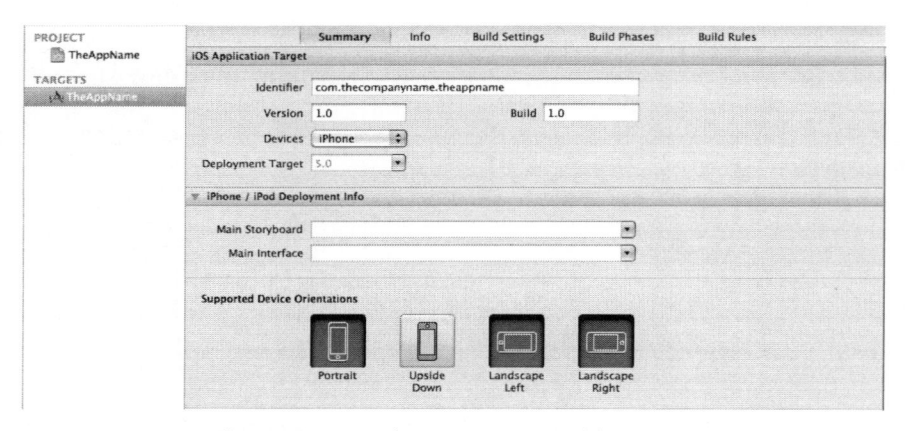

Figure 38-14: Xcode 4 – Summary Build Options

2. Select the project name under the "Targets" section.
3. In the summary tab, ensure the proper information has been entered including identifier (which is the app's Bundle ID), the version that must match the version specified in iTunes Connect, and the build that can be the same as the version number, the supported devices, and deployment target. Once these fields are verified to be correct, select the "Build Settings" tab.

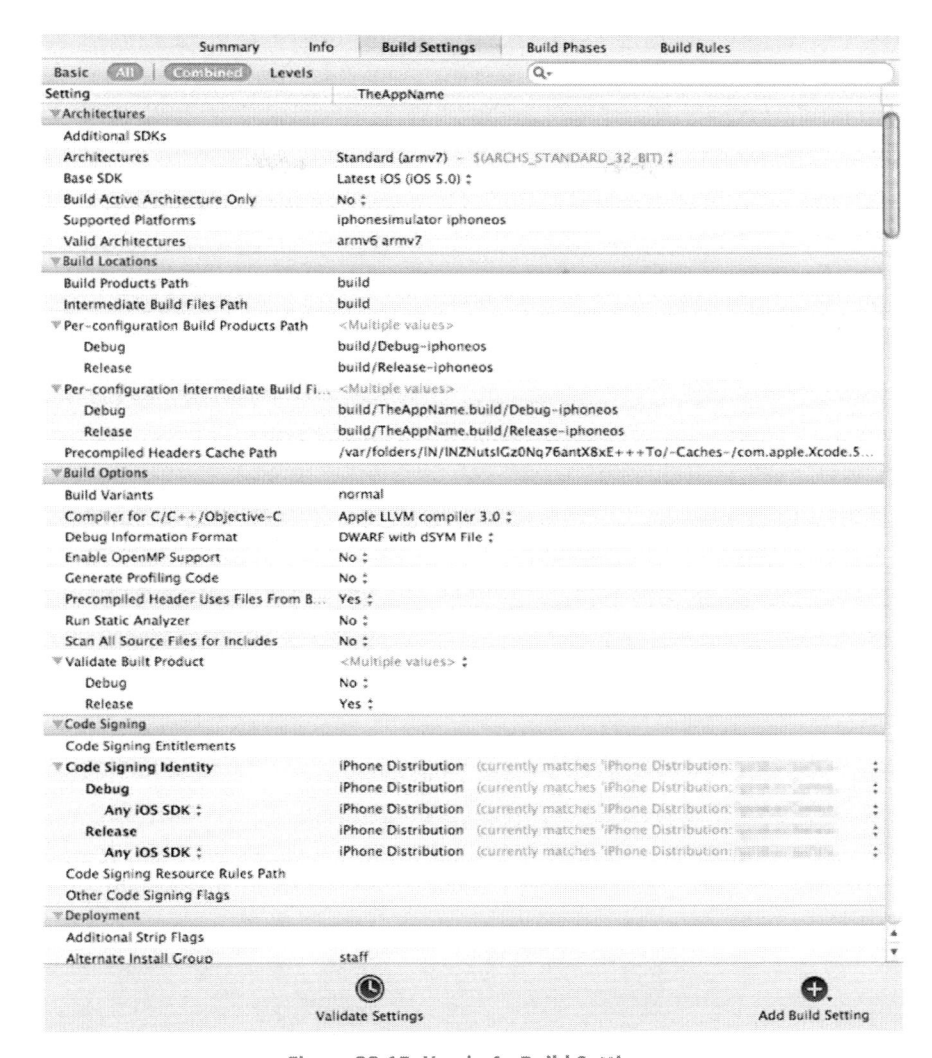

Figure 38-15: Xcode 4 - Build Settings

4. Verify that the build settings are correct and verify that in the "Code Signing" section, the "Code Signing Identity" field and all child nodes are set to "iPhone Distribution."

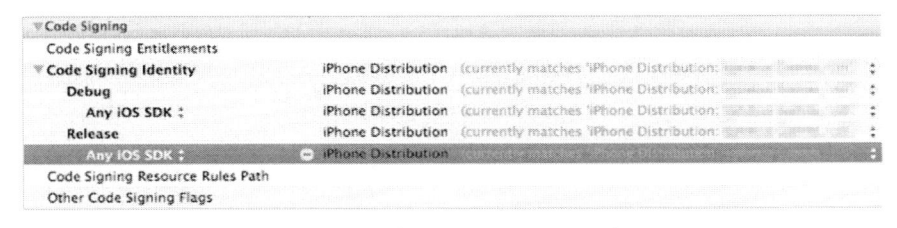

Figure 38-16: Xcode 4 - Build Settings - Code Signing

Chapter 38: Preparing For Submission   429

5. Check if your settings are valid by selecting the "Validate Settings" at the bottom.

Once the build settings are correct, the app can now be built and submitted to Apple for review. Details on this process are covered in the next chapter.

Submitting Apps Using Xcode

It is possible for developers to submit their apps to Apple for review and distribution on the App Store by using Xcode 4 in addition to the standalone Application Loader tool. This is an extremely useful feature of Xcode 4 that makes submitting Xcode projects with a few mouse clicks. It is necessary that the app has been prepared for upload as described in Chapter 38 and has been flagged as "Waiting for Upload" in iTunes Connect.

39.1 Walkthrough on Submitting Apps with Xcode

Once the necessary prerequisites have been completed, as discussed in the previous chapter, it is possible to submit the app directly to Apple using Xcode 4.

1. Open the Xcode project and ensure the Identifier (Bundle ID), version, devices, and deployment target are correct on the Targets level, as shown in Figure 39-1.

Figure 39-1: Xcode 4 – Target Level Summary Settings

2. To ensure that the proper distribution code-signing certificate is being used, check the "Build Settings" section of the Targets properties. Also confirm that the build settings are correct, as described in Chapter 38, Preparing for Upload.

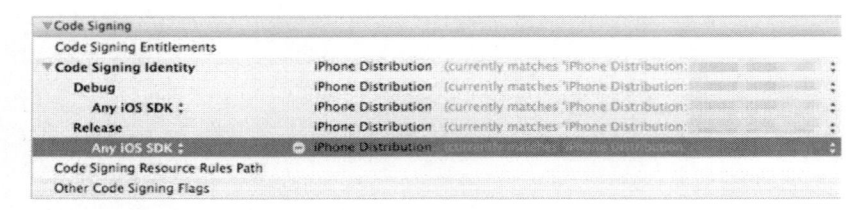

Figure 39-2: Xcode 4 – Target Level Build Settings

3. Select "iOS Device" as the target destination for the build.

Figure 39-3: Xcode 4 – Setting the Build Destination

4. From the top menu in Xcode, select Product > Archive shown in Figure 39-4.

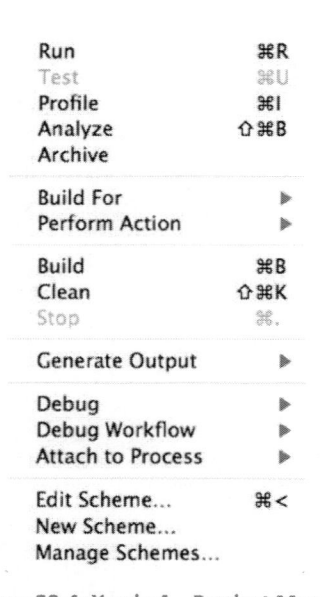

Figure 39-4: Xcode 4 – Product Menu

5. This will create a `.xarchive` file and open the "Archives" section of Xcode's Organizer.

432   Chapter 39: Submitting Apps Using Xcode

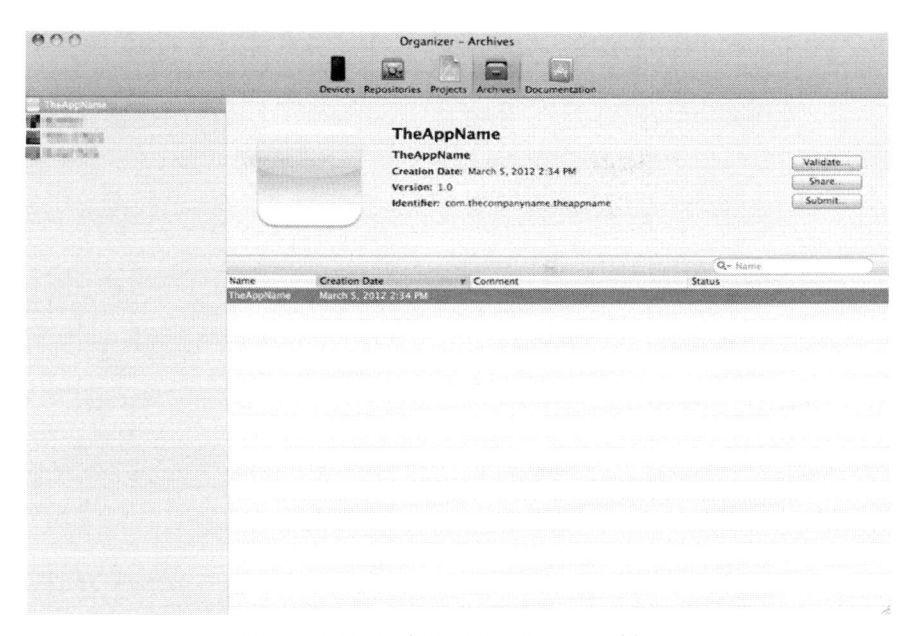

Figure 39-5: Xcode 4 – Organizer – Archives

6. Select "Submit."

7. Enter the credentials for the Team Admin account and press the "Next" button.

Figure 39-6: Xcode 4 – Submitting an App Using Xcode 4

Chapter 39: Submitting Apps Using Xcode   433

8. Xcode will attempt to validate the project and notify you of any issues. If there are none, the app will be uploaded to Apple.

Figure 39-7: Xcode 4 – Successful App Submission

This concludes the chapter on uploading apps using Xcode 4. Once the app is successfully uploaded, the status of the app will change to "Upload Received" and eventually "Waiting for Review" in iTunes Connect and Apple will send out email notifications of a successful upload.

39.2 Viewing Crash Reports

After the app is uploaded and for sale, it's always a good idea to check up on how it's performing on end-user devices. This real-world "in the field" app health information is viewed in what's called Crash Reports within iTunes Connect.

To view crash reports in iTunes Connect, go to the app summary page then click on the version of the app you wish to view crash reports for. Once on the detail page click on the "Crash Reports" link on the right as seen in Figure 39-8.

Links

Version Summary
Crash Reports
Binary Details
Status History
Customer Reviews

Figure 39-8: Crash Reports

When you are in the Crash Reports page, click on "Refresh" to generate the most recent crash report data. After the data is generated select the iOS version tab at the top to display the data as seen in Figure 39-9.

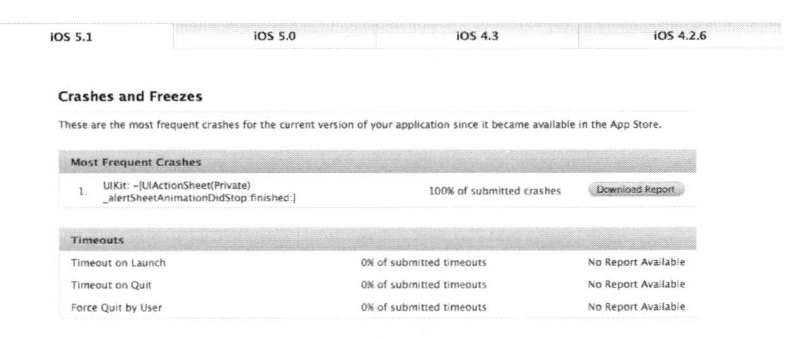

Figure 39-9: Crash Report - Part 1

Once the operating system version is selected you should see a breakdown of the most recent crashes for that version of your app. As you can see the crashes are broken down into "Most Frequent Crashes" and "Timeouts".

There is also a list of Memory issues for the app and a graphic depicting "Crashes vs. Memory". This graphic shows the ratio of crashes and timeouts to memory issues as seen in Figure 39-10.

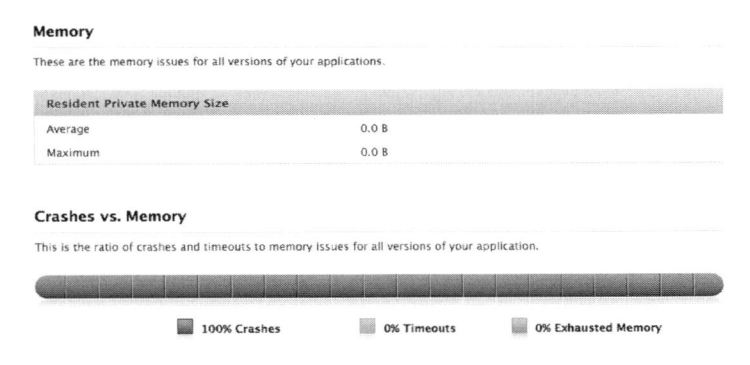

Figure 39-10: Crash Report - Part 2

39.3 How to add a new version of your app in iTunes Connect

There are two big reasons for uploading a revised and updated app to the App Store:

#1: Marketing. Every time an app is uploaded, it's featured very briefly again on the New App section of Apple. Even if very little is changed, doing an update is well known to spike downloads.

#2: Bad reviews. If an old version is crashing too often, or simply has issues making end-users unhappy, there is a way to correct this. The good news is that by uploading a new version of an app marks those negative reviews under the old version number, allowing new and better reviews to be made under the new version number. Ultimately, those new version number reviews will push out the old undesirable version.

To add a new version of your app in iTunes Connect go to the app's page, and under the Versions section, click on "Add Version" as shown in Figure 39-11.

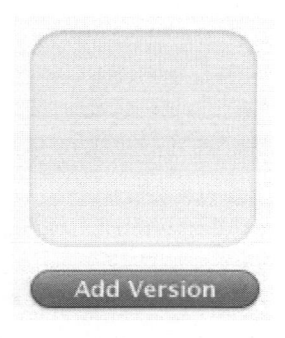

Figure 39-11: Add new version

The following is required to create a new version of your app:

- Version Number
- What's New in this version (to be displayed in iTunes)

Once you have entered the required information the new version will be created.

When you add a new version of your app you can change your keywords, name, enable iAds, and add In-App Purchases and Game Center Leaderboards/Achievements

After you are satisfied with the changes, then upload the binary following the same steps you took to initially release the original version of the app.

With the new app published, users will have the opportunity to download the updates and will once again be happy with your product.

Submitting Apps Using Application Loader

Application Loader is a development tool provided by Apple that is used by developers to submit apps for distribution on the App Store. This standalone tool comes bundled with Xcode and requires Mac OS X 10.6.8 or higher to run. Application Loader is now the required method for app submission for most of the third-party development tools discussed in this book.

In addition to submitting iOS apps to the App Store, Application Loader can be used to submit Mac OS X apps to the Mac App Store; new in-app purchases, and apps to be used for testing new hardware prototypes, such as a Bluetooth headset. This chapter will discuss uploading a new iOS app for review and distribution on the App Store. For more information regarding Application Loader, the Application Loader manual can be found at:
https://itunesconnect.apple.com/docs/UsingApplicationLoader.pdf.

40.1 Walkthrough on Submitting Apps with Application Loader

It is important to review Chapter 38, Preparing for Submission, to ensure that the app being uploaded will be approved. Once the app is in the "Waiting for Upload" state in iTunes Connect, proceed with the following steps.

1. Open Application Loader and log in with your iTunes Connect team admin account. The default location for Application Loader is /Developer/Applications/Utilities/Application Loader.app.

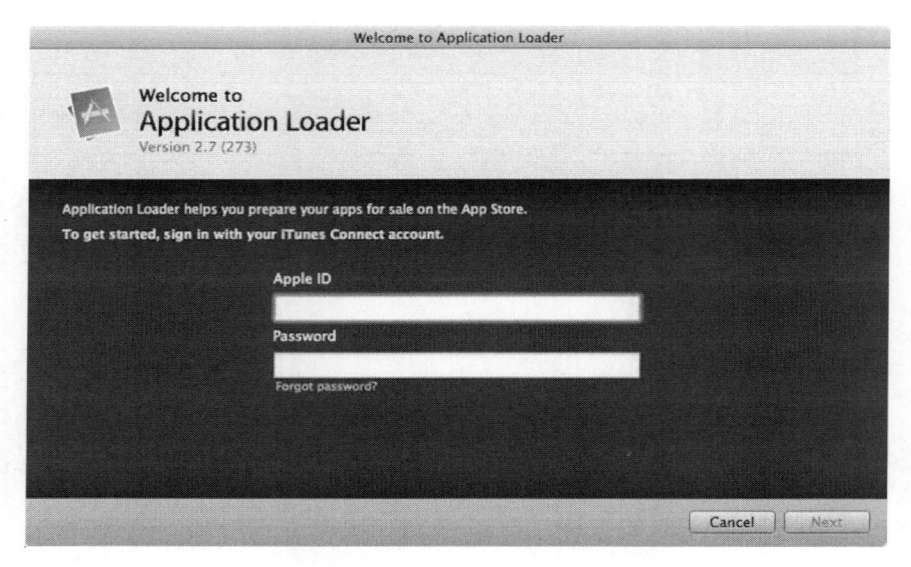

Figure 40-1: Application Loader – Welcome Screen

2. Once logged in, select "Deliver Your App."

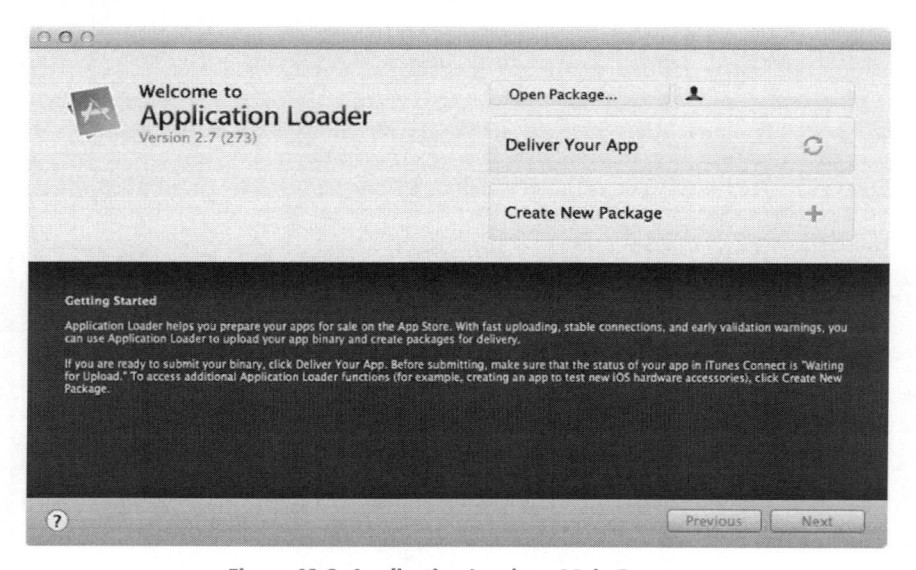

Figure 40-2: Application Loader – Main Screen

This will bring you to a screen that has a drop down menu containing all of your apps with the status "Waiting for Upload."

438   Chapter 40: Submitting Apps Using Application Loader

Figure 40-3: Application Loader – Choose an Application

3. Select the app you wish to upload from the list and press the "Next" button as seen above in Figure 40-3.

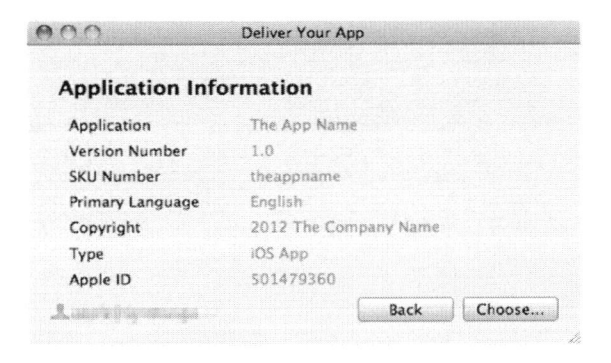

Figure 40-4: Application Loader – Application Information

4. Verify that the details are correct and press the "Choose..." button. Select the binary that is being submitted and select "Open."

5. The file will be checked for critical items, such as the icon file. If no preliminary issues are found, it will be uploaded to Apple for review.

This concludes the section on uploading apps for distribution on the App Store. Once the app has been uploaded, a confirmation email will be sent out and the status of the app in iTunes Connect will change to "Waiting for Review." Once the app is reviewed, the developer will be notified of acceptance or rejection. If the app is accepted, it will become available on the App Store. If the app is rejected, the developer will be notified with the reason and will be directed to the Resolution Center within iTunes Connect that will provide instructions on how to remedy the issues.

40.2 Viewing Crash Reports

As described in the previous chapter, after the app is uploaded and for sale, it's always a good idea to check up on how it's performing on end-user devices. This real-world "in the field" app health information is viewed in what's called Crash Reports within iTunes Connect.

To view crash reports in iTunes Connect, go to the app's summary page then click on the version of the app you wish to view crash reports for. Once on the detail page click on the "Crash Reports" link on the right as seen in Figure 40-5.

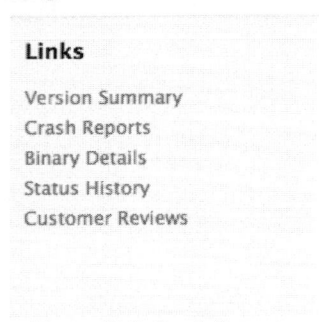

Figure 40-5: Crash Reports

When you are in the Crash Reports page, click on "Refresh" to generate the most recent crash report data. After the data is generated select the iOS version tab at the top to display the data as seen in Figure 40-6.

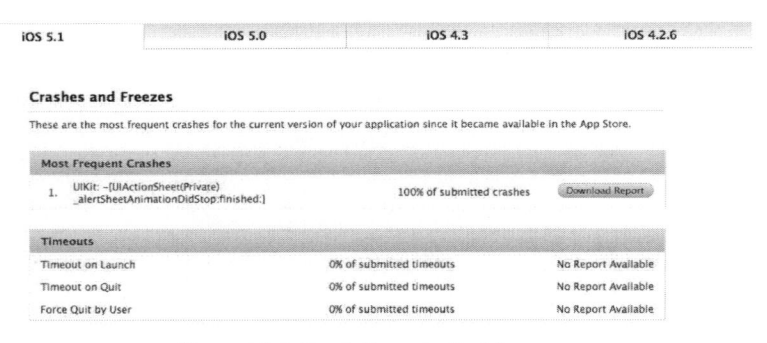

Figure 40-6: Crash Report - Part 1

Once the operating system version is selected you should see a breakdown of the most recent crashes for that version of your app. As you can see the crashes are broken down into "Most Frequent Crashes" and "Timeouts".

There is also a list of Memory issues for the app and a graphic depicting "Crashes vs. Memory". This graphic shows the ratio of crashes and timeouts to memory issues as seen in Figure 40-7.

Memory

These are the memory issues for all versions of your applications.

Resident Private Memory Size	
Average	0.0 B
Maximum	0.0 B

Crashes vs. Memory

This is the ratio of crashes and timeouts to memory issues for all versions of your application.

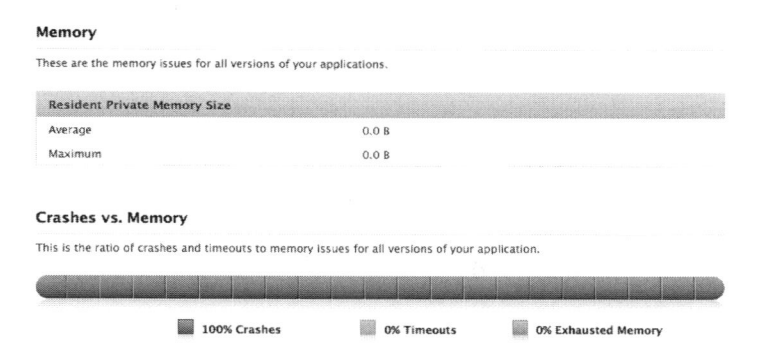

■ 100% Crashes   ■ 0% Timeouts   ■ 0% Exhausted Memory

Figure 40-7: Crash Report - Part 2

40.3 How to Add a New App Version in iTunes Connect

There are two big reasons for uploading a revised and updated app to the App Store:

#1: Marketing. Every time an app is uploaded, it's featured very briefly again on the New App section of Apple. Even if very little is changed, doing an update is well known to spike downloads.

#2: Bad reviews. If an old version is crashing too often, or simply has issues making end-users unhappy, there is a way to correct this. The good news is that by uploading a new version of an app marks those negative reviews under the old version number, allowing new and better reviews to be made under the new version number. Ultimately, those new version number reviews will push out the old undesirable version.

To add a new version of your app in iTunes Connect go to the app's page and under the Versions section click on "Add Version" as shown in Figure 40-8.

Figure 40-8: Add new version

Again, the following is required to create a new version of your app:

- Version Number
- What's New in this version (to be displayed in iTunes)

Once you have entered the required information the new version will be created.

When you add a new version of your app you can change your keywords, name, enable iAds, and add In-App Purchases and Game Center Leaderboards/Achievements

After you are satisfied with the changes, then upload the binary following the same steps you took to initially release the original version of the app.

With the new app published, users will have the opportunity to download the updates and will once again be happy with your product.

Bonus Chapter of Gotchas!

A "Gotcha" in reference to computer systems refers to highly unintuitive or otherwise wildly unexpected results, usually discovered after many painfully wasted hours of attempting to follow the instructions from counter-intuitive documentation.

This chapter will help you to avoid many general pitfalls – or Gotchas! – that might be encountered while going through an otherwise sophisticated process of what can be broken down and simplified into a routine.

Let the hard lessons learned by the contributors of this collection be your warnings to avoid.

41.1 Team Admins: There Can Be Only One!

Unfortunately, as of this writing, Apple developers cannot have more than one Team Admin contact under their iTunes Connect account. This means, among other things, that you cannot have multiple employees uploading apps at the same time on your behalf. Developers are currently limited to a single dedicated workstation for all app uploads. This may not sound like a big deal, but if part of your marketing strategy is to quickly upload many apps at once in an ongoing manner (as is often done), or you are simply a larger organization with lots of projects going on at any one time, then be sure to plan for the fact that multi-uploads simply aren't possible at this time.

41.2 App Transfers: No Can Do!

Unfortunately, as of this writing, you cannot transfer any app from one developer account to another - despite what you may have heard from people falsely assuming otherwise.

This currently means:

1. YOU CANNOT GAIN CONTROL OF YOUR APP, IF YOU PAY A DEVELOPER TO UPLOAD IT UNDER THEIR ACCOUNT.
2. You cannot buy an app from another developer and move it into your account.
3. You cannot transfer any app from one developer's account into any third-party account.

Hypothetical scenario: If you've hired a developer to produce your app, and you did not create and use your own App Store developer account to upload and distribute your app (under your own account), then you may be effectively "held hostage" by the developer who uploaded the app to the App Store on their own account. ENSURE THAT HOW THIS IS GOING TO BE HANDLED IS IN YOUR AGREEMENT WITH THE DEVELOPER UP FRONT. Without an agreement specifying this detail, try to imagine the potential for horror stories in this regard of high and unexpected after-the-fact developer "monthly maintenance fees" for hosting your app, which under some wild circumstances someone may go so far as to call extortion: So do not let this happen to you, and either hope the developer does not go bankrupt or sell out one day, and either get this in writing in advance or preferably use your own developer account for full control of your apps.

Currently, as of this writing, the only Apple-supported way to "transfer" an app from one developer account, into to another account, is to:

1. Delete the app from iTunes, thus losing all stats, download history, sales trend data, app rankings, etc.
2. Re-uploading the app to iTunes, under the new developer account. Warning: This will completely destroy any momentum of the app's upward trends, all of its rankings, the ability to send customers updates, etc.!
3. In other words, it's impossible.

ACTUAL EMAIL EXAMPLE:

Please include the line below in follow-up emails for this request.
Follow-up: 123
Re: iOS Developer Program

Hello [Your_name],
Thank you for taking the time to speak to me today regarding the iOS Developer Program.
Below are the detailed instructions about how to remove an app from sale on the App Store and information related to changing a CFBundle Identifier.

To remove an app from sale on the App Store:
1. The Team Agent should log in to iTunes Connect:
<http://itunesconnect.apple.com>
2. Click on the "Rights and Pricing" button from the App Summary Page.
3. Click on the "Deselect All" button to uncheck all App Store territories.
4. Click on the "Save Changes" button.

After removing all assigned territories from your app in the "Rights and Pricing" section, the app's status will change to "Developer Removed from Sale" and your app will not be seen on the App Store within 24 hours.

If you are selling your app to another developer for their own distribution and need to remove it from your iTunes Connect account, it's recommended that you use App Delete so the App Name will be freed up for their use.

> To do this, after completing the above steps, please select the "Delete Application" button. Deleting your app will not allow you to re-use your SKU or App Name and you will not be able to restore your app once deleted.
>
> Once the app has been removed from sale on the App Store, the new company may then upload the binary to Apple via the usual app submission process.
>
> If the app will be selling under the original app name for the new company, the new company Team Agent will need to ensure that the CFBundleIdentifier for the new app submission differs from the originally submitted app, as multiple apps on the App Store cannot have the same CF Bundle Identifier.
>
> In the original Xcode project, a developer is able to change the CF Bundle Identifier for the app in the Properties Pane of the Target Info Window. The field name is called "Identifier". Alternatively, a developer can alter the CF Bundle Identifier by changing the key into their projects Info.plist file.
>
> I hope that this information is helpful to you.
>
> Best regards,
> [Someone_at_Apple]

Ouch! That might even be funny, if it wasn't so painful. In other words, you cannot do it: All download history, ability to send app updates to prior customers, sales data, app popularity rating, etc., are all lost forever.

To reiterate: as of this writing, despite many false assumptions by developers, there is absolutely no true app transfer process whatsoever.

41.3 Don't Scam the Ranking System

Apple has finally acknowledged the growing problem of developers paying inexpensive foreign companies to effectively scam, or defraud, the ratings and reviews of their apps. This is accomplished by performing thousands of downloads of your app in a very short period of time, using automated scripts and bots. These are so-called app "download services," "ranking services," or similar. This affects the downloaded velocity rating (a measure of how many downloads of the app in a certain amount of time), and simulates a "smash hit" – which can cause the app to quickly shoot up to the top of the charts. This "velocity rating" is used by Apple to rank your app's popularity ranking slot. Not only is this use of automated scripts and bots unethical, it can now officially result in the removal of your app – and according to a recently issued Apple warning – the potential deletion of your entire developer account - and the full removal of all your apps from the App Store!

41.4 Don't Spam the App Store

Avoid uploading tons of apps over and over that have little or no actual functionality – such as the ultra-generic and non-functional "business card app" - and or rude or otherwise inappropriate apps, like the traditional "Fart" soundboard apps. While it's true those types of apps were once permitted, and you may see many thousands of them on the App Store, the fact is Apple has quit accepting any more of them. For example, if an app's only function is to play sound clips as a generic soundboard with no other functionality, then it probably will not be permitted anymore. Warning: If Apple notices you are spamming the store with junkware, they've been known to ban entire developer accounts and all apps associated with it.

41.5 Sexual Content is Not Permitted

With the exception of legitimately sexual educational apps, which is a very hard sell to the Apple app reviewers: any "titillating" apps, which contain any ***implied pornography***; as in, images or sounds; sexual innuendos; or targeted marketing implying sex in any way, such as the old so-called "bikini apps" which were all banned long ago, are all generally no longer permitted.

41.6 Make an App Name Reservation

Before starting development, have a lawyer scan for trademark conflicts and lock in your app name first! Before investing in all of that hard work, be sure to advance-reserve the perfect app name. Do not start working on graphics, such as splash screens, which you may have to redo later if you find out after-the-fact the name or a similar one was already taken. In fact, this is so important: please review this book's dedicated chapter covering exactly how to do this!

41.7 Resolving Rejection

In the unlucky event your app is rejected, do not fret, as you will receive an email containing information on how to go to the ***iTunes Connect Resolution Center*** for the reasons why. Simply login to https://itunesconnect.apple.com and select Resolution Center. You will have the opportunity to correct what is wrong and will be able to resubmit the app immediately for another review attempt.

Another possibility is to request an "appeal" at Apple for a secondary re-review of your app, if you believe it follows the guidelines and was merely misunderstood:

https://developer.apple.com/appstore/guidelines.html

Test before upload: To save time, test the app on an actual device, to make sure that your app can handle unexpected errors or outright crashes and still recover gracefully.

41.8 Easter Egg Apps, Misleading Descriptions, etc.

Avoid app descriptions which contain misleading, hidden, implied but missing, or any false advertising features - which either do not work in the as-is version of the app as described, or are hidden and activated through some other means and not described. For example, "lite" apps that describe features only contained in the paid version, and "Easter egg" apps, which contain hidden features. Promotional "Bundled" apps, such as those, which contain multiple functions of many of your apps combined into a single app for a cheaper price, are also something to avoid wasting time being rejected.

41.9 Following Apple Guidelines

Make sure that your app follows all of Apple's Privacy Policies and Human Interface Guidelines, among others. If not, it is a definite rejection. Other definite rejections: Likeness of any Celebrity or Famous Person, disparaging apps, mimicking or resembling previously banned apps. Also, apps which appear confusingly similar to or contain any Apple product, software feature, name, logo, etc., will be rejected on the spot. In addition, any app deemed to show Apple in a negative light, in any way whatsoever, will be rejected. For example, if your app pretends that the iPhone screen is broken to scare the user as a joke, then the app will be most likely be rejected. Here's the official Apple informational page on these issues:

https://developer.apple.com/appstore/guidelines.html

Don't circumvent Apple getting its cut: Apps with a "Buy Now" button, which circumvents the iTunes "In App" purchase feature or otherwise attempts to avoid Apple receiving its share of the revenue, are highly discouraged and usually rejected. However, it is permitted to generally link to an outside website to purchase online accounts and subscriptions, as long as they also offer similar functionality using the Apple in-app purchase feature – as long as Apple gets it cut.

Know Your Pricing: All apps have a pricing restriction, in that they must be offered for the same or lowest price on iTunes as any other ported version for other platforms. For example, if you sell the Android version for $0.99 on Google Marketplace, you must also sell the iOS version for the same $0.99 or less on the App Store. *[A note from the lawyer: As an aside, this is not illegal price fixing because the developer remains free to set a price. Apple simply requires that the best price or a lower one apply to iOS apps.]*

41.10 Expedited Reviews

You can now officially request Apple review your app immediately, without waiting for the usual 1-2 week queue. Warning: Expedited reviews are limited to approximately 2 or 3 requests per overall developer account. It is not known for certain how long this takes to reset for a developer to be permitted to request another expedited review, but the general consensus is it could be as long as up to a full year. You can request an expedited review by sending an email, along with your company name with its Organization ID (shown in the Developer Portal), and the app's name with its App ID (shown in iTunes Connect), sent from your Team Admin email address, to the following Apple email address: appreview@apple.com

41.11 Device Limits, Use Safari, Keyword Restrictions

Developer Test Device Limits: Each developer account has a limit of 100 developer-testing devices. This means, for example, there will not be enough device activations on your account to offer any kind of open internet-wide testing of your app. Do not try to create more accounts just to circumvent this.

Always Use Apple Safari: For the Apple iTunes Connect and Developer Websites, using Apple's Safari is highly recommended to avoid painful data-loss and other glitches.

Keyword Restrictions: In addition to the obvious inability to use trademarks of a third party within your keywords list (as discussed in prior chapters of this book), Developers cannot use competing mobile platforms in their search keywords, e.g. "Android" will get your app rejected.

41.12 UDID Myths (aka "The good old days")

Many iPhone books and websites ramble on and on about using the Apple device Unique Device Identifier (UDID) to track end-users for marketing and other usage data. This means that the other book you are looking at is outdated. Recently, Apple had a lockdown and removed UDID tracking or usage by developers due to abuse and bad press. Do not transmit or use the device UDID. Other methods may exist, such as salting a hash for transmitting an alternate scrambled and irreversible device ID for uniqueness reasons. Work with your advertising network, and read the latest Apple guidelines, for the latest information in this ever-changing privacy issue area.

41.13 Reinstalling Xcode if needed

Several common conditions that may require a complete Xcode re-install, may include:

First, if another third-party tool or plug-in that is required for a project is being used which is incompatible or has not been update yet for the new version of Xcode (e.g. Unity 3D).

On the other hand, you may experience issues that are more generic as corruption of the installation or of an upgrade that has gone horribly wrong. For example, if you try to upload, and receive an error similar to the following:

"com.apple.transporter.util.StreamUtil.readBytes(Ljava/io/InputStream;)[B"

This might indicate Xcode 4.x, or some other component, has become corrupted.

If that is the case, you will need to completely uninstall Xcode. Load up Terminal, and then enter the following super user command (you will need the admin/root password):

```
sudo /Developer/Library/uninstall-devtools —mode=all
```

After the uninstall script is complete, try rebooting and re-installing Xcode 4.x.

41.14 Illegitimate Content

There are many free and open-source code-bases, images, graphics, sounds, and other content, which you may be tempted to pull from the Internet and then modify for your use, ***but use caution***: Many free and or overly cheap sources have invalid licenses, are scams, or may even contain outright illegally pirated content ***sold under the guise of being legitimate!*** While this book is not providing any legal advice whatsoever, note that, in theory, even if you have a "receipt" from paying for so-called "pirated" or otherwise unlicensed content in your app, when it comes to intellectual property rights and or copyright law, ***paid receipts MAY NOT HELP your defense!*** Please be sure to validate the license, its legality, and the credibility of the sourced creative, of any code, image, sound, or other content that you use or otherwise incorporate into any project. Just because you see content on a "Sharing" site does not mean the real copyright holder knows it's there or gave permission! When in doubt, ask a lawyer. If you see amazing free or bulk-rate content offerings, just remember the old adage: *Deals that seem "too good to be true", probably are.* While imaginary and theoretical, this is all entirely possible.

41.15 Overview of Gotchas!

Avoiding these pitfalls will save countless hours, or even days, of frustration. Common sense and a quick review of the Apple Human Interface Guidelines will also help to avoid the costly rejection and resubmission process. Time is money. Delays cost everyone.

SECTION VI

AFTER THE UPLOAD

This final section will guide you through several suggestions on sticking with your business plan, and provide recommendations on how to best promote your new app through the use of social media.

The goal of having your app discovered and downloaded by thousands of users can be obtained with a broad marketing strategy that begins way before the app is released to the App Store. Through research and analytic analysis of your competition and current popular culture trends, you can enter the market with a better understanding of how to position your app on the road to succeed where your competition are still attempting to start their engines.

In addition to social media (which is rapidly losing its attraction due to over-saturation and marketing abuse), it is recommended that you also look into working with your local newspapers or print publications to help you issue a press release or solicit an interview. Getting your efforts noticed in any "traditional" news media outlet will help spread the awareness of your app, increasing its likelihood of being picked up by various media outlets: and then by bloggers, news relay websites, etcetera, who can all help combined to spread the news of your app.

If things do not go well with the app and your marketing seems to yield no positive results, then it may be time to review your business plan and take stock of your situation. It is not that you did not work hard enough or that your app was poorly written. You have to take stock of the reality that every app may not be profitable. With so many apps on Apple's iTunes App Store, the competition is quite fierce. **Be sure to check out some of the analytics, sales analysis tools, ranking tracking websites, and other marketing products listed in this book's Appendices.**

The closing chapters of this tome's final Section will outline the specifics for such post-release marketing and analysis.

Post Upload Marketing

42.1   What now?

Now is the time to review your post upload marketing plan, created in Chapter 1. Use everything within your budget, properly coordinated according to your plan: This may include a ***simultaneous*** launch of all tactics at once onto the competition! The App Store is known for being extremely trendy. Time is of the essence. If the app has already been released before reading this, then you may already be running out of time! Go! Go! Go!

Hopefully, you are not in a panicked situation, and are following your app-release schedule and overall plan. Everything should be timed properly with your strategic app-release plan: press releases, advertising campaigns, social media, traditional marketing, in-app banners, everything you can possibly afford, to promote your app – keeping your business plan in mind as you craft and carry out your well-orchestrated marketing strategy.

Post upload marketing is the portion of your marketing plan that takes place ***IMMEDIATELY*** after the app is released. Even just a few lost can be devastating to the app's potential. There are both paid and free ways to promote your app once it has been released. For your reference, a list is provided of the most popular marketing tools and websites within **Appendix D: App Review Websites** and **Appendix E: Marketing Tools and Websites** found at the end of this book. Be sure to incorporate as much as budgeted in your plan, most likely to be timed all at once.

By using some of these services you will get a nice boost of awareness and Internet "chatter" about your app and its impact on the Apple mobile device user community. And yes, you can make "free" apps that will prove to be very profitable. In fact, as stated elsewhere, some of the "Top Ranked: Revenue" generating apps are indeed "free" downloads!

Many app developers are lured to leverage their "free" app download numbers with an in-app advertiser, such as iAds, in order to generate revenue. There are several lines of thought on in-app advertising as many customers may find the ads irritating and, unless you have several productive revenue streams, the payout is relatively low. You will need to find a good balance so that your customers will not find the ads annoying and will continue to use your app, perhaps offering an in-app purchase to remove them, or let end-user purchase the "full paid" version from the App Store. Run your analytics and adjust your offerings accordingly to find the right blend.

Remember, the secret to having a successful app on the App Store is being noticed. If people cannot find your app, then they will likely never go out of their way to look for it.

42.2 Post-upload Marketing review

Remember the target audience of the app, as per your business plan, and craft your marketing to their demographic. Make sure your plan includes a mix of traditional and non-traditional marketing channels. The following are some common ideas for you to consider when making your marketing plan.

42.2.1 Utilize Social Media

Even though their commercial value is starting to come into question with corporate abuse and global over-saturation, social media, blog posts, and relevant online forums, still provide wonderful and inexpensive tools that many seem to forget about when developing a marketing plan for their app. Contingent upon your target audience, these outlets may prove to be the most fruitful avenues to recoup your development investment. Do not underestimate the power that a positive review from a popular site can bring to the download of your app – especially if it is "free" or if you provide a "lite" version with slightly less features than the "paid" version.

By making the most of the use of social media outlets, (e.g. Facebook, Twitter, etc.), it is very easy to have conversations directly with the customers of your product that provides a means to vaulting the experience of using your app to a very personal level. This is the absolute, unspoken great power of the iPhone and all mobile devices, as people are always connected to each other via their online communities and will use the device to virtually and instantly interact with others. ***The fantastic integration of Facebook and Twitter into Apple's latest iOS 6 is a powerful statement to the topic.*** These social sites keep very detailed information on how their customers utilize their services. If you are able to tap into any of this data (via a subscription) then you can take the informational power of the giants into marketing your own apps.

42.2.2 Review your competitors

See what is working for your competitors, and what is not. Attempt your best in order to learn from their mistakes and success as you craft your own marketing campaign. Use the Internet to search for articles about your target audience and the apps they use. Add notes to your plan.

Many industry and consumer publications, in addition to general news websites, often produce a review of their Top 100 / Top 50 / Top 10 apps that will make things easier tracking down your top competitors. Often times the lower the number of apps, the more the reviews are written to a particular audience, such as college students, travelling road warriors or foodies. If you find a review of apps that deals with just your targeted audience, then you have struck gold. Just remember, the competition is fierce and there will be other developers out there who are likely performing the same home work as you.

42.2.3 Review your own App Page in iTunes Connect

Do you have a compelling icon? This will likely be the first impression that you can make upon a potential customer. Take the time to make (or contract out) a stylish and compelling icon that may capture the essence of your app. For this reason, the app screen shots, and this one icon desktop graphic, should be a reasonable portion of your graphics budget.

Take the time to just look through the artwork of the top selling apps on the App Store and see if you can find some artistic inspiration: but not plagiarism. Often that artistic inspiration may come when you are least expecting it.

Screenshots are important, but believe it or not, they do not have to actually be exact screenshots of the app itself. They can be modified with an overlay of text or borders to grab the users' attention. This is similar to a call-out, or splash on a book cover. If possible, show several screenshots of the different capabilities of your app. Your perspective customer will like to window shop a bit before deciding upon a purchase or download.

Read the reviews of your app. Find the positive and capitalize on enhancing those features in future upgrade releases of your app. Find the negative reviews and thoroughly validate any negative claims. If comfortable, reach out to the reviewer and see if they are willing to go into more detail about the issues they may have with your app. Submitting fast updates for your app to the App Store to quickly address any complaints is highly recommended.

The personal touch will go a long way towards winning the hearts and minds of your customers.

42.2.4 Promotional Codes

One of the most highly guarded items every developer has: the limited promo codes given to you by Apple on iTunes Connect. Give out promo codes sparingly, only for people to download a free version or your app who you think may review it on a well-known website or other news media outlet. Do not just place your precious few codes on random marketing pieces or websites. If you use multiple marketing campaigns, be sure that each marketing channel utilizes a different promo code so that you can easily trace what method is driving traffic to your app. Actively change your promo codes monthly or quarterly in order to keep driving demand for your app and to help sustain the downloads and any revenue.

42.2.5 Set an App Release Date

Use a set release date opposed to Apple's automatic release so you can prepare and generate marketing "buzz" using some of the methods described within this section. Think about releasing more than one app on the same day, in order to create a buzz about a suite of apps instead of just one.

Setting the Availability Date of your initial app release is configured within the Rights and Pricing screen of the iTunes Connect portal.

By setting the release date, you remain in control of how to best position your app for your coordinated marketing campaigns.

42.2.6 Use Your Own Website

The ability to host your own website has significantly reduced in price over the past few years. Many hosting providers now offer monthly package deals for domain name registration, an SSL certificate, and a Joomla or WordPress website for as little as the price of a single value meal at a fast food restaurant. Take advantage of popular web hosting services provided by companies such as, by way of example only, HostGator (www.hostgator.com) and 1&1 (www.1and1.com) to further spread the word of your apps, and to potentially provide a host for any online data access your app may require.

Use your company/app website to update interested parties with news about the new app(s) and their release date – in advance of its release! You can also link back to pages with information on your app, such as user guides, and screenshots. You can also use the website to offer a promo code for a free download of your app, but again: limit your offers to only a few, in order to hold your remaining precious limited number of promo codes for reviewers and news media outlets, etcetera.

Be sure that your website supports feeds for interested parties to subscribe, watch and follow. There are some great web tools that you can easily implement on your website that offers the interface to tap into the massive ocean of social media. Again, this is another area your ad network sales representative can earn their pay: by giving you suggestions on which solutions work best with their ad platform.

> ***TIP:*** *Upon your app's approval, either before or upon its release date,* ***Apple will send you a very special email with what some developers call the "magic" URL link.*** *This "magic" link is actually just a very special website URL redirector link, which when clicked in a web browser, triggers the locally installed iTunes to activate and bring the end-user directly to your app's location in the over-crowded App Store! Be sure to prominently display this special direct app download link on your website! Make sure it is clearly visible on your home page, as well as on other relevant pages you are using to market the app. If you have an email service associated with your website, add the link to your outgoing marketing emails, to your signature line on your business email, and add the links to all your social media pages. Everywhere!*

42.2.7 Video Impact

Internet video is very popular. Register a YouTube site for your company (whether an individual or bona-fide company), dedicated for the app, and produce some promotional videos to share with the rest of the world. These can be anything from educational user guides to short films of your app in action. Films are great for highlighting games and game play hints. Like the film industry, you can even produce a video trailer of coming attractions currently under development and a promised release date. Videos used to show functionality, ease of use, and common tasks make great content for the YouTube channel. Moreover, you can even have these videos accessible from your own website by uploading the video to your YouTube channel and then linking it to your own website.

Make sure to share the video through your social media channels. It may serve your marketing needs best to create a series of videos to release every week, or every few days, to help generate a sensational media buzz for the big app release date. Again, be sure to add the direct "magic" link to your app on the App Store, the moment Apple emails it to you.

42.3 Press Releases

Press releases are a great way to "get the word out" about your app. Used properly, they will funnel your target audience back to the app page on your website, or to your YouTube channel featuring a trailer of your app. Make sure that each press release takes you to a page that is specific to the release. Your audience will lose interest if the press release redirects them to an irrelevant webpage. The press release is specific for the app, so make sure your potential customers can locate it easily. Again, the press release service will help you with the release as they want you to be successful: in the hopes that you will want to use their services again in the future. Some may charge a reasonable fee, others may help for free –be sure to ask!

If used correctly: Press releases can be an incredibly powerful marketing tool in and of themselves; although, they are usually combined with other aspects of an overall marketing strategy. **For example, you can set a press release to automatically go out to the news wires, at the same time your app is automatically made available to the public using the iTunes Connect control panel.** It's usually worth the money to pay for professional news writers to help create the press release in a professional manner, so it's more likely to be picked up by professional news reporters. You can, of course, write your own news press release and work to submit it to various outlet venues.

However, it's important to correctly conform to the writing style, perspective, word usage, release formatting, and other technical aspects of the press release, so the big news wires will even see your story. Otherwise, if your news press release isn't conforming to industry standards, it may very well be simply deleted or ignored by news reporters. This is why so many of the professional news press release service companies will help you compose, SEO optimize, release, and distribute a professionally tweaked press release for you – for a fee.

By way of example only, PRWeb (http://www.prweb.com), in addition to WebWire (www.webwire.com) and Press King (www.pressking.com) are just a few businesses that offer these types of services for fast and easy distribution to some of the most influential media outlets. Think of it as hiring your own Public Relations firm for your marketing campaign. *More of these similar types of services are mentioned in this book's Appendices.*

42.4   Advertising Campaigns

Try to honestly describe what your advertisement is selling, and what you hope to achieve through the investment in your campaign. Advertising is more than just fliers, brochures, banner ads, and videos. At the core of every advertising campaign is the opportunity for you to connect your concept personally to your customers. An advertising campaign is used to make that human connection and to help position your product above all others. Advertising is about selling trust. If people trust you, then they will give you money in exchange for the trustworthy goods you produce.

Nothing says you need to run only one campaign. You can have multiple media campaigns running simultaneously with each geared towards a specific demographic of your target audience.

An ad campaign can be configured through several of the ad networks mentioned in Chapter 3: Getting Started with In-App Advertising. To generate revenue, mobile ad networks allow the self-creation of campaigns to run paid-advertising promotions that display advertisements, such as banner ads, to a target audience within the apps that are using their service. This is an opportunity for your own app's banner ad campaign, to be displayed directly within a competitor's app. How great is that!

When running a campaign containing in-app banner ads, make sure they are eye-catching and visually stunning as there are several options and variation you can incorporate into your app. If the in-app banner is not captivating, then the user may not click it at all. The in-app banner you select is just as important is the landing page where it delivers the user when they touch upon it. Be sure that your banner ads are relevant to the landing page and craft your ad campaign specific to the advertising page to which you are delivering the user.

42.5 Sticking to your Business Plan

At the beginning of this book, especially in Chapter 1, it was recommended to consider and develop a business plan for entering into the app development market. No matter your app development background or experience level, the creation of such a plan is important for you to gauge your measure of success, or failure.

Two of the many important reasons to have an app business plan and to stick to it, as reiterated from Chapter 1, may include:

1. On the positive side, say the app becomes successful and downloaded a few thousand times with good reviews. This can cause anyone to become so excited, that they'll fail to perform many of the strategic steps recommended in order to maximize their exposure - which could have, perhaps, brought the app to a higher status, which may have resulted in the app being downloaded a million times instead!
2. On the other hand, perhaps the app does not do so well no matter what is done, and it's simply not going to be successful for whatever reason. The business plan (which should contain, among other things, when to halt the investment and cut losses in the event of failure), can be absolutely instrumental in overcoming anyone's burning desire to continue to burn through resources and money trying to make the app successful – up to their bankruptcy.

In other words, whether it's letting go and cutting losses, or sticking to the post-release checklist in the event of success to maximize revenue, it is hard for anyone in a time of crisis to think clearly - that's just human nature. Either way, the business plan will greatly help to remove the emotion from what must be a pre-calculated and logical process.

If you have yet to create a business plan, then just sit down and do it NOW! Plow through it! Get it done! It should be one of the most enjoyable parts of the entire app planning process: dreaming of success, and writing down those thoughts of success on paper. The process will ultimately make you, your apps, and your company, much stronger and more competitive.

From the beginning of this book, you were asked to take an honest approach to laying out your goals and your own customized roadmap to success for realizing those goals, you should have assigned a cost value to measure your success or failure – in advance.

In addition to the native services offered to track the sales and downloads of your app via the iTunes Connect portal, there are many companies that offer free and paid analytical data services on your app sales from the App Store and related ad campaigns. Some of these service providers are listed in **Appendix F: Useful Services & Websites** at the end of this book. The analytical services will help you to monitor your investment and the rate at which you may be burning through your cash and credit reserves. These tools were designed for you to track your success, but they may also determine the grim reality of looming failure – and when to refer back to the business plan for a solid number that was previously decided, logically, without emotion, to cut losses and try again with another app.

If the numbers of your investment of time, money and energy are not adding up to a profitable endeavor, as previously determined, then it may be time to drop the support of your unprofitable apps and transfer those efforts to a new and different app project where you will start the process all over again. A successful business knows when to absorb the loss of investment and change to other potential streams of revenue. A business can only operate for so long without producing a profit. In the absolute worst-case scenario: If you were honest with yourself in the beginning, if all Apps fail, then you are better prepared mentally to know when to close shop and to begin road mapping the next big thing.

On a more positive note, business markets change – and none change faster than the trendy App Store sales market! You will always have the need to constantly adapt your business plan in order to address the ever changing or the entirely new and emerging mobile market opportunities. In other words, some of the most successful Apps in the store were the result of fast response times and good timing in general for new opportunities as they presented themselves to attentive developers. Being on guard, and aware of the market and competition at all times, will greatly increase the chances for success.

AFTERWORD

"For all sad words of tongue or pen, the saddest are these "it might have been."

-John Greenleaf Whittier, "Maud Muller" (1856)

Just remember that the saddest of words is, "it might have been." These are the very words that people often speak when they had a dream and elected not follow it. The phrase is the lament of missed opportunity, of windows closed forever and of roads not taken. Each of us have seen countless individuals who have had an amazing concept or an awesome idea who, but for one reason or another, were unable to follow-up and make the most of on that vision. We at Unknown.com, Inc. want to help you enthusiastically recognize and fanatically foster your desire of developing apps for the use on the greatest mobile computing platform, Apple's iOS and iPhone, iPad and iPod touch devices. That is why we decided to publish this easy to follow roadmap so that you may discover and find the drive to find a life that is full of meaningful purpose.

Many of us here at Unknown.com, Inc. are following our passion and had to sacrifice to make the time to do so. If you are not already following a dream, then you are most likely trapped inside a repeating cycle of grinding out a living working for someone else. Just the fact that you are reading this book means that you have already made some initial steps of breaking that recycling sequence and towards making some changes in your life to find and live the dream you are chasing. In order to catch and live that dream you have to make the critical investment of time. Time, once lost, is the one thing you can never recover. Time carries a perceived range of value by different people who bill by the hour, but in truth, the value of your time is priceless. Today, with so many demands and requests placed upon your time, you must prioritize and choose what to best do with the limited resources a 24-hour day provides. All we can ask is that you make the time to explore those things that interest you. If you find yourself too busy with the obligations of work, family, and other elements of your life, then you have to make a dedicated, conscience effort to make the time to know yourself and have fun while learning and progressing your dream.

If you find yourself starting to take some steps outside of your normal comfort zone of our familiar daily grind and experiencing some anxiety, then this is good. That feeling is fear. Fear is one of the most powerful human emotions that both protects you from danger and keeps you from trying new things that may be fun. If you are feeling a bit insecure about learning some new programming tools, starting your own software development business or facing negative reviews of your app, then you need to embrace and face those fears to put them behind you immediately. We at Unknown.com, Inc. do not want you to fall as a victim to the crippling fear of the unknown. Do not give into the fear. If you quit now at the thought of failure of ridicule, then you will always be left with the lingering question of what might have been. Besides, you may find that living with the sad thought of what might have been will likely prove more difficult than conquering those initial fears and striving to follow your dream.

From the beginning, the foundation of this 'Tome of Knowledge' arose as a collection of closely guarded secrets gathered in an organized binder as training material for our developers to rapidly ramp-up to speed on the intricacies required for creating and publishing apps for iOS Apple devices. If it were not for the beer-infused vision of a dedicated intern who inspired the crafting of this book, then our revelation, publishing and sharing of this work would have never existed. Reflecting back upon the production of this book and the many steadfast hours consumed by our development team in its publication, we at Unknown.com, Inc. consider our invested time and effort as our sacrifice for investing in your improvement. We hope that our roadmap serves as a guide for your journey of realizing those dreams of establishing your own business and producing professional and popular apps.

Appendix A: Apple Contact Info Quick Reference

Apple Developer Support Phone Numbers
For iOS, OSX, and Safari: (800) 633-2152
For iBookstore: (877) 206-2092

eMail Support
Legal, usually DMCA or Trademark issues
AppStoreNotices@apple.com

App review process inquires (e.g. to submit one of your few per year allowed app review expedite requests)
appreview@apple.com

General OSX, iOS, Safari Developer support
devprograms@apple.com

General iBookstore support
iBookstore@apple.com

iTunes Connect general help
itunesconnect@apple.com

iTunes Connect accounting help
itunesappreporting@apple.com

iAds support
iAdSupport@apple.com

Direct Website Logins (Shortcuts)
Direct iTunes Connect login
https://itunesconnect.apple.com

Direct iOS Developer login
https://developer.apple.com/devcenter/ios

Direct download link for developer tools without using Mac (OSX) App Store
https://developer.apple.com/downloads/index.action

Consolidated App Guidelines pages
https://developer.apple.com/appstore/guidelines.html

Appendix B: Apple iOS 6 New Features

Publication of this book was held for months, pending Apple's release of their iOS 6.0 operating system – on September 19th, 2012 – and our subsequent release of Non-Disclosure Agreements and ability to finalize and release this book as an authorized iOS developer on that same day!

The new Apple iOS 6 allows for the development of iPhone, iPod touch, and iPad apps to take advantage of a fresh collection of Application Programming Interfaces (APIs) as well as several modified versions of existing APIs. These new APIs will establish an innovative platform for the development of a new breed of apps taking advantage of these features. Expect to see an explosion of new and updated apps to be published to the App Store over the next several months. Out of the hundreds of new features, some favorite new major additions, specifically from iOS 5.1.x to iOS 6.0, are described below:

Map Kit
With version 6.0 of iOS Apple has replaced Google Maps with their own newly created Maps engine. Routing apps can be created using this API to give users specific types of directions, like hiking trails, bike paths, or subway routes to work.

Social Integration (Facebook)
Developers can now add Facebook support to their app with this new API. With it developers can allow users to post content to their wall and add an option to share media with their friends.

Passbook and Pass Kit
With the new addition of the Passbook app that manages boarding passes, movie tickets, retail coupons, loyalty cards, gift cards; The Pass Kit APIs are used to interface your app with this new system. This includes displaying, adding and updating passes in the Passbook, as well as, sending them via email or setting items to appear at certain locations such as when a user reaches their terminal.

Reminders and Event Kit
Developers can now access and share to-do lists from within their app. The ability to create and modify reminders and set time and location-based alarms is also added. You also have the ability to assign properties such as priorities and due dates.

In-App Purchase
Apple has launched In-App Purchase Content Hosting in ITunes Connect. This means that apps with non-consumable purchases can have the content hosted by Apple as long as it is approved prior to the In-App Purchase being submitted.

Game Center and Game Kit
The latest version of iOS 6 comes with new Game Center features. Now developers can turn high scores and achievements into challenges. Game Groups have been added as well which enable sharing of leaderboards and achievements across iOS and Mac versions of apps.

Camera
New to the Camera in iOS 6 are APIs that let the developer control exposure, focus and region of interest. There is also new face detection APIs that allow accessing and displaying faces, as well as hardware-enabled video stabilization.

iCloud
New iCloud Storage APIs, higher storage limits, and faster updates for Key Value Store are new in iOS 6 for Apple's iCloud.

Webkit and Safari
With the release of iOS 6, Safari now has the ability to create audio for interactive web apps using the new Web Audio API. In addition, it can now use advanced color and pixel effects with CSS filters, as well as, upload videos and images from the Photo Library.

Appendix C: Additional Reading

If other important Apple iOS documentation is skimmed (or left unread entirely), one very important document that should be carefully re-reviewed is Apple's iOS Human Interface Guidelines. Plus, it's freely provided by Apple to all iOS developers!

https://developer.apple.com/appstore/guidelines.html

Besides Apple's developer documentation, and this Tome of Knowledge, there are other noteworthy works: "Appreciate your competitors; they are your stimulus for change."

- http://parinaya.tripod.com/proverbs.html
- http://vellaturi.homestead.com/files/Golden_quotes.htm
- http://www.calicutnet.com/variety/quotes/competitors.htm

In literally the final days of post-editing (after our already-completed unique work, "Producing iOS 6 Apps: The Ultimate Roadmap for Both Non-Programmers and Existing Developers" was finished), *the decision was then made to take our own advice on competitive analysis*, and to overnight a pile of books with similar subject matters - to help determine the proper angles to market and advertise in an already brutally competitive iOS book market. Neither the author contributors nor publisher recommend any particular work for any particular use; but they're listed here as a matter of convenience for good reason. Kudos to the few uniquely high-quality competitors out of the pile as listed below; the very brief comments under each are our own thoughts on why they caught our eye:

The Business of iPhone and iPad App Development: Making and Marketing Apps that Succeed
(Apress | ISBN-10: 1430233001 | ISBN-13: 978-1430233008)
Our opinion: In stark contrast to our attorney's sections revolving around avoiding legal trouble, their attorney's sections focus more on how to protect your intellectual property - and using the court system to enforce your legal rights.

App Savvy: Turning Ideas into iPad & iPhone Apps Customers Really Want
(O'Reilly Media | ISBN-10: 1449389767 | ISBN-13: 978-1449389765)
Our opinion: Fascinating interviews with successful and influential people as intertwined throughout the relevant subjects.

App Empire: Make Money, Have a Life, and Let Technology Work for You
(Wiley | ISBN-10: 111810787X | ISBN-13: 978-1118107874)
Our opinion: This captivating book has qualities usually only ever seen from professional motivational speakers.

Appendix D: App Review Websites

The app review business has become a huge industry. Most of these common app review websites take payment in exchange for an expedited review process, thereby substantially reducing wait-time. The savvy developer will sign up for as many review websites ***at the same time for maximum impact*** as humanly possible. Even if they do not accept payment, or if paying for every single one of them simply isn't in the budget, then try to utilize as many as possible. Obviously, pick all of the sites that cater towards the particular app's focus (e.g. game or not). Neither the author contributors nor publisher recommend any particular site for any particular use; but they're listed here as a matter of convenience for good reason. To help get started, some of the more popular ones are listed here in alphabetical order for convenience and future reference:

http://www.148apps.com
http://www.alliphoneappsreview.com
http://www.appadvice.com
http://www.appoftheday.com
http://www.appsmile.com
http://www.appstoreapps.com
http://www.appaddict.net
http://www.appboy.com
http://www.appchatter.com
http://www.appcraver.com
http://www.appgamer.net
http://www.appmodo.com
http://www.appolicious.com
http://www.appsafari.com
http://www.appsfire.com
http://www.appshopper.com
http://www.appstorm.net
http://www.appvee.com

http://www.chomp.com
http://www.crazymikesapps.com
http://www.dailyappshow.com
http://www.giggleapps.com
http://www.ikidapps.com
http://www.iphoneappreview.com
http://www.iphoneapplicationlist.com
http://www.iphoneappcafe.com
http://www.iusethisapp.com
http://www.modojo.com
http://www.slapapp.com
http://www.slidetoplay.com
http://www.theiphoneappreview.com
http://www.toucharcade.com
http://www.touchgen.com
http://www.touchmyapps.com
http://www.whatsoniphone.com
http://www.yappler.com

Appendix E: Marketing Tools & Websites

The app sales reporting and marketing research business has also become a big industry. Neither the author contributors nor publisher recommend any particular site for any particular use; but they're listed here as a matter of convenience for good reason. To help get started, some of the more popular ones are listed here in alphabetical order for convenience and future reference:

http://www.appfigures.com
App Store reporting, rankings tracking, emailed reports. App integration for tremendous detail.

http://www.appnannie.com
Track app analytics, rankings, reviews, sales, etc.

http://www.appstorehq.com
App Store Keywords analysis. Maximize every precious character of the app's 100-byte limitation.

http://www.distimo.com
App Store monitoring, revenue reports, competitive analysis. Reports on different app stores.

http://www.ideaswarm.com
App Store sales trends and profit reports using a powerful downloadable software tool called AppViz. Basically, it logs into the developer account, downloads all of the sales reports, and makes sense of it with easy-to-read charts (a team favorite for the past few years).

http://www.google.com/analytics/features/mobile.html
Very powerful. Works well with Google AdSense, etc.

http://www.prweb.com
PRWeb is a well-known and very popular press release news service, which can also provide good advice with its news writing assistance services.

http://www.prlog.org
PRLog is a free and fully automated press release site with good search engine optimization and rankings, for those who already know how to write professional news articles.

http://www.webwire.com
Web Wire is another popular press-release news service.

http://www.pressking.com
Press King is another popular press-release news service.

http://www.tapmetrics.com
Owned by Millennial Media.

WARNING REGARDING APP ANALYTICS:

Analytics are important: knowing what end-users are doing within the app (e.g., what they are clicking on) helps the perceptive developer to know where to tweak their app for maximum interest and therefore revenue. However, due to some bad press in the past, and more recent privacy concerns, Apple has clamped down on the usage or tracking of end-user activity within an app (e.g. without the end-user's permission). Be sure to follow Apple's latest guidelines and requirements, as buried deep in the developer program contracts.

Appendix F: Useful Services & Websites

The following is a list of very popular and helpful sites providing iOS developers with various unique resources, depending on the need. Neither the author contributors nor publisher recommend any particular site for any particular use; but they're listed here as a matter of convenience for good reason. To help get started, some of the more popular ones are listed here in alphabetical order for convenience and future reference.

http://www.adobe.com
Adobe Flash Runtimes. Content delivery with powerful DRM fully integrated; AIR SDK APIs.

http://www.buzztouch.com
Online guided training for making apps, and generates project source code using easy learning wizards.

http://www.canappi.com
Generate an mdsl file from either a Balsamiq or Xcode Interface Builder file. Mobile DSL (m|dsl) files are for cross-platform development.

http://www.groupzap.com
Brainstorm on projects with groups online using an HTML5 compatible web browser.

http://www.moodle.com
Server-based Open Source Course Management for educational content deployment; and it is even SCORM compliant.

http://www.rubymotion.com
Ruby on Rails is very popular. Build Ruby for iOS native Apps.

http://www.theappcode.com
Create an entire app without coding, using Bitzio's revenue-focused App Builder "EveryoneApps".

http://www.zinecity.org
Create your own social-oriented content and "publish" it to their viewer app.

OUTSOURCERS:

If you are looking for an independent contractor to write part of an app, or an artist for some graphics or icons, or maybe a musician for sound effects, then look no further!

http://www.elance.com
Find a contractor to outsource your work (graphics, writing, coding, etc.).

http://www.odesk.com
Find a contractor to outsource your work (graphics, writing, coding, etc.).

http://www.vworker.com
One of the oldest and most respected for contracting programmers, providing services similar to "escrow/arbitration" for conflict resolution in case of dispute. Previously called "Rent-A-Coder";

Contractors bid on your "jobs" to do all aspects of development (graphics, writing, coding, etc.). This has been a favorite for years.

CODE REPOSITORIES:

Need to store a more complex project in a single place, or maybe work with a team of people on a single project - at the same time? No problem! Many solutions exist for this requirement, such as Subversion and Git. Some of this functionality is integrated directly into Xcode. Examples include:

http://www.beanstalkapp.com
Collaboration, code management, etc.

http://www.gittiapp.com
A Git client for Mac OS X.

http://www.github.com
Collaboration, code hosting, change tracking, etc.

http://www.versionsapp.com
A Subversion client for Mac OS X.

http://www.unfuddle.com
Both Git & Subversion hosting, tracking, etc.

http://www.zennaware.com
A subversion client for Mac OS X.

Appendix G: Graphics, Images & Icons

After analyzing the competition and feeling the excitement of finding a niche for an app, creating an outline of what's to be created *before* plowing forward into it blind will save tons of time and help avoid discouragement or later burnout.

It is highly unlikely a developer will have graphical artistic capabilities; and likewise it's equally unlikely a graphics artist will have programming capabilities. The whole "right-brain vs. left-brain" issue comes into play. Accordingly, the following is a brief list of extremely popular sites and creative solutions to this "brain hemisphere" issue. Never fear, Graphics are near!

Also, it is a common strategy to create "mockups" or essentially sketches of the app's layout, and basic look and feel, prior to beginning development of the app. It is also the most exciting part of development, because an app can be designed exceptionally fast and efficient with the right tools!

Neither the author contributors nor publisher recommend any particular site for any particular use; but they're listed here as a matter of convenience for good reason. To help get started, some of the more popular ones are listed here in alphabetical order for convenience and future reference.

http://www.omnigroup.com
Multiple sets of tools for both Mac and Windows! Be sure to check out OmniGraffle and track down some Apple-themed stencil plugins for your mockup brainstorming, specifically:

http://www.graffletopia.com
Apple iPhone Stencils for OmniGraffle, which include popular graphical elements (e.g. buttons, slider bars, input boxes, etc.), as supported by the Apple Human Interface Guidelines.

http://www.balsamiq.com
Fantastic online and offline options for collaborating on a single app design. Not limited to just "apps", this amazingly flexible solution for all kinds of software development, are a favorite among many software designers. Balsamiq files can be read by www.canappi.com, and output to Mobile DSL (m|dsl) files for cross-platform development.

http://www.mocabilly.com
A professional prototyping tool for designing iPhone apps, with interesting features, avoids the need for stencils.

INDEX

A

Admob · 16
Adobe Flash Builder · 319
 Distribution · 327
 Hello World · 320
 Installation · 319
Adobe Flash Professional · 331
 Distribution · 341
 Hello World · 334
 Installation · 331
AdWhirl · 20
App Upload
 Application Loader · 437
 Preparation · 413
 Xcode · 431

B

Business Decisions
 Account Type · 38
 Expired Copyright Usage · 31
 Fair Use, From Our Lawyer · 33–35
 Legal Disclaimer · 23
 Likeness Laws · 33
 Open Source · 29
 Searching for Existing Trademarks · 27
 Third-Party Content · 24

C

Cocos2d · 343
 Distribution · 362
 Hello World · 346
 Installation · 343
 Sprite Sheet · 350
Corona SDK · 365
 Advertising · 375
 Distribution · 375
 Hello World · 367
 Installation · 367
 Simulator · 374

D

Developer Portal
 App ID · 50
 Bundle Identifier · 51
 Bundle Seed ID · 51
 Contracts · 47
 Development Certificate · 57
 Distribution Certificate · 419
 iOS Dev Center · 50
 iOS Provisioning Portal · 50
 Keychain Access · 57
 Manual Provisioning Profiles · 63
 Team Admin · 57
 UDID (Unique Device ID) · 60
 WWDR Intermediate Certificate · 59, 420
Device Resolutions · 416–17

F

Foundation Framework
 NSArray · 134
 NSCalendar · 133
 NSData · 131
 NSDate · 132
 NSDateFormatter · 133
 NSDictionary · 137
 NSFileManager · 182
 NSFormatter · 133
 NSLog · 105
 NSMainNibFile · 151
 NSMutableArray · 135

NSMutableSet · 136
NSMutableString · 132
NSNumber · 131
NSNumberFormatter · 133
NSObject · 122
NSRange · 132
NSSet · 135
NSString · 132
NSTimer · 132
NSUserDefaults · 134
NSValue · 131
NSXMLParser · 184
Frameworks
    Address Book · 195
    AVFoundation · 197
    Core Animation · 198
    Core Audio · 198
    Core Data · 190
    Core Graphics · 199
        CGPoint · 147
        CGRect · 147
        CGSize · 147
    Core Location · 219
        Compass · 220
        GPS · 219
    Core Motion
        Gyroscope · 220
    Event Kit · 197
    Media Player · 198
    OpenAL · 198

G

GameSalad · 377
    Distribution · 383
    Hello World · 377
    Installation · 377
Gotchas · 443
    App Developer Transfers · 443
    App Rank · 445
    App Review Expedited · 447
    Apple's Cut · 447
    Easter Eggs · 446

Guidelines · 447
Inappropriate Content · 446
Pirated Content · 449
Spam · 446
Team Admin Limit · 443
UDID · 448
Xcode Reinstallation · 448
Greystripe · 17

H

How to Use this Book
    The Sections · xxxii

I

iAds · 16
Interface Builder
    IBAction · 145
    IBOutlet · 145
    Inspector Pane · 143
    Inspectors
        Attribute · 143
        Connection · 144
        Identity · 143
        Size · 144
Internet Crime Complaint Center (IC3) · xxv
iOS 6 New Features · 465
iOS Dev Options
    Adobe Flash Builder · *See* Adobe Flash Builder
    Adobe Flash Professional · *See* Adobe Flash Professional
    Alternatives to Xcode · 75, 79
    App Store vs Mac App Store · 75
    Apple's Xcode · *See* Xcode
    Cocos2D · *See* Cocos2D
    Corona SDK · *See* Corona SDK
    Cross-Platform Development · 75
    GameSalad · *See* GameSalad
    Marmalade · *See* Marmalade
    MonoTouch · *See* MonoTouch
    MoSync · *See* MoSync

PhoneGap · *See PhoneGap*
Quick Reference Chart · 83
ShiVa3D · *See ShiVa3D*
Titanium Studio · *See Titanium Studio*
iOS Developer Program
   Annual Fee · 41
   Apple ID · 39
   Doing Business As (DBA) Restrictions · 39
   Signing Up · 41
iTunes Connect
   Add New App · 52
   App Crash Logs · 434
   Bank Info · 46
   Contact Info · 45
   Game Center · 426
   iAds · 427
   In-App Purchases · 426
   Metadata · *See also Metadata*
   Metadata (app price, icon, etc) · 54
   Signing Up (After Dev Program) · 43
   Tax Info · 46

J

Jumptap · 17

M

Marketing
   Advertising Campaigns · 457
   Press Releases · 457
   Social Media · 454
   Tools · 471
   Website Reviews · 469
Marmalade · 305
   Hello World · 309
   Installation · 305
      Mac OS X · 305
      MS Windows · 307
   Metadata · *See also iTunes Connect*
   Bundle Display Name · 414
   Icon Files · 415

Locating the Metadata · 413
Millennial Media · 16
MobClix · 20
Model View Controller · 139
Monetization Techniques
   Ad Aggregation Advantages · 19
   Advertising Aggregators · 19
   Advertising Networks · 16
   Different Revenue Options · 11
   Free vs Paid vs GROSSING! · 13
   In-App Advertising · 15
   In-App Purchases · 12
   Market Research & Analysis · 11
   Paid Apps · 13
   Penalty for switching free/paid · 11
MonoTouch · 293
   Distribution · 301
   Hello World · 295
   Installation · 294
   Interface Builder · 297
MoSync · 397
   Distribution · 405
   Hello World · 399
   Installation · 397
   Layout · 403
   Native UI Project · 400

O

Objective-C
   Attributes · 114
      assign, copy, retain · 114
      getter, setter · 115
      nonatomic, atomic · 115
      readwrite, readonly · 115
   Cocoa · 107
   Cocoa Touch · 107
   Data Management · 181
      NSFileManager · 182
      read · 181
      SQLite · 186
      write · 182
      XML · 183

dealloc, destructor · 119
Instance Variables · 110
Memory Management · 120
    autorelease · 120
    Garbage Collection · 121
    Manual Retain Release, Automatic Reference Counting · 120
    release · 120
    retain · 120
Methods · 110
NeXTSTEP · 107
Properties · 114

P

PhoneGap · 287
    Build Services · 292
    Distribution · 292
    Hello World · 288
    Installation · 287
Planning for App Success
    App Store vs Mac App Store · 3
    Following Apple's Guidelines · 9
    Following your Plan for Success · 458
    Return On Investment (ROI) · 4
    SBA's 10 Easy Steps · 6
    Small Business Administration (SBA) · 5
    Turn Your App Idea into a Plan · 3

S

ShiVa3D · 275
    AIModel · 278
    Authoring Tool · 283
    Dynamics Controller · 282
    Hello World · 277
    Installation · 275
    Material · 280
    Model · 279
    onInit handler · 279
    Scene · 278
Smaato · 17

T

Titanium Studio · 389
    Distribution · 394
    Hello World · 390
    Installation · 389
Touch Events
    touchesBegan · 201
    touchesCancelled · 201
    touchesEnded · 202
    touchesMoved · 202

U

UIKit Framework
    Accelerometer · 221
    addSubview, insertSubview, removeFromSuperview · 148
    UIActionSheet · 163
    UIAlertView · 161
    UIApplication · 150
    UIApplicationDelegate · 151
    UIButton · 164
    UIImageView · 157
    UILocalNotification · 151
    UINavigationController · 169
    UIPickerView · 158
    UISegmentedControl · 164
    UITabBarController · 171
    UITableView · 166
    UITableViewCell · 167
    UITestField · 164
    UITouch · 203
    UIView · 139
    UIViewController · 149
    UIWebView · 160
Unity 3D · 257, *See Unity 3D*
    GameObject · 262
    Installation · 257
    OnCollisionEnter · 266
    Physic Material · 264
    Publishing Builds · 270

Rigidbody · 264
Scripting · 265
Walkthrough · 261

V

View Lifecycle
    viewDidLoad, viewWillAppear, viewDidAppear, viewWillDisappear, viewDidDisappear, viewDidUnload · 149

X

Xcode · 87
    breakpoints · 154
    Debug View · 97
    Editors
        Assistant · 93
        Standard · 93
        Version · 94
    Installing Xcode · 90
    Instruments · 223
        Activity Monitor · 231
        Allocations · 224
        Automation · 237
        Core Animation · 246
        Custom · 251
        Energy Diagnostics · 244
        File Activity · 244
        Leaks · 229
        OpenGL ES Analysis · 249
        OpenGL ES Driver · 247
        System Usage · 245
        Threads · 243
        Time Profiler · 232
        Zombies · 241
    Interface Builder, XIB · 140
    iOS Simulator · 89
    Navigators
        Breakpoint · 97, 154
        Debug · 96
        Issues · 95, 154
        Log · 97, 154
        Project · 94
        Search · 95
        Symbol · 94
    Organizer · 100
    Utilities View · 98

> **This book is also available in electronic (eBook) format.**

All UnknownCom Inc. works, including but not limited to its printed books and eBooks, are subject to worldwide copyright protection. All Rights Reserved.

Please see the inside front cover of this book for its ISBN's, LCCN, additional important notices, and important disclaimer information.

Reminders: This book and its eBook are © 2012 UnknownCom Inc. All Rights Reserved. Note the "UnknownCom Inc." abbreviation of the copyright owner's/author's/publisher's full legal name "Unknown.com, Inc." was required for typesetting purposes (© 2012 Unknown.com, Inc. All Rights Reserved). All information in this book is provided on an "as is" basis, without warranty of any kind. The source code described in this book is available and licensed to readers, under a modified Free BSD License, on an "as is" basis without warranty of any kind, at:

http://apps.unknown.com.

Thank you!

Printed in Great Britain
by Amazon.co.uk, Ltd.,
Marston Gate.